Introduction to Quantitative Hydrology

This textbook serves as an introductory quantitative course on the fundamental elements of the hydraulic cycle. It enhances students' understanding by discussing the latest advancements in hydrological science, covering both experimental and computational techniques.

This textbook is self-contained, requiring no prior knowledge, and includes numerous illustrations to clarify scientific concepts. Complex mathematical treatments are minimized, focusing on clear, step-by-step examples and guides that utilize scientific calculators and spreadsheets. Where appropriate, chapters include assignments that reinforce the textbook's role in academic settings. A virtual laboratory section is also provided, featuring experiments and example datasets for student analysis. Additionally, the text outlines the equipment needed to set up a physical laboratory, making it practical for educators to implement.

Targeted at first-year college students, this book supports early career exploration in fields such as natural resources, earth sciences, and civil and environmental engineering. Offering this course early allows students to make informed decisions about their academic and career paths before they reach their senior year, providing them with ample time to pursue specialized interests.

Introduction to Quantitative Hydrology

Aly I. El-Kadi

 CRC Press
Taylor & Francis Group
Boca Raton London New York

CRC Press is an imprint of the
Taylor & Francis Group, an **informa** business

Cover image: Redrawn and modified from Shutterstock: author burakbl

First edition published 2026
by CRC Press
4 Park Square, Milton Park, Abingdon, Oxon, OX14 4RN

and by CRC Press
2385 NW Executive Center Drive, Suite 320, Boca Raton FL 33431

CRC Press is an imprint of the Taylor & Francis Group, an informa business

British Library Cataloguing-in-Publication Data
A catalogue record for this book is available from the British Library

ISBN: 978-1-032-95912-2 (hbk)
ISBN: 978-1-032-96493-5 (pbk)
ISBN: 978-1-003-58714-9 (ebk)

DOI: 10.1201/9781003587149

Typeset in Times
by codeMantra

The book is dedicated to my loving family: my wife Faten and our children Shereen, Aladdin, and Enjy and her husband Omar Noor.

Contents

Preface

Interest in water-related knowledge has significantly intensified due to climate change, resulting in frequent droughts, floods, and widespread concerns about water quality. Serious water-related conflicts are prevalent both within and between countries. Consequently, the employment market for water-related jobs is flourishing, with significant demand for technical expertise in these fields. Such employment encompasses a wide range of jobs related to the management, conservation, and distribution of water resources. These jobs can be found in both private and public agencies and typically involve a combination of office and fieldwork. A background in hydrology or water science is a primary requirement for employment in these fields. Hydrology is a quantitatively oriented subject and is usually taught at the senior/graduate level in universities, as students must have prerequisite skills in mathematics, chemistry, and physics.

This textbook is designed as an introductory quantitative course for first-year college students to provide them with an early start in pursuing a career in this exciting field. Relevant university disciplines include natural resources, earth sciences, and civil and environmental engineering. Offering this course to such students aids them in making informed career choices early on, rather than waiting until their senior year when they may have limited time until graduation. By starting early, students can acquire essential knowledge in water-related courses and develop expertise. In addition to hydrology, relevant topics include mathematics, physics, chemistry, biology, and numerical methods. Typically, universities offer a wide range of these courses at both graduate and undergraduate levels to support the primary field of study in hydrology. With an early start, students can address all their additional needs to pursue careers in various government agencies, industries, and academia.

To meet the needs of the target students, this textbook is self-contained, and no prior background knowledge is necessary. It includes all the required information regarding the description of various water processes and relevant mathematical formulations. Ample illustrations are used to facilitate understanding of scientific concepts. While the quantitative nature of the subject is emphasized, complex mathematical treatments are avoided. Additionally, the text contains numerous examples and step-by-step guides utilizing scientific calculators and spreadsheets. These computations cover rainfall analysis, open-channel and subsurface flows, and contamination travel times and spread, among other subjects. Assignments are integrated with each chapter to support the use of the publication as a textbook.

The text also includes a virtual laboratory section with several experiments and relevant example datasets for students to analyze the results. A description of the necessary equipment for establishing a physical laboratory is conveniently included.

To further enhance students' background, the textbook discusses recent advancements in hydrological science, encompassing both experimental and computational approaches. The coverage is limited to introducing the concepts and example applications, recognizing that in-depth coverage is reserved for advanced students, as it presupposes a high-level background, which may not be suitable for first-year undergraduates without the necessary prerequisites. Advanced topics covered include numerical modeling, geostatistics, stochastic analysis, open-source programming in R and Python, geophysical techniques, geographic information systems (GIS), laboratory water quality analysis, and least squares analysis. Among these, the text provides further details on stochastic analysis and least-squares techniques, complete with example applications and spreadsheet-based implementations.

Acknowledgments

This publication was significantly improved by the critical review of an earlier draft by my colleagues Ripendra Awal, Ahmad Elshall, Ali Fares, Xiaolong (Leo) Geng, Michael Knight, Olkeba Leta, Scott Rowland, and Yin-Phan Tsang, whose contributions are deeply appreciated. Funding to publish this textbook as open access was provided by the University of Hawaii's School of Ocean and Earth Science and Technology (Chip Fletcher, Dean), the Department of Earth Sciences (Garrett Apuzen-Ito, Chair), and the Water Resources Research Center (Tao Yan, Director). I am grateful for their support in making this textbook freely available to students and other interested readers.

About the Author

Aly I. El-Kadi earned his Ph.D. in Groundwater Hydrology from Cornell University's School of Civil and Environmental Engineering in Ithaca, New York. He currently serves as Professor Emeritus in the Earth Sciences Department and the Water Resources Research Center (WRRC) at the University of Hawaii (UH). His teaching and academic focus were on hydrology and groundwater modeling. His research addressed critical areas such as the impact of climate change on water resource sustainability, numerical modeling of groundwater systems, and the simulation of multiphase flow and hydrocarbon transport. His work addressed the growing pressure on Hawaii's coastal aquifers due to population growth, economic development, and climate change. He also contributed to research in American Samoa and Jeju Island, Korea. He is the author or co-author of approximately 170 publications, including two edited books. His career highlights include serving as the Associate Director at WRRC, which focuses on water resource challenges in Hawaii, American Samoa, and other Pacific islands.

Introduction

1

1.1 HYDROLOGY AND HYDROLOGISTS

The term "hydrology," which pertains to the science of water, has its origins in the Greek words "hydro," meaning "water," and the scientific suffix "-logy," denoting "study" or "science." This field encompasses various aspects of the water cycle, describing the storage and movement of water as it circulates between the ocean, the atmosphere, and the Earth. Natural processes primarily involve precipitation, evaporation, transpiration, and surface and subsurface water flow and storage. Water exists in multiple states: liquid, gas (vapor), and solid (ice). Precipitation can take the form of liquid water or snow, while evaporation and transpiration result in water vapor being released into the atmosphere from water bodies and plants, respectively. Water flows on the Earth's surface, through streams, and penetrates the ground, saturating porous materials to create subsurface water domains.

Water resources are developed for human use by directly accessing surface water from sources like lakes and rivers or by drilling wells to extract subsurface water. However, human activities related to water use and environmental practices can disrupt the delicate ecological balance through overuse and the introduction of contaminants. Climate-related issues are compounded by the emission of gaseous pollutants that contribute to global warming, leading to climate change, rising sea levels, and extreme weather events such as floods and prolonged droughts.

Hydrology encompasses various branches, including ecohydrology (which explores interactions between organisms and the hydrological cycle), hydrogeology (focused on subsurface water presence and movement), hydrometeorology (studying the exchange of water and energy between the land and the lower atmosphere), and surface hydrology (examining processes at or near the Earth's surface).

Analytical hydrology, the subject of this text, is a branch of hydrology that employs mathematical and statistical methods to analyze different aspects of the hydrological cycle. As a quantitative discipline, hydrology relies heavily on assessment and analysis tools crucial for resource management, risk forecasting, and environmental remediation. Mathematical equations and statistical techniques play a vital role in quantifying and interpreting hydrological processes, facilitating informed decision-making regarding water-related issues.

At the college level, hydrology programs are offered within natural, environmental, and earth sciences, as well as in physical geography and civil engineering. A solid

DOI: 10.1201/9781003587149-1

understanding of hydrology is essential not only for hydrologists but also for professionals in related fields such as civil engineering (focused on designing and building water
infrastructure like dams, reservoirs, and sewage systems), irrigation (aimed at efficient
water use in agriculture and system design), and sustainable resource management in
food and energy production.

The primary objective of hydrologists is to comprehend and predict water behavior
in the environment. In pursuit of this goal, hydrologists apply scientific knowledge and
mathematical principles to address specific issues related to water quality and availability.
They also tackle problems such as flooding and structural damage to the ground. Assessing
water availability involves finding sustainable water supplies for various purposes, including
domestic use, energy production, irrigation, and industrial demands. Hydrologists are also
involved in reducing the influx of contaminants and sediment into open waters. Furthermore,
their work extends to pollution cleanup and the safe disposal of hazardous wastes.

Hydrologists may pursue both fieldwork and office tasks. In the field, they oversee
data collection, direct field teams, and operate equipment. Office-based duties include
interpreting hydrological data and conducting analyses related to project objectives.
Computer software is an essential tool for data analysis and the preparation of project
reports. Prediction software, known as simulation models, is often used to assess future
scenarios, such as different cleanup strategies for contaminated sites. Besides a strong
hydrology background, effective writing and communication skills are crucial for fulfilling these and other project responsibilities.

As students progress in their careers, they should select a specific branch of hydrology and focus on courses that align with their chosen path. The choice between a bachelor's and a graduate degree will influence the tracks and courses to pursue. Doctoral
degrees are well-suited for those pursuing academic or research-based careers, while
bachelor's and master's degrees are generally suitable for careers in consulting firms
or government agencies. Numerous employment opportunities exist in various US federal agencies, including the US Geological Survey, the US Environmental Protection
Agency, and the National Oceanic and Atmospheric Administration. State-level agencies, water supply providers, and local water management bodies also offer employment
opportunities. For example, in the State of Hawaii, these include the Honolulu Board
of Water Supply, the Department of Land and Natural Resources, the Commission on
Water Resource Management, and the State Office of Environmental Quality Control.

1.2 WATER AND CIVILIZATION, CULTURE, AND RELIGION

Civilizations throughout history have relied heavily on water sources, particularly
major rivers, to establish and sustain themselves. Prominent civilizations such as
Mesopotamia, Egypt, China, and the Harappan civilization (modern-day India and
Pakistan) thrived because of their proximity to rivers and their ability to effectively
manage water.

Throughout history, water management practices have evolved in response to the needs of civilizations. From the development of irrigation systems and water purification techniques to the construction of aqueducts and subterranean water structures, ancient peoples demonstrated ingenuity in addressing water challenges. Their innovations continue to inspire modern technology, offering valuable lessons for sustainable water use and climate resilience today.

The Mesopotamians, whose name literally means "between rivers" (referring to the Tigris and Euphrates), adopted innovative techniques to protect themselves from floods and irrigate their fields. They constructed canals, dug large storage basins, and built up riverbanks to guard against flooding. These early engineering feats allowed them to support agriculture in a region with an unpredictable water supply (Finkelstein, 1962; Guerin-Yodice, 2021).

Similarly, the success of ancient Egypt was tied to the Nile River. Egyptian farmers were among the first to practice large-scale agriculture, cultivating crops like wheat and barley. They developed an irrigation system that involved forming basins with earthen banks and using channels to direct floodwaters. The rich soil left behind by receding floodwaters made the land fertile for farming. Egyptians also built nilometers—stone columns marked with water levels to monitor floods and droughts, ensuring the efficient use of the Nile's resources for agriculture and transportation (Shaw, 2003; Romeo, 2016).

Ancient Chinese civilizations also demonstrated advanced water management. They implemented comprehensive water supply systems, including pipelines and water purification methods in cities like Yangcheng. China's water management practices, including aqueducts and water impounds, were compared to those in ancient Greece, both of which significantly influenced global hydraulic engineering (Du and Chen, 2007; Zheng and Angelakis, 2018).

The Indus or Harappan civilization also demonstrated advanced water management techniques. The civilization (~3000–1500 BC) built hydraulic structures and wastewater disposal systems that reflect an early understanding of the water cycle. They constructed dams, reservoirs, and spillways for efficient water distribution and were one of the earliest societies to develop sophisticated systems for wastewater treatment (Khan, 2014). The Mauryan Empire, which followed the Harappans, further refined water management techniques with innovations in dams, reservoirs, and rainfall measurement.

Ancient civilizations also utilized aqueducts to transport water over long distances. The Romans perfected the aqueduct system, constructing networks that stretched hundreds of kilometers. They used gravity and the natural slope of the land to channel water through pipes, tunnels, and bridges, supplying cities with water for drinking, bathing, and irrigation. The Roman aqueduct system, with 11 aqueducts supplying Rome alone, set a high standard for water distribution. Some Roman aqueducts, like the Aqua Virgo, still function today, supplying water to modern-day Rome (USGS, 2018; Britannica, 2023a; National Geographic, 2023).

One of the most remarkable ancient water structures, however, is not river-based. The Nazca culture of Peru built the puquios—a series of spiral-shaped holes designed to funnel wind into underground canals, forcing water from deep subsurface reservoirs to the surface. These structures, built over 1,500 years ago, are still functioning today, providing water in one of the driest regions on Earth (Cartwright, 2014; BBC, 2016; Jasim, 2023).

Another impressive ancient water management system is the acequia, built by the Moors in Spain's Sierra Nevada between the 8th and 10th centuries. This 3,000-km irrigation network is still used to capture snowmelt and replenish underground aquifers. Similar systems are found in New Mexico and Colorado in the United States. The acequia not only sustains agriculture but also supports local ecosystems and reduces soil erosion, showing how ancient methods can offer solutions for modern challenges such as climate change (Fernald et al., 2007, 2015; National Geographic, 2019a; BBC, 2022).

Water treatment technology has ancient roots as well. The earliest water treatment methods, dating back to the 15th century BC, were documented in Sanskrit writings and Egyptian tombs. These included techniques such as boiling water, heating it under the sun, and filtering it through sand or gravel. Notably, the ancient Egyptians used a combination of aluminum and iron sulfates to remove suspended solids from water, a precursor to modern chemical water treatment processes (Khoury and Gallisdorfer, 2020; AWT, 2023).

Water has been integral to human life, culture, and religion throughout history. It is not only essential for survival but also symbolizes life, purification, and hope across many ancient and contemporary traditions. In ancient China, water was central to both social and spiritual life. Ceremonial bathing played a significant role in early Chinese religion and medicine, symbolizing purification and renewal (Salguero, 2017). In Buddhism, water is metaphorically associated with adaptability and harmony with nature, a concept that has resonated in Eastern philosophy for over 2,500 years. In Hinduism, water is revered as a life-giving force from which all forms emerge. The Upanishads, ancient Hindu scriptures, frequently refer to water as a metaphor for purity and wisdom. This is reflected in the Vedic tradition, which holds water as one of the five essential elements of nature (Sharan and Pathak, 2017). Water's role in binding and sustaining life is central to the Hindu worldview, with sacred rivers like the Ganges serving as symbols of spiritual purification and rejuvenation.

Similarly, in ancient Egypt, water, specifically the Nile River, was deeply intertwined with both daily life and religious beliefs. The Egyptians divided their world into the fertile "black land" of the Nile Valley and the barren "red land" of the desert. This division influenced their cosmology, with the river representing life and abundance, while the desert symbolized death and chaos. Deities like Hapy, the god of the Nile's flooding, were worshiped for their role in providing the water necessary for agriculture and survival. The Nile was also central to Egypt's technological advancements, including irrigation and boat building, and even played a role in constructing the pyramids.

In sub-Saharan Africa, water carries immense cultural and spiritual significance. Among the Bantu people, water is considered the birthplace of creation, a symbol of life's origin. The Bamileke people of Cameroon use water in rituals, such as a father blessing his daughter with water infused with fefe leaves on her wedding day. African proverbs emphasize the spiritual importance of water, often likening it to the divine. Water is used in rituals for purification, atonement, and renewal, and is central to social interactions and relationships.

Native Hawaiians also held a deep reverence for water, or wai. It was considered sacred and essential for both physical and spiritual well-being. The Hawaiian word for wealth, waiwai, literally means "abundance of water," highlighting its value. Hawaiians managed water resources by organizing their communities into ahupua'a, land divisions that included mountain, plain, and sea areas (e.g., Smith and Pai, 1992). Streams originating in the mountains were used to water crops, with excess water flowing back into the streams to be used by others downstream. This practice of sharing water resources was codified in Hawaiian law, which prioritized the equal distribution of water.

Across these diverse cultures and religions, water serves as a unifying element that symbolizes life, purity, and interconnectedness. Whether through ritual purification, religious ceremonies, or the everyday use of water for sustenance, it is clear that water has played and continues to play a profound role in shaping human spirituality and culture. The reverence for water across these traditions underscores its universal significance and the need for its careful management and protection.

Water has always played a crucial role in human religion. Many religions, including Christianity, Islam, and Hinduism, hold water as sacred. In the Bible, water is described as a symbol of God's guidance and care: "The Lord will guide you continually, giving you water when you are dry..." (Isaiah 58:11). The Quran refers to water as the source of life: "... And We created from water every living thing" (Quran 21:30). Water is also significant in purification rituals in various religions, and religious texts, such as the Bible and Quran, contain references to the water cycle long before scientific knowledge about evaporation, precipitation, and hydrology was established.

In the Bible, rainwater saturates the Earth: "For just as the rain and the snow pour down from heaven And do not return there until they saturate the earth, making it produce and sprout, Giving seed to the sower and bread to the eater" (Isaiah 55:10). Similarly, in the Quran, rainwater is the source of groundwater: "We send down rain from the sky in a measured amount, causing it to soak into the earth. And We are surely able to take it away" (Quran 23:18). And water also rises from springs: "See you not, that Allah sends down rain from the sky, and causes it to penetrate the earth, sends it as water-springs and afterward thereby produces crops of different colors..." (Quran 39:21).

Water sustainability is secured if people's use of the water in aquifers is within the rate at which rainfall replenishes them. Otherwise, the aquifers can be drained by excessive pumping, rendering wells useless. In the Quran, there is a reference to the fact that groundwater can diminish, making it unusable as a source: "Or its (a garden's) water may sink into the earth, and then you will not be able to seek it out" (18:41). And "Say, 'Consider this: if your water were to sink into the earth, then who else could bring you flowing water?'" (Quran 67:30). Reasons for water to sink deep into the earth are related to overuse and droughts that decrease groundwater recharge from rain. Spring water originates as rainfall that infiltrates the soil and permeates underground aquifers. Springs form when water flows from these aquifers to the surface, influenced by various geological factors. Depending on the geological conditions, different types of springs can form. For example, artesian springs occur when water under pressure in a confined aquifer finds an exit point, while fissure springs form due to cracks in the confining layers of rock.

Springs and spring water have long captivated human imagination and are deeply embedded in both religious texts and ancient beliefs. In the past, it was thought that springs were formed by ocean water moving inland to the surface, creating streams. However, religious texts such as the Quran had already provided a more accurate explanation, attributing springs to rainwater percolating underground: "See you not, that Allah sends down rain from the sky and causes it to penetrate the earth, sends it as water-springs…" (Quran 39:21). Both the Quran and the Bible emphasize the significance of springs in their teachings, using them as symbols of spiritual sustenance and divine provision. The Bible uses springs in a metaphorical sense, such as "you will joyously draw water from the springs of salvation" (Isaiah 12:3).

Springs are particularly important in Islamic tradition, where they are seen as signs of divine mercy and sustenance. Prophet Abraham's prayer for his descendants in a barren valley was answered with the gushing of the Spring of Zamzam, which brought life to the area (Quran 14:37). In another instance, Prophet Moses demonstrated divine power by striking a rock, which caused twelve springs to flow, providing water for each of the twelve tribes of Israel (Quran 2:60; Exodus 17:1–7). Water is also significant in the Quranic account of Noah's flood, where gushing springs combined with torrential rain to cause the flood (Quran 54:11–12). Similarly, Prophet Job was healed through the refreshing waters of a spring, which alleviated his suffering (Quran 38:41–42).

Religious teachings also promote the conservation and protection of water resources. In Hinduism, food conservation is linked to water sustainability, as wasting food is discouraged. Similarly, Islam considers wastefulness a sin, emphasizing the need for responsible resource management (Quran 7:31). Therefore, people of faith, who make up the majority of the global population, are expected to be especially conscious of conservation and environmental protection, not just as a civic responsibility but also as a religious duty. For such individuals, there should generally be no conflict between religious beliefs and the application of the best available science for environmental stewardship.

1.3 HISTORICAL MILESTONES IN MODERN HYDROLOGICAL SCIENCE

Historical milestones in the advancement of modern hydrological science are documented in works by Bras (1999), AboutCivil (2017), and Rosbjerg and Rodda (2019). As will be discussed in Chapter 2, water science primarily revolves around the water or hydrologic cycle, which describes the continuous movement of water between land, ocean, and the atmosphere. Hydrological research encompasses the development and analysis of field and laboratory experiments related to various water flow and storage processes within this cycle. These experiments are supported by theoretical studies that use specific hypotheses to advance our understanding of these processes and their interactions through the development of mathematical formulations.

For instance, evaporation, the process responsible for water loss to the atmosphere from water bodies as vapor, can be linked to measurable meteorological parameters, including air temperature, humidity, and wind speed. The relationship between the amount of evaporation and these parameters is based on observations that evaporation increases with rising temperature and wind speed while decreasing with increased air humidity, which reduces the air's capacity to accept more vapor. Experimental data can then be used to establish a functional relationship between evaporation and these parameters. Such a functional relationship is valuable as these parameters are relatively easier to measure compared to evaporation amounts.

Table 1.1 summarizes information covering advances in hydrology over five centuries, up to the end of the twentieth century. Achieving a comprehensive understanding of the hydrologic cycle appears to have evolved over approximately 300 years. This journey began with the work of Leonardo da Vinci, who developed field techniques for measuring channel flows. A comprehensive understanding of the cycle was not reached until around 1800, thanks to John Dalton's contributions. During these years, significant research was dedicated to linking infiltration to precipitation, improving streamflow measurements, and establishing that rainfall serves as a significant source for springs and stream flows. Other notable achievements included detailed evaluations of specific processes, such as the assessment of water flow in pipes and streams, the evaluation of surface water runoff, and the measurement of evaporation.

TABLE 1.1 Milestones for advances in modern hydrological research (partly based on Bras, 1999; AboutCivil, 2017; Rosbjerg and Rodda, 2019)

YEAR	SCIENTISTS	MILESTONE
1452–1519	Leonardo da Vinci	Completed physical experiments, such as measuring stream velocity.
1575	Bernard Palissy	Hypothesized that springs and streams originate from rain.
1674	Pierre Perrault	Measured rainfall, runoff, and drainage in an area of the Seine River and concluded that rainfall was enough to support springs and rivers.
1686	Edme Mariotte	Supported the findings of Perrault by completing experiments relating infiltration to rainfall and developing better streamflow measurements.
1700	Edmond Halley	Published the results on evaporation measurements.
1738	Daniel Bernoulli	Published the equation for frictionless pipe flow.
1802	John Dalton	Became the first to give a complete and correct description of the water cycle based on reliable observations.
1840	Gotthilf Hagen and Jean Poiseuille	Described laminar pipe flow.
1776	Antoine Chézy	Described turbulent river flow.

(Continued)

TABLE 1.1 (*Continued*)   Milestones for advances in modern hydrological research (partly based on Bras, 1999; AboutCivil, 2017; Rosbjerg and Rodda, 2019)

YEAR	SCIENTISTS	MILESTONE
1856	Henry Darcy	Developed the fundamental equation for groundwater flow.
1850	Thomas Mulvany	Addressed measuring flow from urban areas.
1883	Wenzel Rippl	Introduced an approach for reservoir design.
1911	William Green and Gustav Ampt	Developed equations for the description of soil infiltration and saturation.
1914	Weston Fuller	Introduced statistical frequency analysis in hydrology.
1931	Lorenzo Richards	Developed the unsaturated flow equation.
1932	Leroy Sherman	Introduced an approach for surface flow assessment.
1935	Charles Theis	Presented an equation for estimating the response of aquifers to well pumping (Theis, 1935).
1948	Howard Penman	Developed an equation for estimating potential evaporation based on measured meteorological variables.
1964	Ven Te Chow	Published encyclopedic Handbook of Applied Hydrology (Chow et al., 1988).
1970s	Jacob Bear	Introduced in-depth analysis of the contaminant dispersion (spreading) phenomena in groundwater (Bear, 1972).
1978	Holcomb Research Institute of Butler University, Indianapolis, Indiana	Establishment of the International Ground Water Modeling Center, which advanced the dissemination and use of groundwater models (IGWMC, 2023).
1979	R. Allen Freeze and John A. Cherry	Published a classic groundwater book (Freeze and Cherry, 1979).
1980s	Lynn Gelhar	Advanced the statistical analyses of groundwater processes (Gelhar, 1993).
1988	M.G. McDonald and A.W. Harbaugh	Developed a groundwater simulation model that became a standard for the analysis of water flow and contaminant transport (McDonald and Harbaugh, 1988).
1988	C.W. Fetter	Published the second edition of his classic Applied Hydrogeology book (Fetter, 1988).

Milestones related to subsurface water flow and related processes included the development of Darcy's law, which assesses the rate of water flow and the subsequent fate and transport of contaminants in porous media. Significant progress was made in subsurface analyses through the development of techniques that addressed the dispersion (or spreading) of contaminants and the statistical representation of various processes, overcoming the lack of knowledge in understanding these processes.

Finally, significant milestones included classic key books, particularly those by Chow et al. (1988), Freeze and Cherry (1979), and Fetter (1988).

Simulation models, which are computer-based software, gained popularity starting in the 1980s as tools for understanding hydrological processes and predicting future water levels and the spread of contaminants. Although there are uncertainties regarding their accuracy, these models have greatly enhanced research efforts, especially in data collection, and serve as efficient tools for comparing various management approaches aimed at conserving and protecting water resources.

1.4 WATER AND HUMAN HEALTH

Water constitutes approximately two-thirds of the healthy human body. A loss of 4% or more of total body water leads to dehydration (NIH, 2023), manifesting symptoms that can vary in severity (MAYO CLINIC, 2023). Mild symptoms may include extreme thirst, excessive sweating, reduced urination, fatigue, headaches, and muscle cramps. Severe symptoms encompass dizziness, infrequent urination with dark-colored urine, rapid heartbeat, extremely dry skin, accelerated breathing, chills or fever, irritability, confusion, and fainting. Left untreated, dehydration can result in more severe complications, including seizures, brain damage, and, in extreme cases, death, which can occur with a loss of 15% of body water.

The National Academy of Medicine (2004) recommends a daily fluid intake of 2.7 L for women and 3.7 L for men, including fluids from all sources, comprising water-rich foods. This equates to approximately nine cups for women and 12 ½ cups for men daily. Consuming an adequate amount of water not only prevents dehydration but also lowers the risk of developing chronic diseases, premature mortality, or exhibiting biological aging beyond one's chronological age (Dmitrieva et al., 2023). These authors reveal that elevated serum sodium levels in the blood are indicative of various abnormal conditions, including dehydration. Lower serum sodium levels are associated with better health outcomes than higher levels. Furthermore, water provides direct benefits to the body, such as lubricating joints and eyes, aiding in digestion, flushing out waste and toxins, and maintaining healthy skin.

More critically, water possesses unique properties (LibreTexts, 2023) that support life and cater to diverse organismic needs. Water profoundly influences cellular functions in the human body, owing to its distinctive characteristics at both the molecular and chemical levels. At the molecular level, water is a simple molecule composed of two small, positively charged hydrogen atoms and one large, negatively charged oxygen atom. When hydrogen atoms bind to oxygen, they create an asymmetrical molecule with a positive charge on one side and a negative charge on the other, known as a polar molecule (Wikibooks, 2023). This polarity enables water to interact favorably with other polar molecules, including itself. This interaction is driven by the attraction between opposite charges, resulting in water molecules forming strong connections or bonds with adjacent polar molecules, including other water molecules. This bonding causes water molecules to adhere together, a property referred to as cohesion (Khan Academy, 2023). Cohesion facilitates water transport in plants, contributes to water's high specific heat capacity and boiling point, and helps animals regulate body temperature (Sciencing, 2023).

Water molecules can also form bonds with other molecules, enveloping both their positive and negative regions and effectively breaking them apart, causing dissolution. This occurs when sugar crystals dissolve in water due to their polar nature. Similarly, water breaks apart ionic molecules composed of oppositely charged particles. For instance, salt dissolves in water because it consists of sodium and chloride ions. Water's capacity to dissolve a wide range of molecules has earned it the label of a universal solvent, making it indispensable for sustaining life. Within the body, water as a solvent facilitates the transportation and utilization of substances, including nutrients. Water-based solutions like blood transport molecules to necessary locations. In this capacity, water aids in oxygen transport for respiration and influences the efficacy of drug delivery to their targets in the body (Riveros-Perez and Riveros, 2017).

Moreover, water plays a crucial structural role by filling cells and maintaining their shape and integrity. The pressure exerted by water inside many human cells counteracts external forces, ensuring that everything within cells maintains the correct molecular-level shape, a critical factor in various biochemical processes. Water also contributes to the formation of cell membranes by creating selective layers that regulate the entry and exit of substances, such as nutrients. Without water, cell membranes would lack structure, impeding the ability to retain vital molecules and exclude harmful ones.

Water also exerts a profound influence on fundamental cellular components, notably DNA and proteins. Proteins provide structure, receive signals, and catalyze chemical reactions in cells, including muscle contraction and nutrient digestion. Without the appropriate shape, proteins cannot perform these vital functions, jeopardizing cell and human survival. Likewise, DNA requires a specific shape for accurate decoding of its instructions, and water molecules align themselves around DNA in an ordered fashion to support its structure (Khesbak et al., 2011). Without this structure, cells cannot faithfully interpret the instructions encoded by DNA, posing threats to growth, reproduction, and overall survival.

In addition to molecular-level processes, water directly participates in various chemical reactions that build and break down components of human cells, which are composed of repetitive units of smaller molecules. Water is necessary for reactions that degrade these molecules, enabling cells to obtain nutrients or utilize segments of larger molecules for other functions. Additionally, water serves as a buffer against the harmful effects of acids and bases (Resource Center, 2023). Finally, water's adaptability allows it to act as both an acid and a base, moderating pH changes caused by acidic or basic substances in the body. Ultimately, this adaptability protects proteins and other molecules within cells.

1.5 WATER AND ENVIRONMENTAL HEALTH

Water is crucial for maintaining the health, productivity, and resilience of ecological systems, including rivers and wetlands. These systems benefit plants, animals, and people alike. A well-functioning ecosystem relies on consistent water flow in various surface water bodies. The direct advantages of healthy ecosystems include support for tourism, agriculture, recreational fishing, and public health. Regrettably, human activities disrupt natural water flows through actions like water diversions, overuse, and the construction of structures such as dams. This interference reduces the available water for various activities and impacts the overall health of the environment.

In the absence of such disruption, natural flow regimes positively affect the health of river systems flowing into floodplains and wetlands. Firstly, river water flow energizes the food web and triggers fish breeding and movement. Young fish utilize floodplain wetlands as a growth environment before returning to the river to complete their life cycle. Secondly, rivers typically deposit sediment on floodplains, enriching soils and providing grazing areas for native animals and livestock. Thirdly, wetlands support aquatic plants that filter water as it moves through the system, slowing down flows and performing nutrient cycling functions. These plants produce flowers and seeds that serve as food and shelter for various insects and animals. Additionally, birds that feed, breed, and move through the surrounding landscape enhance plant pollination and pest control. Some of these birds utilize wetlands as stopovers during their long migrations, especially during inundation periods.

The Amazon Basin serves as an example of where human activities have increasingly disrupted the delicate balance of the forest's highly complex ecology. Deforestation has accelerated, and mineral discoveries have attracted new settlers and corporations to the region. Deforestation has been particularly notable in areas of Brazil, Colombia, and Ecuador. The consequences are severe because the forest efficiently absorbs carbon dioxide, contributing to global warming and the greenhouse effect. Human activities may also reduce the region's evapotranspiration, disrupting the hydrologic cycle by increasing surface runoff and decreasing groundwater recharge. Deforestation poses a threat to the unique gene pool of the Amazon Rainforest, which contains many undiscovered organisms. This jeopardizes biodiversity and risks the potential loss of unknown and untapped resources, such as pharmaceuticals. Lastly, these disruptions threaten the survival and well-being of the region's indigenous peoples, who play an integral role in the rainforest's ecosystem.

1.6 WATER DISASTERS AND CONFLICTS

Throughout its extensive history, the world has endured numerous water-related disasters and conflicts. These include drought-related devastations, such as the Great Bengal Famine of 1769–1773, attributed to a failed monsoon season in 1769 (Kumar, 1983). The lack of sufficient rainfall to support crops resulted in the complete failure of two separate rice harvests, leading to the tragic loss of approximately 10 million lives. Similarly, the Doji Bara Famine of 1789–1792 in India stemmed from prolonged droughts caused by a four-year consecutive monsoon failure (Grove, 2007), resulting in the deaths of more than 11 million people. A comparable situation occurred with the Chalisa Famine across the Indian subcontinent between 1783 and

1784 (Grove, 2007), where the death toll surpassed 11 million. Lastly, also attributed to droughts, the Northern China Famine of 1876–1879 caused a series of crop failures (Owlcation, 2023), estimated to have claimed the lives of 9.5–13 million people (Edgerton-Tarple, 2008).

More recently, states in the American West, including Arizona, California, Colorado, Nevada, and New Mexico, have fallen victim to a decades-long drought, classified as a "megadrought" (The Guardian, 2022). Williams et al. (2022) labeled this as the most severe drought in 1,200 years, exacerbated by increased water consumption during the same period.

Major historical water disasters are also linked to flooding, exemplified by the Bengal Famine of 1943–1944, which resulted in the loss of three to five million lives (e.g., Mansergh and Lumby, 1973; Greenough, 1982). The famine was caused by powerful cyclones, tsunamis, flooding, and rice crop diseases, with wartime colonial policies from the British contributing to the disaster. Flooding also triggered the Chinese Famine of 1906–1907 (Historic Mysteries, 2021), initiated by heavy rains during the 1906 season that resulted in extensive flooding over large areas, causing approximately 25 million casualties.

More recent floods include the catastrophic one in Pakistan (CNN, 2022), which claimed about 1,500 lives and displaced over 30 million people. Some locations experienced up to eight times more rain than usual, and the rain was accompanied by the melting of glaciers, exacerbating the severity of the floods.

Another category of water-related problems involves transboundary water conflicts, which occur when two or more countries share the same body of water (Petersen-Perlman et al., 2017). These conflicts often revolve around the construction of dams and other projects that control or divert rivers by upstream countries. Petersen-Perlman et al.'s study identified 286 surface water basins and 592 aquifers that cross international boundaries.

A comprehensive compilation and analysis of conflict cases is provided by Climate Diplomacy (2023a). This information includes the effects of climate change and environmental change on fragility and conflict risks. Examples of such conflicts include the Tigris and Euphrates Rivers, where Turkey, Syria, and Iraq share the resources (Climate Diplomacy, 2023b). Water conflicts have also arisen in the Mekong Basin, where dams in China potentially reduce water shares for downstream countries, specifically Vietnam, Laos, Cambodia, and Thailand (Climate Diplomacy, 2023c; Foreign Policy, 2023). The construction of a dam has led to conflict between Egypt and Sudan on one side and Ethiopia on the other regarding the Grand Ethiopian Renaissance Dam (Roussi, 2019; Climate Diplomacy, 2023d).

1.7 WATER RESOURCE MANAGEMENT

Water resource management is defined as the administration of water resources under established policies and regulations. Decisions are usually aimed at controlling water resources to maximize their efficient use for sustainability. In this regard, management practices should encompass planning, development, and optimal distribution of these resources. Ideally, the management plan should consider all competing demands for water and strive to allocate water equitably to satisfy all uses and demands. However, achieving these objectives can be exceedingly challenging due to several factors:

- **Limited Understanding**: There is a lack of accurate understanding of the scientific processes governing water flow and storage.
- **Information Limitations**: Accurate information about the system, such as the nature and characteristics of water sources, is often limited.
- **Use Conflicts**: Conflicts arising from existing and potential uses of water resources pose significant challenges.
- **Valuation Difficulties**: In some cases, assigning a monetary value to certain benefits, such as those resulting from environmental quality improvements, is difficult.

In practice, decisions depend on a multitude of issues that go beyond simple demand allocations. Final decisions often necessitate compromises that may not lead to truly optimal solutions. The following sections will discuss key issues that are highly significant in designing effective water resource management strategies.

1.7.1 Water Use and Competing Demands

Freshwater represents only about three percent of the water on Earth, and the majority of that is locked up in ice caps and glaciers. Only about 0.6% of the world's entire freshwater supply is usable, with most of it located in the subsurface, presenting associated management difficulties and extra development costs. In some cases, water availability is reduced due to water quality problems caused by illegal waste disposal and excessive chemical usage by humans, such as in agricultural and industrial practices. Climate change can exacerbate water shortages by altering precipitation patterns and increasing water scarcity in specific regions. As discussed earlier, competing demands within and across national borders have led to conflicts.

Due to such scarcity, freshwater resources often face competing demands from various sectors, including agriculture, industry, domestic use, and ecosystems. Agriculture, as the primary source of food production, is a major consumer of freshwater, primarily for irrigation. Industrial freshwater use encompasses water consumption in various industrial processes and activities, such as manufacturing, cooling, cleaning, and other purposes. Growing populations require more water for drinking, sanitation, and other essential services. Natural ecosystems also depend on water for their health and biodiversity. It is crucial to balance all these water needs with conservation efforts.

To address this challenge, integrated water resource management, efficient water-use technologies, and sustainable policies are essential to ensure equitable access and protect ecosystems while meeting the needs of various sectors.

Table 1.2 provides percentages of major freshwater uses in the US. Approximately 90% of water use is allocated to thermoelectric power, irrigation, and public supply. Public water supply serves domestic, commercial, and public services, including public pools, parks, firefighting, water and wastewater treatment, and municipal buildings. It is important to note that these percentages can vary widely between different states. For instance, agricultural use dominates in major agricultural centers like the State of California, where the percentage rises from the US average of 32% to more than 60% (Johnson and Cody, 2015).

Water use varies by location, season, and, in some cases, local regulations. As an example, the average daily household water use in the US is estimated at 82 gallons per person, translating to an average of over 300 gallons per family. Roughly 70% of this usage occurs indoors (EPA, 2023a). Eighty-two gallons per person per day is exceptionally high compared to water-scarce countries, such as some parts of Africa, where individuals are limited to five gallons in certain locations.

Table 1.3 presents percentages for household water uses, with values that are comparable for the top uses. The estimated leakage rate of 12% is significant and of great concern.

TABLE 1.2   Percentages of US freshwater withdrawal (EPA, 2022a)

ITEM	PERCENTAGE USE (%)
Thermoelectric power	45
Irrigation	32
Public supply	12
Industry	5
Others	6

TABLE 1.3   Percentages of household uses (EPA, 2022a)

ITEM	PERCENTAGE USE (%)
Toilet	24
Shower	20
Faucet	19
Clothes washer	17
Leak	12
Other	8

1.7.2 Water Sustainability

Publications such as Bredehoeft (2007) and Sophocleous (2000) have introduced different definitions and approaches to assess the sustainability of drinking water and other resources. For example, Sophocleous defines sustainable development as the practice that "meets the needs of the present without compromising the ability of future generations to meet their own needs." In this context, optimal or sustainable water use requires setting upper limits on water withdrawal to avoid compromising the source, which can be a stream, a lake, a groundwater aquifer, or a combination of these. For instance, it is commonly suggested to establish a maximum pumping threshold that does not exceed the sustainable yield of an aquifer. Sustainable yield is defined as the maximum amount that can be withdrawn from an aquifer without negatively affecting the source. If exceeded, water levels can significantly decline, leading to water quality deterioration, such as increasing salinity due to saltwater intrusion in near-ocean aquifers.

However, estimating or practically utilizing a sustainable yield is often challenging due to strict assumptions in assessment approaches and the lack of reliable information. For instance, although water use in the State of Hawaii, USA, is subject to regulations regarding sustainable yield, studies have shown declining freshwater resources and rising salinity in some locations (Oki, 2005). This problem is further complicated by the need to address sustainability concerns related to other factors, such as the impacts of human activities on ecological systems (Elshall et al., 2022). Usage should also consider ecological factors, such as setting minimum values for stream flows and ocean discharges, which affect the well-being of fish, other species, and coral reefs. Examples related to designing minimum stream flows include setting in-stream flow standards in Hawaii to protect waterways, considering hydrology, in-stream, and non-in-stream uses (CWRM, 2023). Additional political, legal, and socio-cultural considerations add another layer of complexity, including cultural, religious, and native water rights. More recently, the concept of personhood of water was introduced in 2017 by New Zealand through groundbreaking legislation granting such status to the Whanganui River (AP News, 2017). The law declares that the river is a living entity, from the mountains to the sea, encompassing all its physical and metaphysical elements.

An approach for assessing global water sustainability was introduced by the World Resources Institute (Gassert et al., 2013; Luo et al., 2015). This approach provides water stress projections for various countries, defined as the ratio between total water withdrawals and available renewable surface water. Total stress is a combination of various individual uses, specifically agriculture, industrial, and domestic. The approach is most suitable for making comparisons among countries for the same year, as well as between decades and various scenarios for the same region.

Luo et al. (2015) identified areas of serious stress, mainly located in the Middle East, North Africa, Asia, and Australia. The USA, Canada, and some South American countries also fall into this category. Stress due to agricultural use dominates in some countries, such as the United States, India, and South Africa, while others, like Turkey and Indonesia, are influenced more by domestic and industrial use.

It is important to note that some countries that share borders and water resources are subject to extremely high water stress, specifically Israel, Syria, Lebanon, and Jordan. The same applies to Turkey, Syria, and Iraq in one case, and India and Pakistan in another. Certainly, water disputes and conflicts have occurred in these regions (see Section 1.6) and are likely to become more concerning under climate change, with related incidents of severe droughts.

1.7.3 Water Conservation

As discussed in Section 1.7.1, water use covers power generation, agriculture, domestic, and industrial activities. The top users worldwide are mostly industrial or developing nations (e.g., Water Footprint Calculator, 2023), where wasteful management practices should be a concern, especially under climate change and persistent droughts.

In the United States, virtually all the Western states are classified as having high water use and population growth rates. The average water use in these states exceeds 100 million gallons per person per day, which is higher than the national average of 82 million gallons per day. Usage is even higher in states like Idaho, Utah, and Arizona, which is mainly attributed to agricultural practices.

The United States' domestic water use percentages are dominated by toilets, showers, faucets, and cloth washers. Among all uses, toilet water consumption is significant, and the use of efficient units can certainly help. The alarming 12% leakage rate can translate to an average of more than ten gallons per person per day, exceeding the total water use of individuals in water-scarce countries. This number is particularly concerning if it translates to 400 million gallons per day for a state like California.

Water conservation efforts include reducing water waste, identifying new water resources, and implementing water reuse. Among these, waste reduction plans are considered primary efforts, achievable by adopting water-efficient practices. These efforts should address the major uses in various categories, specifically in energy, agriculture, and domestic use, through integrated collaboration between governments and individuals. Water savings can be achieved directly by reducing water consumption and indirectly by reducing related uses, such as food and energy.

The study by Hoekstra and Mekonnen (2012) reported that 92% of global freshwater use is attributed to the agriculture industry. Nearly one-third of all freshwater is used for animal products, such as meat, dairy, and eggs (Gerbens-Leenes et al., 2013). Hence, conservation measures should prioritize these activities, especially concerning irrigation.

In general, farmers tend to over-irrigate to ensure that plants are adequately hydrated. The main challenge is to develop irrigation scheduling systems that precisely indicate when crops need watering and how much water they require. An effective irrigation system needs to be sensitive and flexible enough to adjust based on changing conditions, such as the onset of rainfall. This approach would primarily rely on careful monitoring of soil moisture and weather conditions, especially regarding cloud cover, evapotranspiration, humidity, and rainfall. Among these, a critical factor is the soil moisture level and the amount of water needed to maintain acceptable moisture levels.

A potential approach would involve using buried sensors that can transmit signals regarding these quantities to an irrigation controller.

It is also imperative that irrigation utilizes efficient methods, such as sprinkler and surface and subsurface drip systems (HydroPoint, 2023). With sprinklers, water droplets are sprayed like raindrops over the landscape through rotating nozzles connected to pipelines. This technique is effective for sandy soils and uneven terrain and can protect crops from extreme frost or temperature changes. It also facilitates the application of fertilizers and pesticides through the sprinkler system. The relatively low rates of water application help prevent soil erosion.

In the drip system, water flows through narrow pipes laid on the ground surface and directly drips through small holes (called emitters) at the plant roots. This allows roots to absorb water directly, minimizing water waste through evaporation, runoff, and wind. The method also inhibits weed growth since water is applied directly to plants. In general, bacterial growth is limited because the area near the plant remains relatively dry.

Subsurface drip systems differ by using drip tubes buried below the soil surface within the plant root zones. This method can reduce water use by 25% compared to aboveground sprinkler systems and works well in irregularly shaped fields and on slopes. Similar to the aboveground system, this method also reduces weed growth and plant diseases.

A more advanced approach to conservation aims to design and implement smart farming, which also aims to increase production quantity and quality (Big Ideas, 2021; IBERDROLA, 2023). These approaches are characterized by integrating agricultural and cattle production practices.

Surface water and stream runoff are another major cause of water waste, as water can end up in the ocean. Reducing such waste can be achieved either at the source by enhancing infiltration or after runoff generation through surface and subsurface water harvesting. Enhanced infiltration replenishes aquifers, reduces runoff, helps prevent soil and stream bank erosion, and saves costs associated with managing runoff water by government agencies. Unfortunately, urban areas have replaced forests and meadows with buildings and pavements, resulting in a drastic decline in infiltration. Rainwater runs off roofs and driveways into the streets and can pick up contaminants such as fertilizer, oil, pesticides, and dirt as it travels through storm drains and ditches to various water bodies. Due to difficulties in controlling and treating these chemicals, polluted runoff is one of the greatest threats to clean water. In addition to other benefits, a reduction in runoff can thus help prevent or reduce water pollution.

Approaches to reduce runoff in urban areas include the use of green roofs, permeable pavements and driveways, strips of greenery along roadsides, and green parking lots. Green roofs, a vegetative layer grown on a rooftop, can also lower building energy costs. On the other hand, permeable pavements are made of either porous material or nonporous blocks spaced to allow water flow. They enable infiltration and reduce construction costs by decreasing concrete volumes and the need for conventional drains. Green parking lots incorporate trees and permeable pavements, among other elements. An example of green parking design is described in San Diego County (2019).

Managing surface water is traditionally done using complex infrastructure like dams and reservoirs, which have the combined benefits of harvesting water and reducing the risk of flooding. The stored water becomes available for various uses when needed, including electricity generation, an added benefit of dams, achieved by utilizing water elevation differences between the upstream and downstream sides of the dam.

Instead of surface water harvesting in reservoirs and other water bodies, runoff control can be achieved by artificially recharging groundwater aquifers. This plan, called aquifer storage and recovery, simultaneously mitigates flooding and enhances aquifer storage. The aquifer storage and recovery option is much superior to surface water storage behind dams, considering construction costs, land space requirements, and virtually nonexistent evaporation, especially for deep aquifers. Among related efforts, Santa Cruz's Recharge Initiative at the University of California is advancing the design of field projects for aquifer storage (PMC, 2023).

In California, the Los Angeles Department of Water and Power has invested $130 million in runoff capture projects, such as the Tujunga Spreading Grounds, covering about 150 acres of dirt basins that average 20 ft deep (Los Angeles County Public Works, 2023). Runoff water is pumped into the area to infiltrate into the aquifer. When fully developed, the agency expects it to provide enough water for 64,000 households a year. Between October 1, 2022, and January 10, 2023, an estimated 11 billion gallons of water recharged the aquifer, enough to serve about 140,000 households for a year. Ongoing enhancements include expanding and combining the spreading basins and installing new intake structures, which will increase the facility's storage and intake capacity.

1.7.4 Water Contamination and Protection Efforts

Accidental releases and incidents of inappropriate or illegal dumping of various contaminants can have severe consequences, as they can ultimately find their way into surface water or groundwater. Examples of such incidents include the improper disposal of used car oil and leftover home chemicals on the ground or into streams. Additionally, certain contaminants may result from the residues or excessive use of chemicals that are legally permitted.

Contaminants can originate from either point or nonpoint sources. Point sources refer to relatively localized contamination, such as contaminants discharged from wastewater treatment facilities, leaking septic systems, chemical and oil spills, as well as instances of illegal dumping. In contrast, nonpoint source pollution arises from diffuse sources, including agricultural runoff and stormwater runoff. Nonpoint source pollution is a major contributor to water pollution in the United States and is challenging to control and regulate effectively.

Contaminants, in this context, refer to any material that can impact water quality. In contrast, the term "pollution" is used to describe situations where water quality becomes unacceptable due to contaminant levels that exceed permissible limits, resulting in adverse effects on both humans and the environment.

1.7.4.1 Surface water contamination

According to the U.S. Environmental Protection Agency, nearly half of U.S. rivers and lakes are polluted, rendering them unfit for swimming, fishing, and drinking (EcoWatch, 2022). Nutrient pollution, which includes nitrates and phosphates, is a significant type of contamination in these water sources. It occurs due to the use of fertilizers on farms with chemicals that dissolve in runoff water. Streams and lakes can become directly contaminated by runoff or through interaction with contaminated aquifers. Furthermore, municipal and industrial waste discharges contribute to the problem, introducing a range of organic and inorganic chemicals.

1.7.4.2 Groundwater contamination

Groundwater becomes polluted as contaminants leach from various sources, including landfills, onsite disposal systems, storage tanks, and surface water bodies. Onsite disposal systems encompass cesspools or septic systems, particularly prevalent in rural areas with limited access to sewer lines. Underground and aboveground storage tanks are necessary for civilian and military uses in airports, military bases, and gas stations. Remediation of groundwater can be challenging, often bordering on impossible, due to the slow-moving nature of water. Site assessment and remediation efforts face complexities in the geological features of the site, including material variability and preferential flow, which hinder attempts to trace the contaminant. Contamination may continue undetected for years or even decades, concealed underground. Discovery is often prompted by the spread of human or ecological diseases linked to pollution. Once polluted, an aquifer may remain unusable for decades or even thousands of years, and groundwater can also transmit contamination as it seeps into streams, lakes, and oceans.

1.7.4.3 Ocean contamination

Eighty percent of ocean pollution originates on land, where contaminants are transported from farms, factories, and cities via streams and rivers, ultimately reaching bays and estuaries and eventually the ocean. Other contaminants take the form of debris, such as plastic, which can enter water bodies through storm drains, sewers, or direct dumping. Additionally, both large and small oil spills and leaks are common sources of ocean contamination.

1.7.4.4 Types of contaminants

As previously discussed, the agricultural sector is the largest consumer of global freshwater resources, primarily for farming and livestock production. Unfortunately, it is also a significant source of water pollution. Contaminants in this context mainly include fertilizers, pesticides, and pathogens such as bacteria and viruses. Nutrient pollution can lead to the development of algal blooms, which produce toxic algae harmful to both people and wildlife.

Wastewater results from household waste and various commercial, industrial, and agricultural activities. Treatment facilities aim to reduce pollutant levels before discharging the treated water back into waterways, such as the ocean or via deep injection wells. However, potential releases to the environment can occur due to aging facilities and their inability to meet increasing demands.

Petroleum products contribute to contamination through oil and gasoline leaks from millions of cars and trucks. While ocean contamination can result from tanker spills, the primary sources are land-based facilities, including factories, farms, and cities.

Radioactive waste is mainly generated by uranium mining, nuclear power plants, and the production and testing of military weapons. Hospitals also use radioactive materials for research and medical purposes. The disposal of radioactive waste presents challenges as its radiation can persist in the environment for thousands of years. For example, the decommissioned Hanford nuclear weapons production site in Washington requires billions of dollars for cleanup and will take many years (GAO, 2022).

Household hazardous waste is generated during household and maintenance activities both inside and outside the home, including cleaning, home improvement, and car maintenance (DTSC, 2022).

1.7.4.5 U.S. federal and state regulations

Government at both the federal and local levels bears the responsibility of protecting human health and the environment. In the USA, at the federal level, the Environmental Protection Agency (EPA) is tasked with ensuring clean air, land, and water for the population (EPA, 2023a). Its specific duties include reducing environmental risks based on the best available scientific information, enacting federal laws to safeguard human health and the environment, providing access to accurate information for communities, individuals, businesses, and state, local, and tribal governments to effectively manage health and environmental risks, and enforcing regulations based on national standards, while assisting states, tribes, and companies in compliance.

An example of the EPA's role relates to the Federal Clean Water Act (CWA; EPA, 2022b). This act establishes the framework for regulating pollutant discharges into U.S. waters and sets quality standards for surface waters. Under the CWA, the EPA has implemented pollution control programs, including the establishment of wastewater standards for industries. Additionally, the agency has developed recommendations for national water quality criteria for pollutants in surface waters. The CWA prohibits the discharge of pollutants from point sources into navigable waters without obtaining a permit.

At the state level, agencies like state departments of health are responsible for locally preventing pollution and promoting the preservation of a clean, healthy, and natural environment. This includes resource conservation and the protection and enhancement of air and water quality (e.g., HDOH, 2023). State agencies also play a part in implementing federal regulations and reporting to the EPA.

HDOH (2023) outlines various actions that individuals and agencies can take to prevent or minimize water pollution, including

- *Reducing plastic consumption through reuse or recycling.*
- *Properly disposing of chemical cleaners, yard chemicals, oils, and non-biodegradable items, as waste poured into sinks or toilets often goes untreated before being released into waterways.*
- *Taking efforts to prevent car oil, antifreeze, or coolant leaks.*
- *Designing landscaping to minimize water use and reduce runoff, as well as limiting or eliminating the use of pesticides and herbicides.*
- *Properly disposing of leftover medicine through pharmacies.*
- *Cleaning storm sewers blocked by litter to prevent trash from entering the water.*
- *Educating neighbors, coworkers, and community groups about water pollution prevention.*

1.8 CLIMATE-RELATED PROBLEMS

1.8.1 Global Warming

Global warming is caused by greenhouse gases, which are wrapped around the Earth, trapping the sun's heat and raising temperatures (National Geographic, 2019a; IPCC, 2023; NASA, 2023a). Evidence shows that human activities have greatly contributed to the problem since the 1800s. The Earth is now about 1.1°C warmer than it was in the late 1800s. Alarmingly, the last decade (2011–2020) was the warmest on record.

Greenhouse gases include carbon dioxide and methane, and major sources of emissions include vehicles, energy and industrial facilities, buildings, agriculture, and waste disposal sites. Carbon dioxide is generated from burning fossil fuels, including coal, oil, and gas. Examples of contributing activities include using gasoline for driving and coal for heating purposes. Carbon dioxide is also generated through clearing land and forests. Methane is emitted from various human-generated and natural sources, including landfills, oil and natural gas systems, agricultural activities, coal mining, stationary and mobile combustion, wastewater treatment, and certain industrial processes (EPA, 2022a).

The rise in temperature seems small, but in reality, it causes catastrophic effects on the Earth as a system, because changes in one area can influence changes in all others. The main consequences of global warming are climate change, sea level rise, and direct and indirect ecological damage. All factors together can lead to a decline in biodiversity and the dominance of invasive species.

1.8.2 Climate Change

Climate change refers to long-term shifts in temperatures and weather patterns. These shifts may be natural, such as through variations in the solar cycle. However, since the 1800s, human activities, as described above, have been the main driver. Details of climate change processes and related serious effects are described in EPA (2023a), NOAA (2021), and UN (2023a). Problems include increases in extreme events, such as severe and prolonged droughts, devastating flooding, and catastrophic storms. Widespread fires occur due to droughts and shifting weather conditions. These problems negatively affect human health and safety and the ability to grow food.

In principle, solutions for climate change are possible, which can protect the environment and deliver economic benefits. Efforts include those by the United Nations (UN), which have developed global frameworks and agreements to guide progress in this regard. The broad categories of action are cutting emissions, mediating climate impacts, and securing the financial resources required for adjustments (UN, 2023a). Reducing emissions requires switching energy systems from fossil fuels to renewable sources, such as solar and wind. As documented in UN (2023a), a growing coalition of countries is committing to net-zero emissions by 2050. In this regard, about half of emission cuts must be in place by 2030 to keep warming below 1.5°C. In addition, fossil fuel production must decline by roughly 6% per year between 2020 and 2030.

More recently, the United Nations Climate Change Conference COP27 was held in Sharm el-Sheikh, Egypt, in November 2022 (UN, 2023b). Of interest is the fact that, for the first time, water was referenced in a COP outcome document, which emphasized the critical role water plays in climate change adaptation. The meeting reaffirmed the commitment to limit global temperature rise to 1.5°C above pre-industrial levels. Efforts to limit the temperature will require reductions in global greenhouse gas emissions of 43% by 2030 relative to the 2019 level. The meeting closed with a breakthrough agreement to provide "loss and damage" funding sourced from developed nations, the major emission contributors, to vulnerable countries greatly harmed by climate-related disasters. The outcome of COP27 also emphasized the need to strengthen universal coverage of early global climate warning systems.

1.8.3 Sea Level Rise

The rise in temperature is causing the melting of glaciers and ice sheets worldwide, adding water to the ocean and causing subsequent sea level rise (EPA, 2022b; NOAA, 2022; NASA, 2023b). Specific data about sea level rise compiled by NOAA (2022) indicated that the average global sea level rose 8–9 in (21–24 cm) since 1880, with an increased acceleration rate in recent years. In many locations along the U.S. coastline, the rate of local sea level rise is greater than the global average due to land subsidence caused by oil and groundwater pumping. The occurrence of high-tide flooding is more frequent and severe. Studies have projected an alarming average sea level rise for the contiguous U.S. of 2.2 m (7.2 ft) by 2100 and 3.9 m (13 ft) by 2150.

The sea level rise contributes to surface (marine) flooding and subsurface inundation (Rotzoll and Fletcher, 2012). Low-lying areas are subjected to surface flooding as water from the higher ocean level moves inland. In addition, subsurface inland flow occurs, causing the water table to rise and contributing to higher flooding levels. As documented by Nicholls (2011), more than 20 million people live below normal high-tide levels, and more than 200 million people are vulnerable to flooding during temporary sea level extremes. Most of these areas are urbanized, and rising seas can damage infrastructure, including roads, bridges, water supply facilities, oil and gas lines, power plants, and sewage treatment plants. Serious water contamination can occur due to flooding of waste disposal and treatment facilities as well as sewer line damages. In some rural areas, flooded on-site disposal systems, such as cesspools and septic systems, can also cause serious groundwater and ocean contamination. The quality of fresh groundwater aquifers will deteriorate due to more pronounced saltwater intrusions. Many of these aquifers are invaluable and necessary for sustaining municipal and agricultural water supplies and supporting natural ecosystems. Damages to coastal ecosystems are especially alarming as they constitute significant resources for recreation and habitat for fish and wildlife.

The sea level rise can also contribute to shoreline erosion and the occurrence of hazards from storms. Deadly and destructive storm surges are associated with sea level rise. Examples include Hurricane Katrina, Superstorm Sandy, and Hurricane Michael, which reached farther inland than normal. The likelihood of frequent high-tide flooding increases with sea level rise, which can be disruptive and lead to expensive mitigation costs.

People living in small island nations and other developing countries are more vulnerable to such problems, especially when drought is combined with sea level rise. For example, in some atolls, such as the Marshall Islands, ground elevation is generally only a few meters above sea level, and mitigation efforts, such as internal migration, are not feasible (UNU, 2020).

REFERENCES

AboutCivil. (2017). *Hydrology*. AboutCivil. Retrieved from https://www.aboutcivil.org/hydrology.html

AP News. (2017). New Zealand river's personhood status offers hope to Māori. Retrieved from https://apnews.com/article/religion-sacred-rivers-new-zealand-86d34a78f5fc662ccd554dd7f578d217

BBC. (2016). *The ancient Peruvian mystery solved from space* https://www.bbc.com/future/article/20160408-the-ancient-peruvian-mystery-solved-from-space.

BBC. (2022). *The Moorish invention that tamed Spain's mountains*. https://www.bbc.com/future/article/20221011-the-moorish-invention-that-tamed-spains-mountains

Bear, J. (1972). *Dynamics of Fluids in Porous Media*. New York: American Elsevier.

Big Ideas. (2021). Smart farm. Retrieved from https://bigideas.ucdavis.edu/sustainable-agriculture-smart-farm

Bras, R. L. (1999). A brief history of hydrology. *Bulletin of the American Meteorological Society,* *80*(6), 1151–1164. https://doi.org/10.1175/1520-0477-80.6.1151

Bredehoeft, J. D. (2007). It is the discharge. *Ground Water, 45*(5), 523. https://doi.org/10.1111/j.1745-6584.2007.00348.x

Britannica. (2023a). *Aqueduct Engineering.* https://www.britannica.com/technology/aqueduct-engineering.

Cartwright. (2014). *Nazca Civilization.* https://www.worldhistory.org/Nazca_Civilization/

Chow, V. T., Maidment, D. R., & Mays, L. W. (1988). *Applied Hydrology.* New York, NY: McGraw-Hill.

Climate Diplomacy. (2023a). Case studies. Retrieved from https://climate-diplomacy.org/case-studies?term=&filter%5bgeolocation%5d%5bcenter%5d=-22.1543751,64.8926452&widget%5bmap%5d%5bzoom%5d=2.211

Climate Diplomacy. (2023b). Turkey, Syria and Iraq: Conflict over the Euphrates-Tigris. Retrieved from https://climate-diplomacy.org/case-studies/turkey-syria-and-iraq-conflict-over-euphrates-tigris

Climate Diplomacy. (2023c). Dam projects and disputes in the Mekong River Basin. Retrieved from https://climate-diplomacy.org/case-studies/dam-projects-and-disputes-mekong-river-basin

Climate Diplomacy. (2023d). Grand Ethiopian Renaissance Dam. Retrieved from https://climate-diplomacy.org/case-studies/disputes-over-grand-ethiopian-renaissance-dam-gerd

CNN. (2022). Over 900 killed by Pakistan monsoon rains and floods, including 326 children. Retrieved from https://www.cnn.com/2022/08/24/asia/pakistan-floods-monsoon-rain-deaths-intl/index.html

CWRM. (2023). Instream flow standards. Retrieved from https://dlnr.hawaii.gov/cwrm/surfacewater/ifs/

Dmitrieva, N. I., Gagarin, A., Liu, D., Wu, C. O., & Boehm, M. (2023). Middle-age high normal serum sodium as a risk factor for accelerated biological aging, chronic diseases, and premature mortality. *EBioMedicine, 87,* 104404. https://doi.org/10.1016/j.ebiom.2022.104404

DTSC. (2022). Managing hazardous waste. Retrieved from https://dtsc.ca.gov/household-hazardous-waste/

Du, P., & Chen, H. (2007). Water supply of the cities in ancient China. *Water Supply 7*(1), 173–181. https://doi.org/10.2166/ws.2007.020.

EcoWatch. (2022). 50% of U.S. lakes and rivers are too polluted for swimming, fishing, drinking. Retrieved from https://www.ecowatch.com/us-lakes-and-rivers-polluted.html

Edgerton-Tarpley, K. (2008). *Tears from Iron: Cultural Responses to Famine in Nineteenth-Century China.* Oakland, CA: University of California Press.

Elshall, A. S., Castilla-Rho, J., El-Kadi, A. I., Holley, C., Mutongwizo, T., Sinclair, D., & Ye, M. (2022). Sustainability of groundwater. In: DellaSala, D. A., Goldstein, M. I. (eds.), *Imperiled: The Encyclopedia of Conservation,* pp. 157–166. Elsevier. https://researchprofiles.canberra.edu.au/en/publications/sustainability-of-groundwater

EPA. (2022a). Global methane initiative. Retrieved from https://www.epa.gov/gmi/importance-methane#:~:text=Methane%20is%20emitted%20from%20a,treatment%2C%20and%20certain%20industrial%20processes

EPA. (2022b). Climate change indicators: Sea level. Retrieved from https://www.epa.gov/climate-indicators/climate-change-indicators-sea-level

EPA. (2023). Our mission and what we do. Retrieved from https://www.epa.gov/aboutepa/our-mission-and-what-we-do

Fernald, A. G., Baker, T. T., & Guldan, S. J. (2007). Hydrological, riparian, and agroecosystem functions of traditional acequia irrigation systems. *Journal of Sustainable Agriculture, 30* (2), 147–71. https://doi.org/10.1300/J064v30n02_09

Fernald, A., Guldan, S., Cibils, A., Gonzales, M., Hurd, B., Lopez, S., Ochoa, C., Ortiz, M., Rivera, J., Rodriguez, S., & Steele, C. (2015). Linked hydrologic and social systems that support resilience of traditional irrigation communities. *Hydrology and Earth System Sciences, 19* (1), 293–307. https://doi.org/10.5194/hess-19-293-2015

Fetter, C. W. (1988). *Applied Hydrogeology* (2nd ed.). New York, NY: Macmillan Publishing Company

Finkelstein, J. J. (1962). Mesopotamia. *Journal of Near Eastern Studies, 21*(2), 73–92. https://www.jstor.org/stable/543884

Foreign Policy. (2023). Science shows Chinese dams are devastating the Mekong. Retrieved from https://foreignpolicy.com/2020/04/22/science-shows-chinese-dams-devastating-mekong-river/

Freeze, R. A., & Cherry, J. A. (1979). *Groundwater.* Englewood Cliffs, NJ: Prentice-Hall.

GAO. (2022). Hanford site cleanup. Retrieved from https://www.gao.gov/assets/gao-22-105809.pdf

Gassert, F., Reig, P., Luo, T., & Maddocks, A. (2013). Aqueduct country and river basin rankings: A weighted aggregation of spatially distinct hydrological indicators. Working paper. Washington, D.C.: World Resources Institute. Retrieved from https://www.wri.org/publication/aqueduct-country-river-basin-rankings

Gelhar, L. W. (1993). *Stochastic Subsurface Hydrology.* Englewood Cliffs, NJ: Wiley.

Gerbens-Leenes, W., Mekonnen, M. M., & Hoekstra, A. Y. (2013). Global water footprint of farm animal products. *Ecological Economics, 86,* 306–318. https://doi.org/10.1016/j.ecolecon.2012.08.002

Greenough, P. R. (1982). *Prosperity and Misery in Modern Bengal: The Famine of 1943–1944.* New York, NY: Oxford University Press.

Grove, R. H. (2007). The Great El Niño of 1789–93 and its global consequences: Reconstructing an extreme climate event in world environmental history. *The Medieval History Journal, 10*(1&2), 75–98. https://doi.org/10.1177/097194580701000203

Guerin-Yodice, S. (2021). Ancient Mesopotamia and the Fertile Crescent (Prehistory – ca 1750 BCE). In: Skjelver, D., Arnold, D., Broedel, H. P., Glasco, S. B., Kim, B. (eds.), *History of Applied Science & Technology: An Open Access Textbook,* pp. xx–xx. Grand Forks: The Digital Press. https://press.rebus.community/historyoftech

HDOH. (2023). Program directory. Retrieved from https://health.hawaii.gov/about/

Historic Mysteries. (2021). The Chinese famines of 1907 and 1959: Natural disasters or man-made? Retrieved from https://www.historicmysteries.com/chinese-famines/

Hoekstra, A. Y., & Mekonnen, M. M. (2012). The water footprint of humanity. *Proceedings of the National Academy of Sciences, 109*(9), 3232–3237. https://doi.org/10.1073/pnas.1109936109

HydroPoint. (2023). Modern methods of irrigation. Retrieved from https://www.hydropoint.com/blog/modern-methods-of-irrigation/

IBERDROLA. (2023). Smart farming, precision agriculture to achieve a more sustainable world. Retrieved from https://www.iberdrola.com/innovation/smart-farming-precision-agriculture#:~:text=Smart%20farming%20is%20about%20using,and%20minimising%20the%20environmental%20impact

International Association of Hydrological Sciences. (2019). *Hydrology History.* IAHS. Retrieved from https://iahs.info/About-IAHS/History-of-IAHS/

IGWMC. (2023). Integrated ground water modeling center. Retrieved from *https://igwmc.princeton.edu/*

IPCC. (2023). Global warming of 1.5°C. Retrieved from https://www.ipcc.ch/sr15/

Jasim, H. (2023). The Nazca culture built these incredible aqueducts in the Peruvian desert 1,500 years ago, and they are still in use today. Retrieved from https://hasanjasim.online

Johnson, R., & Cody, B. A. (2015). California agricultural production and irrigated water use. Retrieved from https://sgp.fas.org/crs/misc/R44093.pdf

Khan Academy. (2023). Cohesion of water. Retrieved from https://www.khanacademy.org/science/biology/water-acids-and-bases/cohesion-and-adhesion/a/cohesion-and-adhesion-in-water

Khan, S. (2014). Sanitation and wastewater technologies in Harappa/Indus valley civilization (ca. 2600–1900 BC). In: Angelakis, A. N., Rose, J. B. (eds.), *Evolution of Sanitation and Wastewater Technologies Through the Centuries*, pp. xx–xx. London: IWA Publishing Alliance House. Retrieved from https://www.academia.edu/26736892/Evolution_of_Sanitation_and_Wastewater_Technologies_through_the_Centuries

Khesbak, H., Savchuk, O., Tsushima, S., & Fahmy, K. (2011). The role of water H-bond imbalances in B-DNA substate transitions and peptide recognition revealed by time-resolved FTIR spectroscopy. *Journal of the American Chemical Society, 133*(15), 5834. https://doi.org/10.1021/ja108863v

Khoury, W., & Gallisdorfer, M. S. (2020). A journey through time: How ancient water systems inspired today's water technologies. Retrieved from https://smartwatermagazine.com

Kumar, D. (ed.) (1983). *The Cambridge Economic History of India*, vol. 2. Cambridge, UK: Cambridge University Press.

LibreTexts. (2023). Precipitation reactions. Retrieved from https://chem.libretexts.org/Bookshelves/Inorganic_Chemistry/Supplemental_Modules_and_Websites_(Inorganic_Chemistry)/Descriptive_Chemistry/Main_Group_Reactions/Reactions_in_Aqueous_Solutions/Precipitation_Reactions

Los Angeles County Public Works. (2023). Stormwater engineering projects. Retrieved from https://pw.lacounty.gov/wrd/Projects/TujungaSG/index.cfm

Luo, T., Young, R., & Reig, P. (2015). Aqueduct projected water stress country rankings. Technical Note. Washington, D.C.: World Resources Institute. Retrieved from https://files.wri.org/d8/s3fs-public/aqueduct-water-stress-country-rankings-technical-note.pdf

Mansergh, N., & Lumby, E. W. R. (eds.). (1973). *The Transfer of Power 1942–7, Vol. IV: The Bengal Famine and the New Viceroyalty, 15 June 1943–31 August 1944*. London: H.M.S.O.

MayoClinic. (2023). Dehydration. Retrieved from https://www.mayoclinic.org/diseases-conditions/dehydration/symptoms-causes/syc-20354086

McDonald, M. G., & Harbaugh, A. W. (1988). *A Modular Three-Dimensional Finite-Difference Groundwater Flow Model*. U.S. Geological Survey. https://doi.org/10.3133/twri06A1

NASA. (2023a). Global warming vs. climate change. Retrieved from https://climate.nasa.gov/global-warming-vs-climate-change/

NASA. (2023b). Climate change: Vital signs. Retrieved from https://climate.nasa.gov/vital-signs/sea-level/

National Academy of Medicine. (2004). *Dietary Reference Intakes for Water, Potassium, Sodium, Chloride, and Sulfate*. The National Academies Press. Retrieved from https://nap.nationalacademies.org/read/10925/chapter/1

National Geographic. (2019a). What is global warming, explained. Retrieved from https://www.nationalgeographic.com/environment/article/global-warming-overview?loggedin=true&rnd=1681072506703

National Geographic. (2019b). Centuries-old irrigation system shows how to manage scarce water. Retrieved from https://www.nationalgeographic.com/environment/article/acequias

National Geographic. (2023). Roman aqueducts. Retrieved from https://education.nationalgeographic.org/resource/roman-aqueducts/

Nicholls, R. J. (2011). Planning for the impacts of sea level rise. *Oceanography, 24,* 144–157. https://doi.org/10.5670/oceanog.2011.34

NIH. (2023). DNA. Retrieved from https://www.cancer.gov/publications/dictionaries/genetics-dictionary/def/dna

NOAA. (2021). Climate change impacts. Retrieved from https://www.noaa.gov/education/resource-collections/climate/climate-change-impacts#:~:text=The%20compounding%20effects%20of%20climate,to%20be%20smothered%20by%20sediment

NOAA. (2022). Climate change: Global sea level. Retrieved from https://www.climate.gov/news-features/understanding-climate/climate-change-global-sea-level

Oki, D. S. (2005). Numerical simulation of the effects of low-permeability valley-fill barriers and the redistribution of ground-water withdrawals in the Pearl Harbor area, Oahu, Hawaii. U.S. Geological Survey Scientific Investigations Report 2005–5253. https://doi.org/10.3133/sir20055253

Owlcation. (2023). The top 10 worst famines in history. Retrieved from https://owlcation.com/humanities/The-Top-10-Worst-Famines-in-History

Petersen-Perlman, J. D., Veilleux, J. C., & Wolf, A. T. (2017). International water conflict and cooperation: Challenges and opportunities. *Water International, 42*(2), 105–120. https://doi.org/10.1080/02508060.2017.1276041

PMC. (2023). The recharge initiative. Retrieved from https://websites.pmc.ucsc.edu/~afisher/RechargeInitiative/index.htm

Romeo, B. (2016). Ancient device for determining taxes discovered in Egypt. Retrieved from https://www.nationalgeographic.com/history/article/160517-nilometer-discovered-ancient-egypt-nile-river-archaeology

Resource Center. (2023). Acids, bases & water chemistry 101: Acids, bases & water science lesson. Retrieved from https://learning-center.homesciencetools.com/article/acid-base-water-science-lesson/

Riveros-Perez, E., & Riveros, R. (2017). Water in the human body: An anesthesiologist's perspective on the connection between physicochemical properties of water and physiologic relevance. *Annals of Medicine and Surgery (Lond), 26*, 1–8. https://doi.org/10.1016/j.amsu.2017.12.007

Rosbjerg, D., & Rodda, J. (2019). IAHS: A brief history of hydrology. *History of Geo- and Space Sciences, 10*, 109–118. https://doi.org/10.5194/hgss-10-109-2019

Sciencing. (2023). How does water stabilize temperature? Retrieved from https://sciencing.com/water-stabilize-temperature-4574008.html

Roussi, A. (2019). Gigantic Nile dam prompts clash between Egypt and Ethiopia. *Nature, 574*, 159–160. https://doi.org/10.1038/d41586-019-02987-6

Rotzoll, K., & Fletcher, C. (2012). Assessment of groundwater inundation by sea level rise. *Nature Climate Change, 3*, 477–481. https://doi.org/10.1038/NCLIMATE1725

Salguero, C. P. (2017). Cultural associations of water in early Chinese and Indian religion and medicine. *Water and Asia, 22*(2), xx–xx. Retrieved from https://www.asianstudies.org

San Diego County. (2019). Green parking lots guidelines. Retrieved from https://www.sandiegocounty.gov/content/dam/sdc/dpw/WATERSHED_PROTECTION_PROGRAM/watershed-pdf/Dev_Sup/GPL_Guidelines_2019.pdf

Sharan, S., & Pathak, V. (2017). Concepts of Panchamahabhut at elemental level. *World Journal of Pharmaceutical and Medical Research, 3*(7), 80–89. Retrieved from https://www.wjpmr.com/download/article/24072017/1501485250.pdf

Shaw, I. (Ed.). (2003). *The Oxford History of Ancient Egypt.* Oxford: Oxford University Press.

Smith, M. K., & Pai, M. (1992). The Ahupua'a concept: Relearning coastal resource management from ancient Hawaiians. *Naga, The WorldFish Center, 15*(2), 11–13. https://ideas.repec.org/a/wfi/wfnaga/27697.html

Sophocleous, M. (2000). From safe yield to sustainable development of water resources: The Kansas experience. *Journal of Hydrology, 235*(1–2), 27–43. https://www.sciencedirect.com/science/article/abs/pii/S0022169400002638?via%3Dihub

The Guardian. (2022). US west 'megadrought' is worst in at least 1,200 years, new study says. Retrieved from https://www.theguardian.com/environment/2022/feb/15/us-west-megadrought-worst-1200-years-study

Theis, C. V. (1935). The relation between the lowering of the piezometric surface and the rate and duration of discharge of well using groundwater storage. *American Geophysical Union Transactions, 16*, 519–524. https://doi.org/10.1029/TR016i002p00519

United Nations. (2023a). What is climate change? Retrieved from https://www.un.org/en/climatechange/what-is-climate-change

United Nations. (2023b). COP 27. Retrieved from https://www.un.org/en/climatechange/cop27

United Nations University. (2020). The Marshall Islands: Connecting climate change and migration. Retrieved from https://ehs.unu.edu/news/news/the-marshall-islands-connecting-climate-change-and-migration.html

USGS. (2018). Aqueducts move water in the past and today. Retrieved from https://www.usgs.gov/special-topics/water-science-school/science/aqueducts-move-water-past-and-today#:~:text=There%20is%20even%20a%20Roman,Some%20content%20may%20have%20restrictions

Water Footprint Calculator. (2023). Water footprint comparisons by country. Retrieved from https://www.watercalculator.org/footprint/water-footprints-by-country/

Wikibooks. (2023). Structural biochemistry/water. Retrieved from https://en.wikibooks.org/wiki/Structural_Biochemistry/Water#:~:text=It%20consists%20of%20an%20oxygen,Water%2C%20the%20universal%20solvent

Williams, A. P., Cook, B. I., & Smerdon, J. E. (2022). Rapid intensification of the emerging southwestern North American megadrought in 2020–2021. *Nature Climate Change, 12*, 232–234. https://doi.org/10.1038/s41558-022-01290-z

Zheng, X. Y., & Angelakis, A. N. (2018). Chinese and Greek ancient urban hydro-technologies: Similarities and differences. *Water Supply, 18*(6), 2208–2223. https://doi.org/10.2166/ws.2018.038

Units and a Mathematical Refresher

2

Hydrology is fundamentally a quantitative discipline, with its practical applications focused on aiding decision-makers in the effective management of water resources. This entails evaluating and quantifying key decision variables, such as the impact of human activities on water resources and the status of contaminants, with or without remediation efforts.

In general, an advanced level of mathematical formulation is required for a more accurate representation of hydrological processes. However, in keeping with this book's intended audience, we typically use mathematical equations with simplifying assumptions. Examples are provided to help students grasp and apply the concepts effectively. Additionally, Appendix A covers the basics of spreadsheets, essential tools for performing repetitive calculations and creating informative graphical representations.

To ensure the accurate application of governing equations, this chapter reviews the fundamentals of units, emphasizing the critical importance of correct unit usage, as errors can have significant consequences. For example, on July 30, 1983, The New York Times reported the headline, "Jets Fuel Ran Out After Metric Conversion Errors." In this incident, an Air Canada Boeing 767 ran out of fuel mid-flight due to errors in calculating the fuel supply, stemming from the airline's transition to different measurement units for the aircraft (The New York Times, 1983). Fortunately, the aircraft was able to glide to a safe landing. Additionally, The Los Angeles Times published the headline, "Mars Probe Lost Due to Simple Math Error," on October 1, 1999. In this case, NASA lost its $125 million Mars Climate Orbiter because spacecraft engineers failed to convert vital data to the correct units before the craft's launch (The Los Angeles Times, 1999).

The chapter also includes a mathematical refresher and definitions of basic statistical terms. Finally, it introduces regression, least squares fitting, and an example application, which are widely used in the field of hydrology.

DOI: 10.1201/9781003587149-2

29

2.2 MEASUREMENT UNITS

2.2.1 Basic Units

The fundamental measurement units encompass length, mass, and time, denoted by the symbols L, M, and T, respectively. These units fall into two categories: the English (or imperial) system and the metric system. Currently, only three countries in the world officially employ the English system: the United States of America, Myanmar, and Liberia. Nevertheless, both Myanmar and Liberia have adopted the metric system alongside the English system and are in the process of transitioning fully to the metric system. The United Kingdom occupies an intermediate position between the two systems, where the metric system is the official standard, but people still commonly use the English system for expressing measurements of length or speed.

For the metric system, the basic units consist of the meter (m), kilogram (kg), and second (s), representing length, mass, and time, respectively. Conversely, the parallel units for the English system are the foot (ft), slug, and second (s), which, respectively, correspond to length, mass, and time. A comprehensive list of these basic units is presented in Table 2.1.

2.2.2 Derived Units

Units for other quantities can be derived by utilizing the fundamental units, as outlined in Table 2.2. The "In words" column elucidates the correlation between each quantity and the corresponding basic units. The final two columns specify the units that should be uniformly employed within the respective system. Nevertheless, conversions within each system are also permissible; for instance, utilizing grams instead of kilograms or days instead of seconds is acceptable, provided that the chosen units remain consistent.

It should be noted that the kilogram is commonly used as a unit of measurement in shopping markets to represent weight, which can be misleading since the kilogram is actually a unit of mass. In reality, weighing scales are calibrated to display the mass of an object in kilograms, which, in the context of physics, is equivalent to measuring the weight in newtons. Therefore, for greater accuracy, a package labeled as 1.0 kg should

TABLE 2.1   Basic units for the two systems

VARIABLES	METRIC	ENGLISH
Length (L)	Meter (m)	Feet (ft)
Mass (M)	Kilogram (km)	Slug
Time (T)	Second (s)	Second (s)

TABLE 2.2   Derived units

VARIABLE	IN WORDS	METRIC SYSTEM	ENGLISH SYSTEM
Area (L^2)	Length squared	m^2	ft^2
Volume (L^3)	Length cubed	m^3	ft^3
Velocity (L/T)	Distance per unit time	m/s	ft/s
Water discharge (L^3/T)	Volume per unit time	m^3/s	ft^3/s
Density (M/L^3)	Mass per unit volume	kg/m^3	Slug/ ft^3
Acceleration	Distance per time squared	m/s^2	ft/s^2
Force or weight (ML/T^2)	Mass times acceleration	Newton (kg m/s^2)	lb (pound)
Pressure ($ML/T^2/ L^2$) $=(M/L\ T^2)$	Force divided by area	Newton/m^2	lb/ ft^2
Viscosity ($MLT/T^2/ M^2)=(L/MT)$	Force-second per square meter	Newton-sec/m^2 (pascal-s)	lb-s/ft^2

ideally be labeled as 1.0 newton or 9.81 kg-m per second squared, with 9.81 representing the acceleration due to gravity.

Conversions within the metric system are detailed in Tables 2.3–2.5. For instance, when converting from kilometers (km) to meters (m), centimeters (cm), or millimeters (mm), you simply multiply by the respective factors: 1,000, 100,000, or 1,000,000, respectively. In scientific notation, these numbers can be written as 10^3, 10^5, and 10^6, while in Excel and other computer software, they are expressed as 1E+3, 1E+5, and 1+6. Conversely, when converting from millimeters to centimeters, meters, or kilometers, you multiply by the reciprocal factors: 0.1, 0.001, or 0.000001, respectively, or 1^{-1}, 1^{-3}, and 1^{-6}, and 1E-1, 1E-3, and E-6. To avoid confusion, it's important to recognize that converting from kilometers to other units should result in larger values, while converting from meters or centimeters to kilometers should yield smaller values.

The conversion between the two systems involves specific conversion factors: 0.453592 or about 0.454 for pounds to kilograms and 0.305 for feet to meters.

In hydrology, one frequently encountered unit for measuring volumes is gallons. The US gallon can be converted to cubic meters by multiplying it by 0.00378541. Alternatively, it can be converted to cubic feet or liters by multiplying it by 0.133681 and 3.78541, respectively. (Note that one cubic meter is equivalent to 1,000L.)

Example:

Given $K=0.25$ ft/s, $A=2,000\,ft^2$, $S^1=0.0001$ (dimensionless), estimate the discharge in m^3/day, by using equation (2.1):

$$Q = KAS \tag{2.1}$$

We can either convert all units to metric or then apply the equation:

$$K = 0.25 \text{ ft/s} = 0.25 \times 0.305 \times 86,400 = 6,588 \text{ m/day}$$

$$A = 2,000 \times 0.305 \times 0.305 = 186.05 \text{ m}^2$$

TABLE 2.3   Conversion factors for length (metric units)

	KM	M	CM	MM
km	1	0.001	0.00001	0.000001
m	1,000	1	0.01	0.001
cm	100,000	100	1	0.1
mm	1,000,000	1,000	10	1

TABLE 2.4   Conversion factors for mass (metric units)

	KG	GM	MILLIGRAM
kg	1	0.001	0.000001
gm	1,000	1	0.001
milligram	1,000,000	1,000	1

TABLE 2.5   Conversion factors for time (metric or English units)

	DAY	HOUR	MINUTE	SECOND
Day	1	0.041667	0.000694	1.15741E-05
Hour	24	1	0.016667	0.000277778
Minute	1,440	60	1	0.016666667
Second	86,400	3,600	60	1

In which "×" is the multiplication operator. Note that an area requires conversion by using a factor of $0.305 \times 0.305 = 0.305^2$ (for example, if an area is a square or rectangular in shape, each side needs to be converted to meters).

So, the answer would be

$$Q = KAS = 6{,}588 \frac{\text{m}}{\text{day}} \times 186.05 \text{ m}^2 \times 0.0001 = 122.6 \frac{\text{m}^3}{\text{day}}$$

Another option would be to calculate Q in ft³/s and then convert the final answer to m³/day:

$$Q = KAS = 0.25 \frac{\text{ft}}{\text{s}} \times 2{,}000 \text{ ft}^2 \times 0.0001 = 0.05 \frac{\text{ft}^3}{\text{s}}$$

Converting to m³/day provides the same answer for Q:

$$Q = 0.05 \times 0.305^3 \times 86{,}400 = 122.6 \frac{\text{m}^3}{\text{day}}$$

Note that Q has units of cubic feet or cubic meters (per day), and hence the conversion factor should be $0.305 \times 0.305 \times 0.305$ or 0.305^3.

Finally, online unit conversion calculators are available, which greatly simplify the effort, e.g., UnitConverter.Net (2023). Note that the difference in the answer utilizing this calculator and the above calculations relates to the precision of the conversion factor from feet to meters, which should be 0.3048, but was rounded here to 0.305.

2.3 MATHEMATICAL REFRESHER

2.3.1 Equations and Functions

Equation (2.2) contains a number of symbols, specifically K, k, ρ, g, and μ, representing what we call variables of interest:

$$K = \frac{k\rho g}{\mu} \qquad (2.2)$$

This and any other equation is written in terms of chosen symbols; any letter can be used for a specific variable, including Latin or Greek letters. The equation states that K is estimated by multiplying k by ρ, then by g, and dividing the product by μ. For example, K would be 9.8 for $k=2$, $\rho=1$, $g=9.8$, and $\mu=2$:

$$K = \frac{2 \times 1 \times 9.8}{2} = 9.8$$

Again, we use $\times$ to indicate multiplication, and it is not a variable's name.

The calculations are facilitated by using a calculator, which is available on cell phones. The equation can be rearranged (as shown in equation 2.3) to estimate other variables in the equation, such as k, given K, ρ, g, and μ:

$$k = \frac{K\mu}{\rho g} \qquad (2.3)$$

In the examples presented in the following chapters, we have included units in the applications of the equations. As previously emphasized, maintaining unit consistency is crucial. For instance, consider equation (2.1): if variable A is measured in square meters, K in meters per day, and S is dimensionless (with no units), then Q in this equation would be expressed in cubic meters per day:

$$Q = AKS = 5\,\text{m}^2 \times 650\frac{\text{m}}{\text{day}} \times 0.001 = 3.25\,\frac{\text{m}^3}{\text{day}}$$

Care should be taken in completing the calculations in an appropriate order. For example, for equation (2.4) in the following form:

$$Q = \frac{1}{3}\left(\frac{V1}{t1} + \frac{V2}{t2} + \frac{V3}{t3}\right) = \frac{1}{3}\left(\frac{4}{2} + \frac{5}{3} + \frac{1}{2}\right) = \frac{1}{3}(2 + 1.67 + 0.5) = \frac{1}{3}(4.17) = 1.39 \quad (2.4)$$

The expressions within the parentheses should be evaluated first, followed by dividing the result by 3. It's worth noting that two symbols are used for each of the variables $V1$, $V2$, and $V3$, which is acceptable.

Some of the equations include terms expressed by using a power function, as shown in equation (2.5):

$$Q = 1.84(L - 0.2H)H^{2/3} = 1.84 \times (3 - 0.2 \times 1.5) \times 1.5^{2/3} = 6.51 \text{ m}^3/\text{s} \quad (2.5)$$

In this equation, the variable H should be raised to the power of 2/3. The calculations shown above are based on the values $L = 3$ m and $H = 1.5$ m. To evaluate $H^{(2/3)}$ using a scientific calculator (Figure 2.1), follow these steps:

1. Input the value 1.5 for H.
2. Press the "x^y" button on the calculator.
3. Input the approximate value 0.67, which represents 2/3 (obtained by dividing 2 by 3):

This will give you the result of $H^{(2/3)}$, which is about 1.3.

In equation (2.6), another expression utilizes the natural logarithmic function (ln). Once again, it is important to evaluate the expressions within the parentheses first.

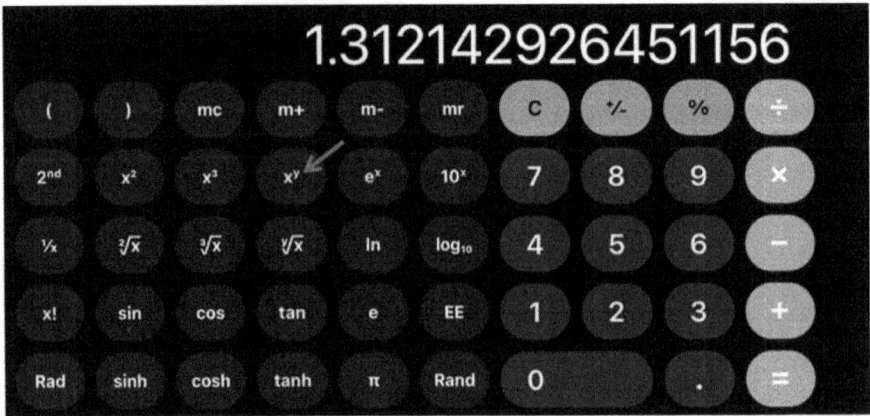

FIGURE 2.1 Using a scientific calculator to calculate power functions.

$$K = \frac{Q}{2\pi b (h2 - h1)} \ln\left(\frac{r2}{r1}\right)$$

$$= \frac{400 \; \dfrac{m^3}{h}}{2 \times 3.14 \times 40 \; m \times (89.6 \; m - 85.3 \; m)} \ln\left(\frac{75 \; m}{25 \; m}\right) = 0.41 \; \frac{m}{h} \qquad (2.6)$$

The argument of ln is r2/r1, which is 75 divided by 25. Therefore, ln(75/25) equals ln(3). To calculate this, you can use a scientific calculator by entering the number 3 and then pressing the ln key, as illustrated in Figure 2.2. The calculated value is approximately 1.1.

Finally, certain equations, often referred to as empirical equations, are formulated with specific units in mind. Consequently, calculations should be performed using these specific units, and the resulting answer should then be converted to the desired units.

2.3.2 Derivatives

In hydrology, a variable such as water level h in a stream or an aquifer can be expressed as a function of distance x, denoted as $h = h(x)$. The derivative measures how this function changes as the value of x changes. In general, x is called an independent variable, while h is called a dependent variable. The derivative of the function $h(x)$ with respect to the variable x is defined by equation (2.7):

$$f'(x) = \lim_{\Delta x \to 0} \frac{h(x + \Delta x) - h(x)}{\Delta x} \qquad (2.7)$$

In this equation, $h(x + \Delta x)$ represents the value at an increment Δx from the original value x. The limit, as Δx approaches zero, represents the rate of change of the function

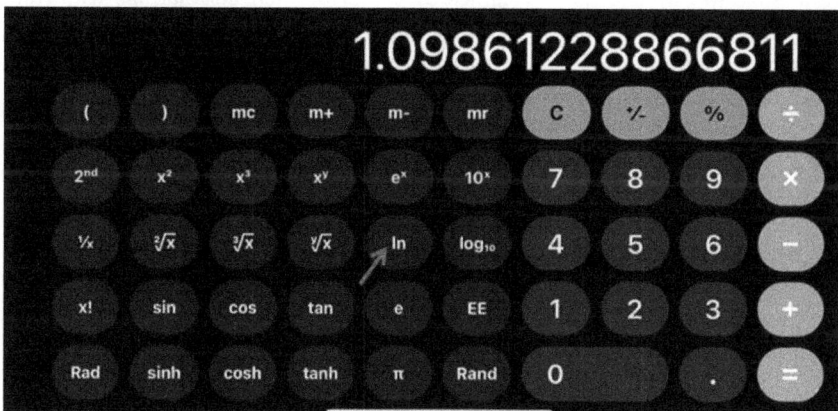

FIGURE 2.2 Using a scientific calculator to calculate the logarithm of a number.

$h(x)$ at any point x, which corresponds to the slope of the tangent line to the curve of the function at that point. This describes how steep the curve is at that specific location.

An example of the use of derivatives in hydrology is Darcy's law (Section 6.3), which describes groundwater flow. The derivative of the function $h(x)$ with respect to distance x determines the amount of water flow through a given area with a certain conductivity (or material's ability to transmit water). A larger derivative, or slope of the tangent, indicates a greater flow rate.

Partial derivatives are used in cases involving functions of more than one variable, such as $h(x,t)$, where t refers to time. In this case, h is a function of both distance x and time t. As shown in equation (2.8), the partial derivative with respect to one of the variables, such as x, is found while keeping the other variable constant, which is t in this case:

$$f'(x) = \lim_{\Delta x \to 0} \frac{h(x + \Delta x, t) - h(x,t)}{\Delta x} \tag{2.8}$$

2.3.3 Definitions of basic statistical terms

Statistical techniques, including frequency and regression analyses, prove invaluable for examining and interpreting hydrological data encompassing rainfall patterns, river flows, and contaminant levels. Frequency analysis stands out as a commonly employed tool for designing dams and other flood protection structures. This design process hinges on assessing the probability of extreme floods and ensuring the dam's capability to withstand such events. To achieve this, historical data pertaining to river flows and rainfall patterns serve as foundational elements for conducting these analyses.

Regression analysis, which involves establishing relationships among two or more hydrologic variables, can help circumvent the need for labor-intensive and costly measurements. For instance, quantifying the volumes of water flowing across the land surface during rainstorms poses significant practical challenges. However, it becomes more feasible to estimate these volumes by exploring their correlations with rainfall intensity and land characteristics. By analyzing historical data, it becomes possible to construct regression models that predict these volumes based on these key variables. These models also prove beneficial for assessing the potential impacts of changes in land use and climate.

The following sections provide definitions and example calculations for some basic statistical terms.

1. **Mean (μ):** The mean describes the average result of a test, survey, or experiment. It is estimated by adding all the values X_i, and dividing the sum by their number N as shown in equation (2.9):

$$\mu = \frac{\Sigma X_i}{N} \tag{2.9}$$

where the Greek symbol Σ refers to the addition process: $\Sigma X_i = X_1 + X_2 + \cdots + X_N$

- Example: Table 2.6 lists a sample of nine values ($N=9$), such as water levels in a stream, in feet. The mean in feet is calculated by adding the nine values and dividing the result by 9, which is 3.75.
2. **Variance (Var) and Standard Deviation (σ)**: These represent a measure of how dispersed the data is in relation to the mean. Data are clustered around the mean for small values of Var and σ. In contrast, larger values indicate that the data are more spread out. Var and σ are respectively estimated by equations (2.10) and (2.11):

$$\mathrm{Var} = \frac{\Sigma(\mu - X_i)^2}{N-1} \qquad (2.10)$$

$$\sigma = \sqrt{\mathrm{Var}} \qquad (2.11)$$

Again, with the Greek symbol Σ referring to the addition process of all the terms.
3. **Coefficient of Variation (CV)**: The coefficient of variation (CV) is a statistical measure of the dispersion of data points in a data series around the mean. CV represents the ratio of the standard deviation to the mean, and it is a useful statistic for comparing the degree of variation from one data series to another, even if the means differ significantly from one to another. CV is estimated by equation (2.12):

$$\mathrm{CV} = \frac{\sigma}{\mu} \qquad (2.12)$$

- Example: Table 2.6 lists the steps used to calculate the variance, the standard deviation, and the coefficient of variation.

TABLE 2.6   Calculating the mean, variance, standard deviation, and coefficient of variation of a sample

X	μ	$(X - \mu)$	$(X - \mu)^2$	$\Sigma (X - \mu)^2$	VAR	σ	CV
3.80	3.75	0.05	0.0025	1.9600	0.2450	0.4950	0.1320
3.85		0.10	0.0100				
4.00		0.25	0.0625				
4.05		0.30	0.0900				
4.55		0.80	0.6400				
2.80		−0.95	0.9025				
3.25		−0.50	0.2500				
3.70		−0.05	0.0025				
3.75		0.00	0.0000				

4. Median: The median is defined as the value of the variable that divides the results in half, or simply the middle value of the data set. The first step is to list the data in increasing order. The estimation will vary based on whether the data set is of has an odd or even number of values.

- Example: Odd set of numbers: For the data listed in Table 2.7, the median is 3.8 ft (the number in the middle).
- Example: Even set of numbers: In Table 2.8, the median value is estimated by adding the two middle numbers, 3.8 and 3.85, and dividing the result by 2, which is 3.825.

5. Mode: The mode is defined as the most common result (or the most frequent value) of a test, survey, or experiment. There is no defined mode value for the data listed in Tables 2.7 and 2.8, considering that there are no repeated values.

TABLE 2.7 Calculating the median of an odd amount of numbers

ORDER	X
1	2.80
2	3.25
3	3.70
4	3.75
5	3.80
6	3.85
7	4.00
8	4.05
9	4.55

TABLE 2.8 Calculating the median of an even amount of numbers

ORDER	X
1	2.80
2	3.25
3	3.70
4	3.75
5	3.80
6	3.85
7	4.00
8	4.05
9	4.55
10	5.00

Spreadsheets, such as Excel, have many statistical functions to directly estimate various quantities. Figures 2.3–2.5 illustrate examples to respectively estimate the mean, the median, and the standard deviation. In the shaded

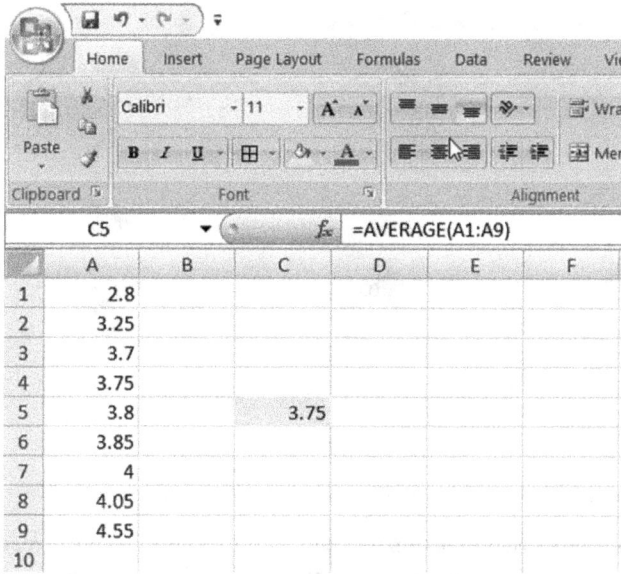

FIGURE 2.3 Calculating the mean using Excel.

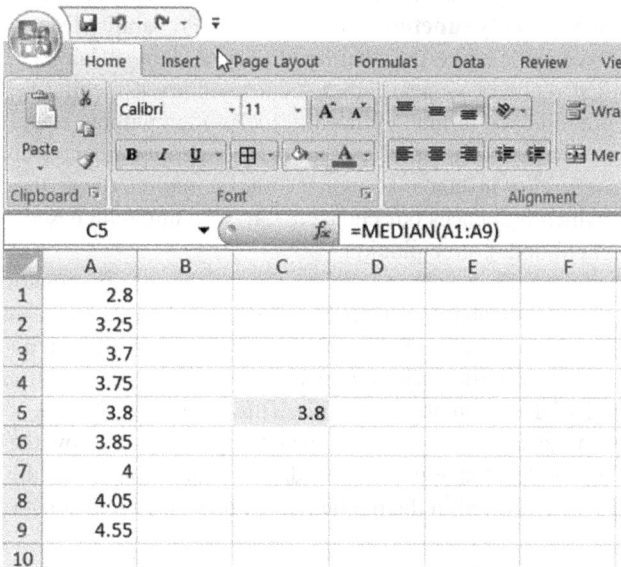

FIGURE 2.4 Calculating the median using Excel.

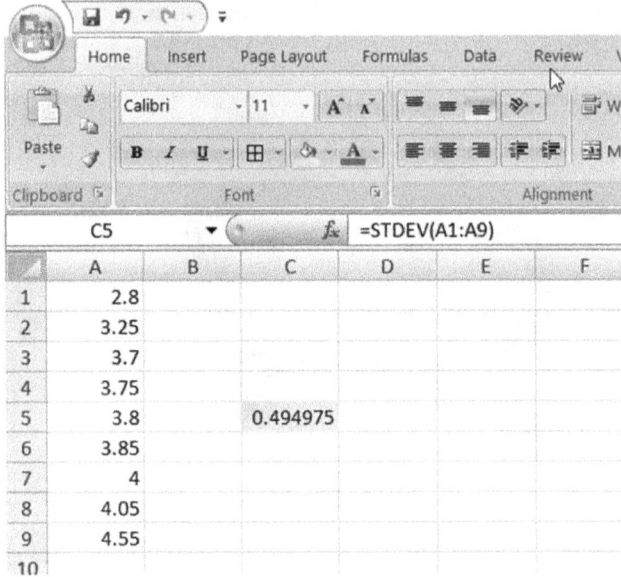

FIGURE 2.5 Calculating the standard deviation using Excel.

cell, the appropriate function is typed in following an equal sign and then the desired range of data values. As seen in the figures, the respective functions are =AVERAGE(A1:A9), =MEDIAN(A1:A9), and =STDEV(A1:A9). There is no need to sort the data for estimating the median. The mode, which is not displayed in the figures, would be evaluated by the function =MODE(A1:A9).

6. **Probability density function (PDF)**: A PDF (Probability Density Function) is employed to describe the likelihood of a random variable falling within a specified range of values, rather than assuming a specific value. We all utilize probabilistic terms in our daily lives, often without even realizing it, such as when planning for rainstorms or making investment decisions in the stock market. For instance, a weather forecast might indicate the probability of rain, which influences our choice to carry an umbrella based on the forecasted likelihood of precipitation.

In addition to its role in data analysis, PDFs can be employed to construct synthetic databases that represent missing data, such as rainfall values over a specific time period. Among various types of PDFs, a normally distributed PDF is particularly attractive because it relies solely on the mean and standard deviation of the variable. This type of distribution is used when there is an equal likelihood of the data being above or below the mean. For a mean value denoted as μ and a standard deviation represented by σ, such a PDF can be expressed in the mathematical form shown in equation (2.13).

$$f(X) = \frac{1}{\sigma\sqrt{2\pi}}e^{\left(-\frac{(X-\mu)^2}{2\sigma^2}\right)}$$

(2.13)

Figure 2.6 illustrates plots of Probability Density Functions (PDFs) for specific mean values ($\mu=3$ and 1.5) and their corresponding standard deviations ($\sigma=1$ and 2). As depicted in the figure, the standard deviation plays a crucial role in determining the spread of the data in relation to the mean

Moving on to Figure 2.7, it demonstrates how PDFs can be employed to estimate the probability of selecting a value X that satisfies particular conditions. For instance, the PDF can be utilized to determine the probability denoted as $P(X<4)$ for a value $X<4$. Alternatively, one might seek to ascertain the probability, denoted as $P(2.5<X<4.2)$, for a value of X falling within the range of 2.5 and 4.2. The solution lies in the shaded areas in Figure 2.7, which can be quantified by integrating the PDF within the respective limits: from $-\infty$ to 4 for the first case, and from 2.5 to 4.2 for the second. It is important to note that the lower limit for the first integral is set to negative infinity, as indicated by the PDF equation provided above, where the PDF reaches a value of zero.

For a more comprehensive understanding and detailed calculation methods for probabilities, particularly for normally distributed PDFs, the Probability Course (2023) offers an in-depth discussion and calculation techniques.

Figure 2.8 illustrates the usage of the probability (PROB) function in Excel. This function calculates the probability of a value, denoted as X, falling within specified upper and lower limits. In the example presented in Figure 2.8, cells A1–A5 represent a range of numeric values for X, while their corresponding probabilities are listed in cells B1–B5. In this case, the lower limit for X is set at 3, and the upper limit is set at 7. The resulting probability for X falling within this range is 0.57.

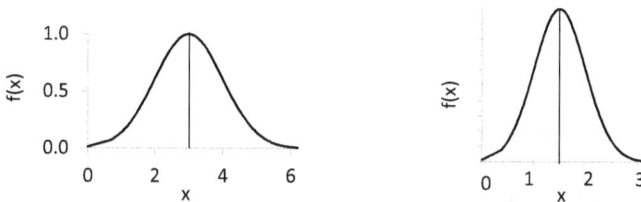

FIGURE 2.6 Comparing PDFs with different means and standard deviations.

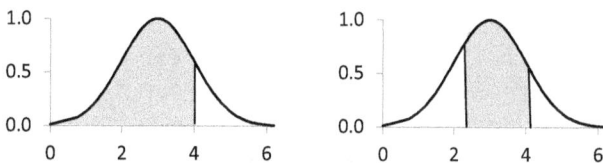

FIGURE 2.7 Shaded areas represent probability P for X that is less than 4 [$P(X<4)$] (left plot), and the probability P for X falling between 2.5 and 4.2 [$P(2.5<X<4.2)$] (right plot).

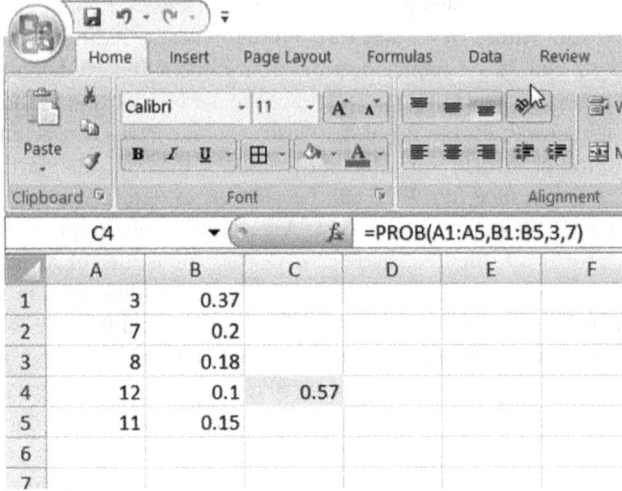

FIGURE 2.8 Using the PROB function in Excel.

It is worth noting that if you omit the upper limit when using the PROB function, it will calculate the probability of the value in the range equaling the specified lower limit.

7. **Creating charts and trendlines in Excel**:

The following are the steps used to plot a data set and the related trendline.

1. Select the data you want to include in the chart by clicking and dragging over the cells.
2. Go to the "Insert" tab in the Excel ribbon.
3. Click on the "Scatter" chart button.
4. Right-click on any data point in the chart and select "Add Trendline" from the context menu.
5. In the "Add Trendline" dialog box, choose the desired type of trendline (linear, exponential, polynomial, etc.).
6. Customize the chart and trendline as needed, such as displaying the equation or R-squared value on the chart.

Figures 2.9 and 2.10 depict charts illustrating an example dataset featuring linear and polynomial trendlines, respectively. These data pertain to the relationship between river stage (water level height) and river discharge. The charts have been customized by including minor gridlines, axis titles, trendline equations, and R-squared values. In the equations, the variables y and x represent discharge and stage, respectively. R-squared, also known as the coefficient of determination, serves as a statistical measure to evaluate the goodness of fit of the trendline. R-squared values range from 0 to 1, with higher values indicating a stronger fit of the line to the data. The trendline equation proves particularly useful for estimating discharge based on a known stage, which is comparatively easier to measure than discharge.

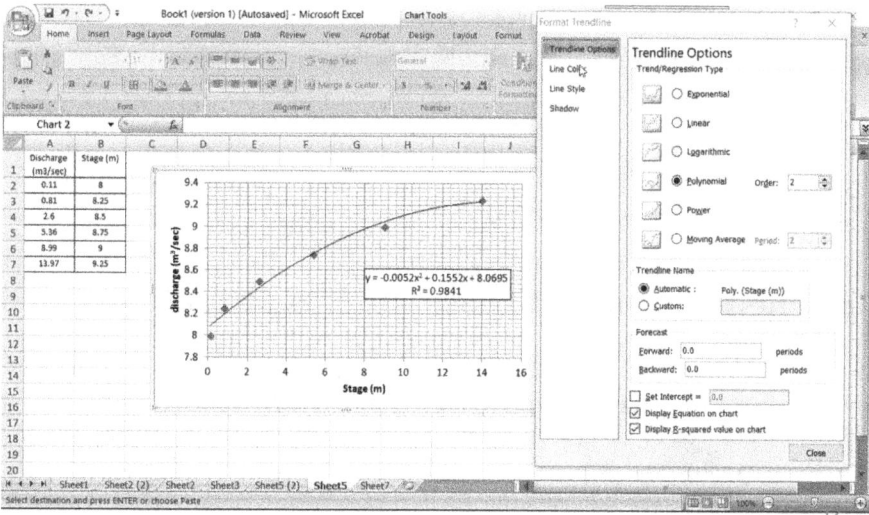

FIGURE 2.9   Plot of data with a polynomial trendline.

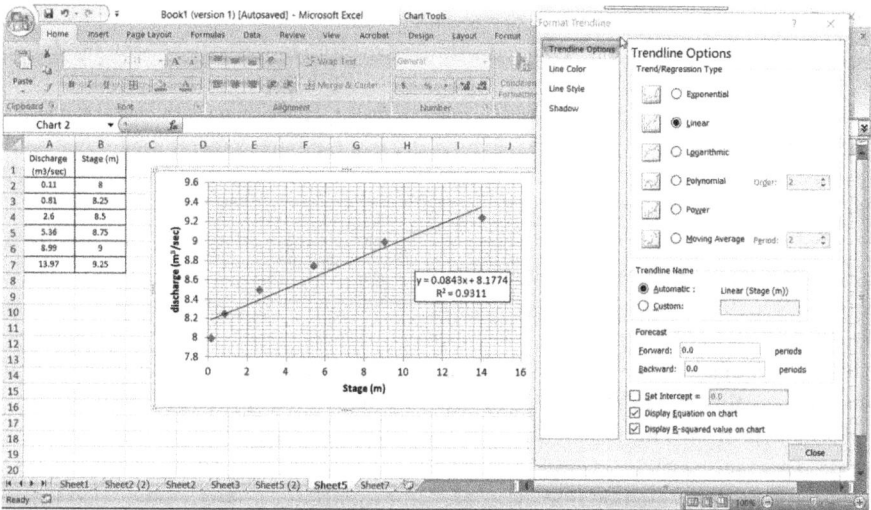

FIGURE 2.10   Plot of data with a linear trendline.

Of the available options, a polynomial trendline yielded the best visual match, boasting an impressive R-squared value of 0.98 (Figure 2.9). Conversely, the linear trendline in Figure 2.10 does not exhibit as strong a fit, with an R-squared value of 0.93. Nevertheless, the linear trendline option may be appealing in situations where linearity is hypothesized. In such cases, a linear trendline can be employed to test this hypothesis. As demonstrated in Figure 2.9, the trendline equation takes the form shown in equation 2.14):

$$y = mx + b \tag{2.14}$$

in which m is the slope of the trendline and b is the y intercept (value of y when $x=0$).

2.3.4 Regression and Least Squares Fitting

Field and laboratory hydrologic tests primarily yield data on hydraulic head or concentration as a function of time and space. As will be discussed in Chapter 6, water levels are measured at various time intervals at specific locations during pumping tests. Subsequently, this data can be analyzed to estimate the hydraulic properties of the aquifer, which are typically represented by hydraulic conductivity and storage coefficient (see Section 6.9 in Chapter 6). In these estimations, fitting techniques are employed, utilizing a line or an arc to match the measured points through visual inspection. However, such an approach is generally inaccurate due to its qualitative nature. An alternative and more accurate approach utilizes least squares fitting techniques, which comprise a set of mathematical methods to define the best-fitting line to the set of data points. These techniques minimize the error of the fit, defined as the sum of the squared differences between the observed data points and the values predicted by a model. In addition to hydrology, the least squares approach is widely used in statistics, engineering, physics, and data analysis, among other fields. Available references include Burden and Faires (2010), Montgomery et al. (2012), and Bates and Watts (2007).

Given a set of measured data represented by x and y, which are known as the independent and dependent variables, respectively, the approach begins by selecting a model, which is a specific formula, either physically or mathematically based, that represents the type of fit required. The physically based model employs equations that establish relationships between x and y using scientific principles, such as Darcy's law, which relates flux (y in this case) to the head gradient (x in this case). Mathematically based models are primarily used for visualization and trend analysis to represent the data in question.

Among various methods, linear regression stands out as one of the simplest and most widely used least squares techniques. This method involves fitting a linear model (a straight line) to the data via equation 2.14 and minimizing the sum of the squared differences between the observed data points and the values predicted by the linear equation (e.g., Montgomery et al., 2012).

Least squares fitting aims to estimate the parameters m and b in this equation to achieve the optimal alignment between the measured data and a straight line. Equations (2.15) and (2.16) provide the expressions for estimating these parameters (e.g., Saylor Academy, 2012):

$$m = \frac{S_{xy}}{S_{xx}} \qquad (2.15)$$

$$b = \bar{y} - m \cdot \bar{x} \qquad (2.16)$$

where $\bar{x}$ and $\bar{y}$ are the average values of x and y, respectively, and S_{xx} and S_{xy} are expressed in equations (2.17) and (2.18), respectively.

$$S_{xx} = \sum x^2 - \frac{1}{N}\left(\sum x\right)^2 \qquad (2.17)$$

$$S_{xy} = \sum xy - \frac{1}{N}(x)\cdot\left(\sum y\right)^2 \qquad (2.18)$$

In this context, N represents the number of data points, while $\sum x$ and $\sum y$ denote the sum values of all x and y data points, respectively. Additionally, $\sum xy$ represents the sum of all products of x and y, and $\sum x^2$ signifies the sum of all squared values of x. The average values of x and y are estimated by dividing the respective values of $\sum x$ and $\sum y$ by the number of data points. As discussed in Section 2.3.3, the parameter R-squared, also known as the coefficient of determination, serves as a statistical measure to evaluate the goodness of fit of the regression line. A value close to 1.0 indicates a good match. Equations (2.19) can be used to estimate R-squared.

$$R^2 = \frac{S_{xy}^2}{S_{xx} \cdot S_{yy}} \qquad (2.19)$$

where S_{xx} and S_{xy} are given by equations (2.15) and (2.16), and S_{yy} is given by

$$S_{yy} = \sum y^2 - \frac{1}{N}\left(\sum y\right)^2 \qquad (2.20)$$

The procedure is detailed in Example 2.1.

Example 2.1: Linear Regression Application

Table 2.9 contains a set of 20 values for both x and y, which are intended for use in the linear regression calculations. Additionally, this table includes the corresponding values of x^2, y^2, and x multiplied by y. In the final row, the summations of x, y, x^2, y^2, and xy are listed. The averages of x and y values are $\bar{x} = 31.9$ and $\bar{y} = 3.69$, respectively. By substituting these quantities into equations (2.17) through (2.18), the values of m and b can be calculated as follows:

$$S_{xx} = 27{,}180 - \frac{1}{20}(329.47)^2 = 6{,}828$$

TABLE 2.9   Least squares calculations for linear regression

X	Y	Y²	X²	XY	ESTIMATED Y	ERROR²
2	0.88	0.35	4	1.77	0.94	0.0037
5	1.18	0.78	25	5.90	1.22	0.0013
9	1.47	1.39	81	13.27	1.58	0.0128
13	1.77	2.18	169	23.01	1.95	0.0325
16	2.06	3.13	256	33.04	2.23	0.0275
17	2.36	4.26	289	40.12	2.32	0.0018
18	2.65	5.57	324	47.79	2.41	0.0580
20	2.95	7.05	400	59.00	2.59	0.1276
26	3.24	8.70	676	84.37	3.14	0.0093
30	3.54	10.53	900	106.19	3.51	0.0009
35	3.83	12.53	1,225	134.22	3.97	0.0195
38	4.13	14.71	1,444	156.93	4.24	0.0132
41	4.42	17.06	1,681	181.42	4.52	0.0100
42	4.72	19.58	1,764	198.23	4.61	0.0117
45	5.01	22.28	2,025	225.66	4.89	0.0150
50	5.31	25.15	2,500	265.49	5.35	0.0013
52	5.60	28.19	2,704	291.45	5.53	0.0049
56	5.90	31.41	3,136	330.38	5.90	0.0000
59	6.19	34.81	3,481	365.49	6.17	0.0003
64	6.49	38.37	4,096	415.34	6.63	0.0199
$\sum x =$ 638	$\sum y =$ 73.7	$\sum y^2 =$ 329.47	$\sum x^2 =$ 27,180	$\sum xy =$ 2977.69		RMS = 0.1362

$$S_{yy} = 329.47 - \frac{1}{20}(73.7)^2 = 57.89$$

$$S_{xy} = 2977.69 - \frac{1}{20}(638.74)\cdot(73.7) = 626.66$$

$$m = \frac{626.66}{6828} = 0.092$$

$$b = 3.69 - 0.092 \cdot 31.9 = 0.76$$

The last column in Table 2.9 displays the squared values of the residuals (or errors), indicating the disparity between each y value and its corresponding estimated value. The bottommost entry in the column represents the sum of these squared

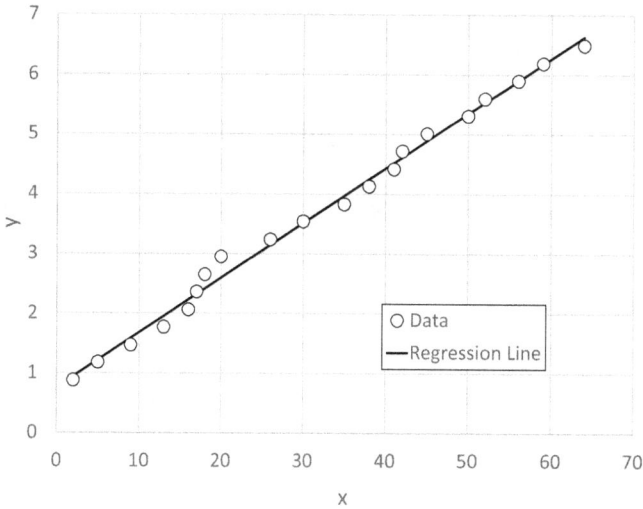

FIGURE 2.11   Comparison between data and a linear regression line.

values, signifying the minimum error. Finally, Figure 2.11 compares the given values of y and the regression line. Visual inspection shows a reasonable match.

Utilizing the data in Table 2.9, the value of R-squared is estimated as follows via equation (2.19):

$$R^2 = \frac{S_{xy}^2}{S_{xx} \cdot S_{yy}} = \frac{626.66^2}{6,828 \times 57.89} = 0.99$$

which indicates a good match.

It is important to recognize that linear regression may not be appropriate for many cases that involve greater complexities, contradicting the assumption of linearity. In such cases, more intricate nonlinear equations are necessary. Similar to linear regression, these relationships can be established through physical or other processes. In instances where explicit physically based equations are unavailable, empirical equations are employed. However, unlike linear regression, these cases often demand more complex mathematical computations to minimize the least square residuals. This is typically achieved through optimization techniques depending on the specific problem and the characteristics of the model.

Here is a brief description of the formulas used in nonlinear regression.

1. Polynomial functions are represented by a polynomial equation, such as the quadratic equation (2.21):

$$y = ax^2 + bx + c \tag{2.21}$$

2. The power law function describes relationships where the dependent variable is a power of the independent variable; it is often expressed by equation (2.22):

$$y = a \cdot x^b \tag{2.22}$$

3. Exponential functions describe relationships where the dependent variable changes exponentially with the independent variable, as shown in equation (2.23):

$$y = a \cdot e^{bx} \tag{2.23}$$

For exponential and power law fitting, the regression method involves transforming the data or the model to make it linear, followed by linear regression on the transformed data, as illustrated in Example 2.1.

4. Logarithmic functions are used when the relationship between variables can be described by the logarithm of one variable being a linear function of another, as shown in equation (2.24):

$$y = a \cdot \ln(x) + b \tag{2.24}$$

As with linear regression, the goal is to estimate the parameters (such as a, b, and c in the equations) that provide the best fit to the available data, which involves minimizing the sum of squared differences between the model's predictions and the actual data points.

Each of the aforementioned formulas has its strengths and weaknesses and is suitable for different types of data and modeling problems. The choice of which least squares fitting technique to use depends on the specific characteristics of the data and the research or analysis goals.

As discussed in Section 2.3.3, Excel allows for the display of trend lines on plots, offering various trending options. These trending options correspond to the same regression formulas mentioned earlier (refer to Figures 2.9 and 2.10). Consequently, the trend lines are essentially equivalent to the least squares analysis explained earlier.

Figure 2.12 illustrates the data for a river rating curve, featuring a polynomial trend line represented by equation (2.21). The fitting parameters for a, b, and c are −0.0052, 0.1552, and 8.0615, respectively. Upon visual inspection, it becomes evident that a linear regression model is not suitable for this type of data. The polynomial line, as depicted in the figure, aligns with the inherent nature of the relationship between discharge and stage. The figure shows that the curve flattens out at higher stages, which is consistent with a smaller change in river cross-sectional area and reduced velocity.

It is important to bear in mind that the application of empirical equations should be restricted to the units used during the regression procedure. For example, in Figure 2.12, only meters and cubic meters per second are permissible. If necessary, unit conversions can be done after the calculations are completed. For example, when provided with a stage measurement in feet, the first step is to convert it to meters before applying the

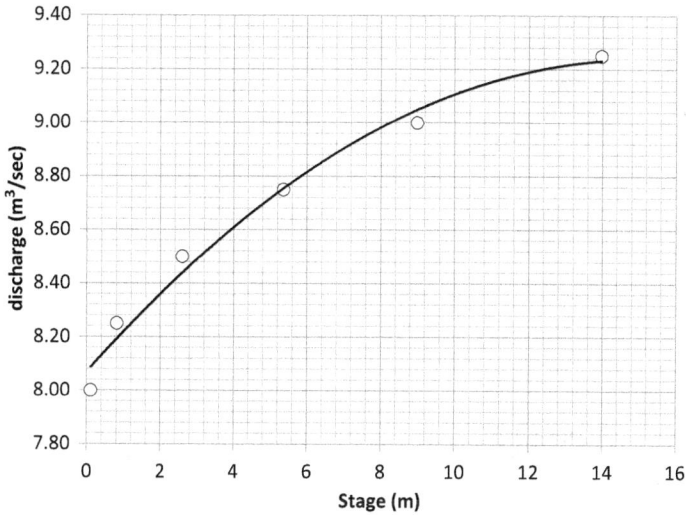

FIGURE 2.12   River duration data and trend (regression) line.

regression equation. Subsequently, the resulting discharge value is converted to cubic feet per second.

A detailed description of the river rating curve is provided in Chapter 5 (Section 5.2.4), which serves to estimate river discharge based on a given river stage (or water level). As mentioned in that section, river discharge relies on several variables, including water slope, river cross-sectional area, wetted perimeter, and a river's friction parameter, most of which are challenging to measure directly. As an alternative, discharge can be determined based on velocity and cross-sectional area. However, directly measuring velocity is a labor-intensive process. Therefore, the empirically based polynomial equation, which only requires the easily measurable river stage, is an appealing option.

The preceding discussion focused on regression for a single dependent variable, denoted as x. However, multiple regression broadens the scope of regression analysis by involving the relationship between the dependent variable y and two or more independent variables, namely x_1, x_2, x_3, ... x_n, where n represents the total number of independent variables. In multiple regression, the objective is to estimate the coefficients associated with each of the independent variables by minimizing the sum of squared errors.

Regression techniques can be further improved through the implementation of various methods. The weighted least squares technique, for instance, assigns distinct weights to individual data points, enhancing the model's accuracy. Robust regression techniques are also valuable in improving results by down-weighting or disregarding outliers in the data. Lastly, total least squares regression proves beneficial in accounting for errors in both the dependent and independent variables.

It is crucial to understand that, as a general rule, the regression equations generated should be used for estimating the value of the dependent variable within the specific range of the independent variable that was used to create the regression line.

Consequently, making extrapolations beyond these ranges is not valid and can lead to inaccuracies.

Commercial software for analyzing pumping tests is readily available, catering to various aquifer and pumping conditions. Some notable options include AQTESOLV (2023), AnsTest (ANS Distributing, 2023), and Pointstar (2023). Additionally, BRGM (2023) has developed a freeware application for this purpose. Least squares regression is a commonly used approach in this software group.

NOTE

1   A dimensionless quantity has no units. For example, here S describes the change of water level per unit distance (or L/L). So there are no units associated with S

REFERENCES

ANS Distributing. (2023). Aquifer test: What it is and how it works. Retrieved from https://ans-dimat.com/aquifer-test.shtml

AQTESOLV. (2023). AQTESOLV. https://www.aqtesolv.com/

Bates, D. M., & Watts, D. G. (2007). *Nonlinear Regression Analysis and Its Applications*. Hoboken, NJ: Wiley..

BRGM. (2023). OUAIP: computer-assisted pumping tests interpretation. Retrieved from https://www.brgm.fr/en/software/ouaip-computer-assisted-pumping-tests-interpretation

Burden, R. L., & Faires, J. D. (2010). *Numerical Analysis*. Boston, MA: Cengage Learning.

Los Angeles Times. (1999). Mars probe lost due to simple math error. Retrieved from https://www.latimes.com/archives/la-xpm-1999-oct-01-mn-17288-story.html

Montgomery, D. C., Peck, E. A., & Vining, G. G. (2012). *Introduction to Linear Regression Analysis*. Hoboken, NJ: Wiley.

New York Times. (1983). Jet's fuel ran out after metric conversion errors. Retrieved from https://www.nytimes.com/1983/07/30/us/jet-s-fuel-ran-out-after-metric-conversion-errors.html

Pointstar. (2023). Aquifer test analysis software. Retrieved from https://pointstar.com/Aquifer/Default.aspx

Probability Course. (2023). 4.2.3 Normal (Gaussian) distribution. Retrieved from https://www.probabilitycourse.com/chapter4/4_2_3_normal.php

*Saylor Academy. (2012). Introductory statistics. Retrieved from https://saylordo*torg.github.io/text_introductory-statistics/s14-04-the-least-squares-regression-1.html#fwk-shafer-ch10_s04_s02_f01

UnitConverter.Net. (2023). Converter express version. Retrieved from https://www.unitconvert-ers.net/

The Hydrologic Cycle

3

3.1 INTRODUCTION

Hydrology, or the science of water, encompasses the comprehensive study of water's storage, distribution, and movement, both above and below the Earth's surface as well as within the atmosphere. Additionally, it explores the influence of human activities on water availability and conditions. A profound understanding of hydrological processes is vital for evaluating global and regional water resources, effectively managing them, and promoting conservation efforts.

The water (or hydrologic) cycle delineates the movement and storage of water across various domains and states. Water is retained in oceans, surface water bodies, the atmosphere, on land surfaces, and beneath the ground (as illustrated in Figure 3.1). As a cyclical process, it can be initiated from any stage within the figure.

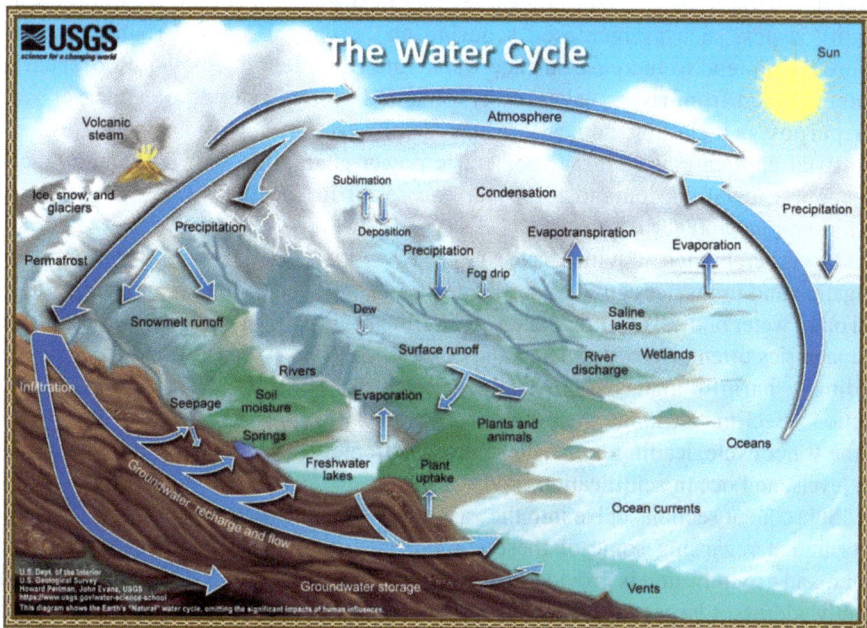

FIGURE 3.1 The water cycle (USGS, 2019).

DOI: 10.1201/9781003587149-3

For instance, the ocean and other open water bodies contribute water vapor to the atmosphere through evaporation. Additionally, vapor is generated via transpiration from plants. On land, these two processes are typically combined and referred to as evapotranspiration. Vapor subsequently condenses to form clouds, which, under suitable conditions, produce precipitation in the form of rain or snow. The liquid precipitation that reaches the ground is divided between surface and subsurface water components. The surface water component includes overland flow (or surface runoff), stream flow, and water stored in lakes, ponds, and other bodies. The subsurface component, known as groundwater, originates through the process of infiltration. Groundwater can exit the system through springs and surface water bodies, such as lakes and the ocean. Water is in a perpetual state of motion within and between these systems, transitioning between liquid, solid (ice), and gas (vapor). Water exhibits movement at various scales, ranging from national and watershed levels to agricultural plots. It also moves at microscales within human bodies, plants, and other organisms.

Human activities exert a significant impact on elements of the water cycle, either by directly altering the processes themselves or through water overuse and the introduction of contaminants into the environment. A comprehensive illustration of these human activities is available in a USGS publication (USGS, 2023), which overlays various human actions onto a water cycle diagram. Given its resolution, it is advisable for the reader to access the diagram directly, as it is not suitable for inclusion here. Specific human actions include river redirection, dam construction, urbanization, deforestation, and wetland drainage. Such actions are likely to disrupt the natural flow patterns of water bodies, leading to sustainability challenges and water quality issues, such as increased scouring and sediment loads. Urbanization and deforestation tend to increase surface water runoff while reducing aquifer recharge. Water sourced from rivers, lakes, reservoirs, and groundwater aquifers serves various purposes, including domestic water supply, irrigation, livestock grazing, and industrial applications like thermoelectric power generation, mining, and aquaculture. The escalating demand for water can result in shortages and disputes, both within nations and across borders.

Moreover, human activities have a noticeable impact on water quality. For instance, irrigation and rainwater can wash fertilizers, pesticides, and sediment into rivers, lakes, and other water bodies. Chemicals can also seep into groundwater aquifers. Power plants and factories often discharge heated and contaminated water into surface water bodies, leading to harmful algal blooms and diseases that threaten habitats. As discussed in Section 1.8, climate change, stemming from human activities, adversely affects aspects of the water cycle, leading to weather-related extremes like droughts and floods, rising sea levels, and ocean acidification.

Subsequent sections delve into the various components of the water cycle. Detailed information regarding some of these components, particularly precipitation, surface water, and groundwater, is provided in the following chapters. This publication predominantly addresses issues related to freshwater resources, and it is essential to underscore the critical importance of these three components. Precipitation serves as the primary source of water, while surface and subsurface resources have a direct impact on people's lives and present significant management challenges.

3.2 THE OCEAN

The water cycle encompasses not only water in a transient state but also various forms of storage, such as the vast bodies of water found in our oceans, which hold more than 96% of the Earth's total water supply. It is estimated that the oceans contribute approximately 90% of the evaporated water that enters the Earth's atmosphere (USGS, 2019). The water within our oceans is saline, characterized by a concentration of dissolved salts of approximately 35,000 milligrams per liter (mg/L). [Concentration is defined as the weight of salt (in milligrams) per unit volume of water (in liters).] Water is considered saline when its concentration exceeds 1,000 mg/L of dissolved salts, while drinking water should ideally have a salinity level not exceeding 250 mg/L.

The oceans play a pivotal role in the water cycle, not only as a primary source of evaporated water but also as a vital resource for food, salt, and minerals. Additionally, the water from oceans is utilized in various industrial processes, cooling applications, and energy generation. Desalination processes allow for the conversion of saltwater into domestic water for arid regions, albeit at significant costs. On the downside, saltwater intrusion can lead to the degradation of freshwater quality in coastal areas, particularly during dry weather conditions when freshwater resources are overdrawn.

Over extended periods and during significant climatic shifts, the volume of water in the oceans can undergo substantial changes. For instance, during the last ice age, glaciers covered large portions of the Earth's landmass, causing ocean water levels to drop by more than 100 m (300 ft) compared to current levels. Conversely, during the last global warming period, sea levels were approximately 5 m (15 ft) higher than they are today. Presently, there are serious concerns regarding sea level rise attributed to global warming, with projections indicating a potential rise of 1 m (3 ft) or more by the end of this century (see Section 1.8.3). Such an increase in sea levels poses the threat of coastal flooding, saltwater intrusion into groundwater sources, and significant ecological repercussions. Additionally, it could lead to the loss of habitable land and damage to essential infrastructure. Indirect groundwater contamination is also a concern, as flooded areas may contain potential sources of harmful substances, such as cesspools or storage tanks.

Tides represent short-term fluctuations that occur daily and are influenced by the gravitational pull of the moon. Tide heights vary depending on location and can reach considerable levels in enclosed bays. For example, the Bay of Fundy in Nova Scotia, Canada, experiences maximum tides of up to 16 m (48 ft) during specific times of the year (Bay of Fundy.Com, 2023). Global warming and resultant climate change have led to larger tidal events, which can result in sporadic coastal damage.

3.3 HILLSIDE HYDROLOGY

Hillslope hydrology focuses on the movement and distribution of water on and just below the surface of hillslopes. The primary objectives of related studies are to understand and evaluate the pathways, storage, and fluxes of water and their interactions with soil, vegetation, and the geological environment. This field is crucial for water resource management, erosion control, and ecosystem health. The following sections will address the main issues related to various features and processes that are part of the hillslope hydrology, as well as the deeper zones.

3.4 EVAPORATION AND EVAPOTRANSPIRATION

Evaporation is the process by which water moves from the Earth's surface into the atmosphere as vapor or gas. This phenomenon occurs when heat breaks the bonds holding water molecules together, transforming them into a vapor phase. The primary source of heat for evaporation is the sun, and this bond-breaking process is similar to what happens when water boils on a stove. Most of the moisture in the atmosphere originates from the evaporation of surface water bodies, including oceans, seas, lakes, and rivers. However, evaporation can also occur from both the surface and subsurface soil, and even from deep subsurface water bodies.

The remaining moisture in the atmosphere comes from plant transpiration, which involves the release of water vapor from plant leaves. After absorbing water from the soil through their roots, plants transport it through their tissues, serving various essential functions for their growth. Transpiration occurs when plant leaves release water vapor into the air through structures called stomata, which are openings in leaves that facilitate growth and enable the exchange of water and carbon dioxide with the surrounding environment. The rate of transpiration can vary widely depending on factors such as weather conditions, plant type, and other environmental variables. Influential factors include soil type, soil moisture levels, sunlight availability and intensity, precipitation, air humidity, temperature, and wind. Transpiration rates tend to be lower when soil moisture is scarce, in fine-textured soils with a higher water retention capacity, at lower temperatures, and in highly humid atmospheric conditions. Notably, transpiration occurs exclusively during daylight hours, as it is dependent on sunlight. Additionally, increased air movement around plants results in higher transpiration rates.

Evapotranspiration is the sum of land-based evaporation and transpiration. Estimating evaporation, transpiration, and evapotranspiration involves various methods, which are covered in textbooks such as Fetter and Kreamer (2022). These methods include the evaporation pan and the lake evaporation nomograph. An evaporation pan (Figure 3.2) is used to estimate potential (or maximum) evaporation rates. This pan is placed on a level wooden base and enclosed by a fence to prevent interference from

FIGURE 3.2 Evaporation pan. Adapted from Evaporation Pan, by Kallgan, licensed under the Creative Commons Attribution 3.0 Unported License. Available at: https://commons. wikimedia.org/wiki/File:Evaporation_Pan.jpg. License details: https://creativecommons.org/ licenses/by/3.0/

animals and insects. The initial water level is set at 5 cm (2 in), and at the end of the day, the pan is refilled. The amount of water needed to return the water level to 2 in is a measure of evaporation. However, this method may not be accurate during heavy rainfall, and corrections are required based on field conditions.

The lake evaporation nomograph is a graphical chart used to estimate potential evaporation. It requires knowledge of mean daily temperature, solar radiation, wind speed, and mean daily dew point temperature (the temperature at which condensation begins).

Transpiration can be measured using phytometers, which are sealed containers filled with soil and plants placed in the field. Humidity in the air within the phytometer is measured to estimate transpiration. However, this method may not precisely replicate natural field conditions due to the sealed nature of the phytometer.

It's imperative to distinguish between actual and potential evapotranspiration. Potential evapotranspiration represents the water loss that would occur if no soil water deficiency exists, while actual evapotranspiration is generally less than the potential value under field conditions. Various approaches can be used to estimate these values, with different degrees of accuracy based on available data. Simple methods may rely on mean monthly air temperature, latitude, and the month of the year, assuming that evapotranspiration is not influenced by vegetation density or maturity. More sophisticated approaches incorporate factors that change with the season. The most detailed schemes utilize climatic data, including vapor pressure, sunshine duration, net radiation, wind speed, and mean temperature, but these calculations may require extensive data that might not always be available.

Field measurements of evapotranspiration can be conducted using lysimeters, which are large containers in the field containing soil and plants. These measurements involve tracking water input from precipitation and irrigation, water output through drainage, and changes in soil water volume. Changes in soil water volume are estimated by measuring soil moisture levels, and evapotranspiration is calculated as the additional volume loss needed to account for the change in soil water volume.

The flux of water vapor can also be directly assessed by using the eddy covariance tower method, a technique employed to measure the exchange of gases between the Earth's surface and the atmosphere (e.g., Aubinet et al., 2012; Harvard Forest, 2021; CSIR, 2023). This method entails the installation of a tall tower equipped with instruments at various heights to measure the vertical gas flux resulting from turbulent air movements. Eddy flux towers are additionally equipped with instructions to measure a range of meteorological variables, including wind velocity, wind direction, relative humidity, and temperature (CSIR, 2023). The covariance tower method is also applied to evaluate carbon dioxide fluxes into the atmosphere, which contribute to global warming.

3.5 PRECIPITATION

Water vapor in the atmosphere condenses into water droplets that can increase in size. Condensation, the opposite of evaporation, is the process in which water vapor returns to its liquid state. This occurs when saturated air cools down, such as on the outer surface of a glass of iced water or on a car's windshield. Particles of dust or smoke in the atmosphere act as surfaces for water vapor to condense on, aiding in the aggregation of water droplets and their growth. When these droplets become sufficiently heavy, they descend to Earth as precipitation in either liquid or frozen forms. In colder clouds, such as those at higher altitudes, the droplets take on an ice form. Depending on the temperature within the cloud and at the Earth's surface, these ice crystals may then fall as snow, hail, or rain.

Another source of atmospheric water is known as fog drip, which refers to water that drips to the ground, usually from trees that have been moistened by drifting fog droplets. Some trees in mountainous regions are efficient enough at this process to supply sufficient water to sustain forests. When assessing the water budget of a river basin or watershed, fog drip is typically combined with precipitation.

It's important to note that precipitation is always fresh, even if it originates from the ocean, because salt does not evaporate with water. However, rainwater can become polluted while in the atmosphere before reaching the Earth's surface. This polluted precipitation is referred to as acid rain, which can increase the acidity of lakes and streams, causing harm to the plants and animals that inhabit these aquatic ecosystems.

3.6 SUBSURFACE AND SURFACE WATER PROCESSES

3.6.1 The Unsaturated Zone

Processes in the unsaturated zone include percolation, which is the downward movement of water through soil layers. Percolation is essential for recharging the deeper groundwater zone, maintaining soil moisture, and supporting plant growth. The rate of percolation depends on several soil characteristics, including texture, structure, porosity, organic matter content, compaction, and moisture level.

Texture refers to the size and distribution of soil particles, while structure denotes the arrangement of these particles into aggregates. Porosity represents the volume of pores or spaces within the soil. Generally, coarser materials, such as sands, allow for faster percolation than clay soils, which have smaller, tightly packed particles. Well-structured soils with good porosity facilitate increased percolation compared to compacted or poorly structured soils. Organic matter improves soil structure, increases water-holding capacity, and reduces percolation. Soil moisture, or the degree of wetting, indicates the volume of soil occupied by water; saturated soils have slower percolation rates due to reduced pore space for additional water. Increased soil compaction affects pore space and negatively impacts percolation.

Additionally, vegetation, root systems, ground slope, and topography play important roles in controlling percolation. Plant roots create channels and pathways that facilitate water movement and can absorb water, reducing the amount that percolates deeper into the soil. Steeper slopes may result in less percolation compared to flat areas due to increased surface runoff. Finally, climate-related factors, such as precipitation patterns, intensity, and duration, significantly impact water movement.

Interflow, also known as subsurface stormflow, refers to the lateral movement of water through the upper unsaturated layers of soil above the main groundwater table during and after precipitation events. Similar to percolation, this process is influenced by factors such as soil type, land cover, topography, and climate. Water pathways include soil pores, fractures, and channels created by roots or soil structure. Interflow contributes to streamflow and maintains base flow in rivers and streams, especially during dry periods. It also plays a role in nutrient transport, soil moisture distribution, erosion, and flooding, particularly in areas with poor drainage or steep topography.

Capillary effects cause the movement of water through soil due to the forces of adhesion, cohesion, and surface tension. Adhesion is the attraction between water molecules and soil particles, allowing water molecules to cling to soil particles and helping water move through soil pores. Cohesion is the attraction between water molecules themselves, helping them stay together as they move through the soil. Surface tension creates a "skin" on the water surface due to the attraction between molecules at the surface, which also helps pull more water molecules up through soil pores, creating a partially water-saturated zone known as the capillary fringe.

Therefore, due to capillary effects, water moves upward or laterally through the tiny spaces between soil particles. The smaller the soil pores, the higher and further the water can move. Capillary effects are vital for maintaining soil moisture, which is essential for plants, especially during dry periods. Water movement contributes to soil drainage and the movement of nutrients within the soil. It is important to note, however, that water in the unsaturated zone is tightly held by soil grains and cannot be easily extracted via wells.

Virtual Experiment 10.6 (Chapter 10) describes an approach to study capillary rise in a soil column. This approach is used to estimate the average size of the soil grains.

3.6.2 Infiltration

Infiltration refers to the movement of water into the subsurface soil and rock following rainfall. The rate of infiltration primarily depends on factors such as the intensity of rainfall, land use and cover, and the characteristics of the underlying subsurface material. Surface features include vegetation that intercepts and stores water. Some of the water that infiltrates the soil remains in the upper soil layers, while some may penetrate deeper, replenishing groundwater aquifers.

3.6.3 Overland and Stream Flow

Overland flow runoff is defined as water flowing over the ground's surface, and its rate depends on the extent of infiltration. For instance, certain soils, such as clayey ones, have a lower water absorption capacity and a slower infiltration rate compared to sandy soils, leading to increased runoff that eventually reaches streams. When the soil becomes fully saturated, additional rainfall contributes to surface runoff. Land cover plays a crucial role in influencing both infiltration and runoff dynamics. Vegetation, for example, can impede runoff and encourage infiltration. In contrast, impervious surfaces commonly found in urban areas, such as buildings, parking lots, and roads, swiftly channel rainfall into storm drains that feed streams or the ocean. Agricultural practices, such as the formation of hard pans, can also alter infiltration patterns, resulting in increased runoff. Moreover, sloping terrain typically exhibits reduced infiltration rates compared to flat land, promoting larger runoff volumes. Water present on the soil surface or within the shallow subsurface can be lost to the atmosphere through evapotranspiration, diminishing opportunities for infiltration.

3.6.4 Saturated or Groundwater Flow

Aquifers are geological formations capable of storing and transmitting water. Infiltrated water recharges the aquifer, which is fully saturated when water completely occupies the voids within rock fractures or between soil particles. The recharge process can vary in speed, depending on subsurface characteristics that influence water flow rates within the unsaturated zone. Ensuring water sustainability relies on managing various water uses within the aquifer's recharge rate that replenishes it. Otherwise, excessive pumping can deplete the aquifer, rendering wells ineffective.

Wells can be drilled into these aquifers to extract water for various purposes. Aquifer water can travel considerable distances or be stored for extended periods, and it may resurface through springs or seep into bodies of water like streams and the ocean. In some cases, aquifer water can gradually seep into streambeds, sustaining flow even during dry periods when there's no direct runoff from precipitation.

3.7 WATERSHEDS

To assess the field elements of the water cycle, studies are conducted across various scales, ranging from small plots to large-scale watersheds. A watershed represents a land area that directs rainfall and snowmelt toward a stream or river, ultimately leading to a water body, such as the ocean. As illustrated in Figure 3.3, multiple watersheds contribute to the drainage of Pearl Harbor on the island of Oahu, Hawaii (Figure 3.3a). Watershed boundaries are determined based on topography, typically following mountain crests, which serve as dividing lines (Figure 3.3b). Within each watershed, numerous creeks and streams transport water from higher elevations to a primary river, eventually emptying into the harbor (Figure 3.3c).

In principle, water from precipitation and surface runoff should remain within a given watershed. However, in some instances, stream water may be redirected across watershed boundaries to meet specific demands. Furthermore, it is theoretically possible to comprehensively account for all other components of the water cycle within a watershed to establish a water budget. Nevertheless, the groundwater component introduces uncertainty because subsurface boundaries may not align precisely with surface watershed boundaries and can shift based on groundwater levels.

Watersheds vary widely in size, with some being relatively small and others spanning thousands of square miles, often containing numerous surface water bodies and significant underlying groundwater aquifers. The largest watershed in the United States is the Mississippi River Watershed, covering an extensive 1.15 million square miles and draining all or parts of 31 U.S. states and two Canadian provinces.

As water traverses these watersheds, it frequently accumulates contaminants from various land uses and sources (Figure 3.3d). These contaminants can adversely affect the ecology of the watershed and, ultimately, the receiving water body, whether it is

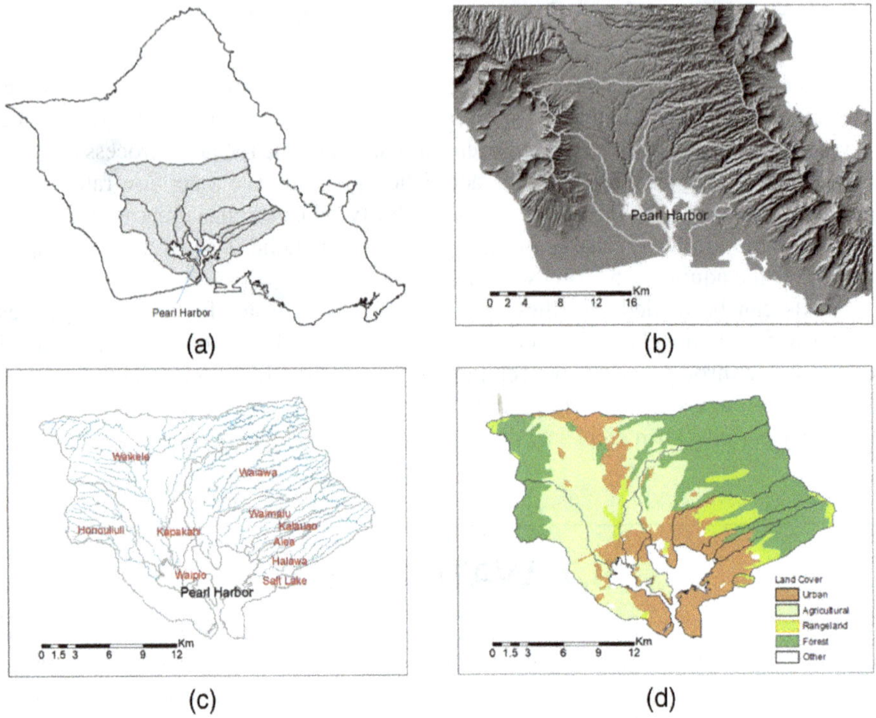

FIGURE 3.3 Watersheds draining into Pearl Harbor (Oahu, Hawaii): (a) Location and boundaries, (b) Boundary delineation, (c) Watersheds and respective streams, (d) land cover type.

a lake, estuary, or ocean. Examples of contaminants include nutrients from forested areas, pesticides and fertilizers from agricultural lands, and nutrients and bacteria from urban areas.

A water budget serves as the fundamental basis for conducting effective water resources analyses and managing water systems. In fact, it can be developed for any system, regardless of its size or scale. Its core principle involves tracking the inflow and outflow of water within the system, along with any resulting changes in storage. However, the accuracy of this assessment heavily relies on the availability of sufficient data to support the use of the budget evaluation scheme.

The publication by Healy et al. (2007) addresses various aspects of the water budget, encompassing the atmosphere, land surface, subsurface, and the exchange of water between these compartments. Additionally, the study delves into issues related to field measurements and the uncertainties inherent in the process. Typically, a water budget is expressed through a mathematical equation that depends on a few underlying assumptions regarding the mechanisms governing water movement and storage. This can be succinctly expressed as follows:

$$\Delta S = \left(P + Q_{\text{in}}\right) - \left(ET + Q_{\text{out}}\right) \tag{3.1}$$

In equation (3.1), P represents precipitation, Q_{in} and Q_{out} denote water flow into and out of the watershed, respectively, ET stands for evapotranspiration, and ΔS represents the change in water storage. In simpler terms, the water storage in equation (3.1) directly accounts for the total inflows ($P + Q_{in}$) minus the total outflows ($ET + Q_{out}$).

This budget can be expressed either in terms of volumes for a fixed time interval, such as cubic meters, or as fluxes, which are volumes per unit time, such as cubic meters per day. Alternatively, it can be formulated in terms of flux densities, expressed as volumes per unit area of land surface per unit time, such as millimeters per day. Equation (3.1) can be refined and customized depending on the specific objectives and scales of a particular study. Precipitation can include water from various sources, such as rain, snow, fog drip, and irrigation. Water flow into or out of the site may involve surface or subsurface flow resulting from both natural and human-related activities, including pumping.

A modified version of equation (3.1) can be applied to estimate groundwater recharge, a crucial component for assessing groundwater resources. This recharge represents the primary source of replenishment for aquifers and is essential for determining the sustainability of water resources. In principle, sustainability requires that water use remains roughly within the bounds of recharge estimates. The equation used for this purpose is as follows:

$$G = P + IR - R - ET - \Delta S \tag{3.2}$$

In equation (3.2), G represents recharge, IR denotes the irrigation amount, and R signifies the direct runoff. P and IR in this equation represent water inflows into the watershed, while R, ET, and ΔS are associated with losses or outflows. The balance represented by G eventually replenishes the aquifer.

The choice of methods for estimating recharge depends on data availability. For instance, to simplify physically based approaches, one can use average field measurements of rainfall and runoff to estimate the fraction of runoff relative to rainfall, which can then be utilized for future predictions. In such cases, runoff can be estimated as a straightforward percentage of rainfall based on available records for a specific watershed. An illustrative example is Shade's study (1995), conducted on the island of Kauai, Hawaii. In this study, data from all streams on the island were employed to estimate mean monthly values for direct runoff and rainfall. These data were further utilized to determine the spatial variability of runoff-to-rainfall ratios across the island. In Shade's study, evapotranspiration was estimated based on the amount of moisture in soil storage from which actual evapotranspiration occurs. The volume of water added to soil storage is the difference between rainfall and direct runoff. When the water in soil storage surpasses its maximum capacity, the excess water recharges the groundwater. Evapotranspiration is subtracted from soil moisture storage at either the maximum potential rate or a lesser actual rate, depending on the available water in storage to meet the demand.

Another example is documented by Rasmussen and Andresen (1959) for the Beaverdam Creek Basin in Maryland, covering the period from April 1950 to March 1952. A summary of this study is provided by Healy et al. (2007). Figure 3.4 includes

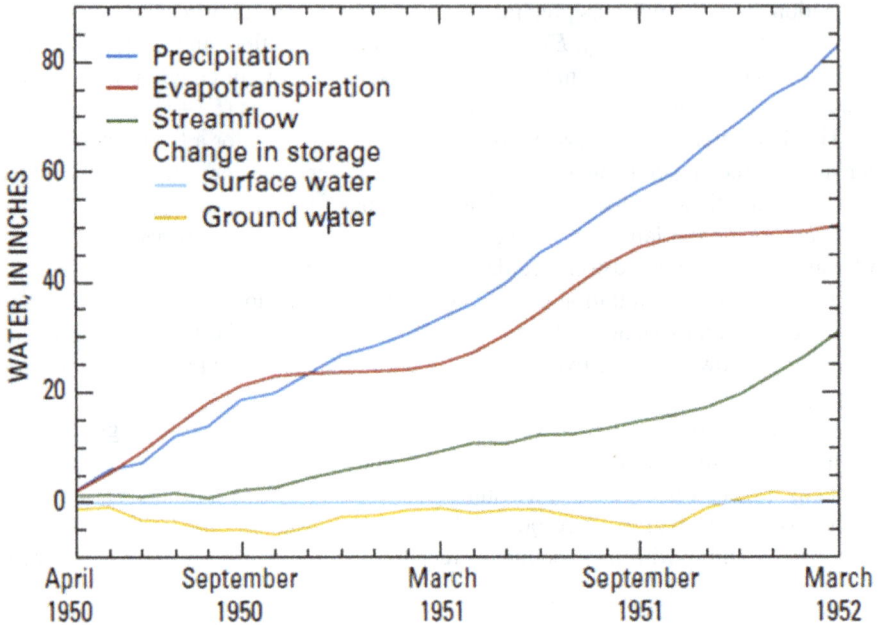

FIGURE 3.4   Water budget for the Beaverdam Creek Basin, Maryland, from April 1950 to March 1952. (From Healy et al., 2007.)

graphs depicting estimates of various elements related to watershed water components. As summarized by Healy et al., at the end of the study period, groundwater recharge was 42.6 in and was divided into 21.5 in of base flow, a 1.7-in increase in groundwater storage, and 19.5 in of evapotranspiration from groundwater. During the summer months, evapotranspiration is the primary draw on groundwater, while for the remainder of the year, base flow is the dominant mechanism of groundwater discharge. Information from such studies is valuable for future planning and for making informed decisions regarding water management.

REFERENCES

Aubinet, M., Vesala, T., & Papale, D. (eds.). (2012). *Eddy Covariance: A Practical Guide to Measurement and Data Analysis*. Springer Atmospheric Sciences, Springer Verlag. https://link.springer.com/book/10.1007/978-94-007-2351-1

Bay of Fundy.Com. (2023). Bay of fundy tides: The highest tides in the World! Retrieved from https://www.bayoffundy.com/about/highest-tides/

CSIR. (2023). Eddy covariance flux towers. https://www.csir.co.za/eddy-covariance-flux-towers

Fetter, C. W., & Kreamer, D. (2022). *Applied Hydrogeology* (5th ed.). Long Grove, IL: Waveland Press.

Harvard Forest. (2021). Eddy flux. Retrieved from https://harvardforest.fas.harvard.edu/research/research-topics/climate-and-carbon-exchange/eddy-flux/?utm_source=chatgpt.com

Healy, R. W., Winter, T. C., LaBaugh, J. W., & Franke, O. L. (2007). Water budgets: Foundations for effective water-resources and environmental management. *U.S. Geological Survey Circular 1308*, 90. https://doi.org/10.3133/cir1308

Kallgan. (2024). Evaporation pan [Photograph]. Wikimedia Commons. Retrieved from https://commons.wikimedia.org/wiki/File:Evaporation_Pan.jpg. Licensed under the Creative Commons Attribution 3.0 Unported License. https://creativecommons.org/licenses/by/3.0/

Rasmussen, W. C., & Andresen, G. E. (1959). *Hydrologic budget of the Beaverdam Creek basin, Maryland (U.S. Geological Survey Water-Supply Paper 1472)*. U.S. Geological Survey. https://doi.org/10.3133/wsp1472

Shade, P. J. (1995). *Water Budget for the Island of Kauai, Hawaii. U.S. Geological Survey, Water-Resources Investigations Report 95–4128*. U.S. Geological Survey. https://doi.org/10.3133/wri954128

U.S. Geological Survey. (2019). The natural water cycle (JPG). Retrieved from https://www.usgs.gov/media/images/natural-water-cycle-jpg

U.S. Geological Survey. (2023). Water cycle. Retrieved from https://www.usgs.gov/special-topics/water-science-school/science/water-cycle

Precipitation

4

4.1 THE PRECIPITATION PROCESS

Evaporated water in the atmosphere condenses to form clouds, which are collections of droplets or small drops of condensed water. While these droplets are large enough to be visible as clouds, they are too small to fall as precipitation because they cannot overcome the updrafts that support the clouds. For precipitation to occur, the droplets must condense on dust, salt, or smoke particles that act as nuclei. When these particles collide, they cause water droplets to grow through additional condensation of water vapor. Droplets will fall out of the cloud as precipitation only if their falling speed exceeds the cloud's updraft speed. However, this process is not very efficient, as it takes millions of cloud droplets to produce a single raindrop. A more efficient mechanism for producing precipitation-sized drops occurs through the rapid growth of ice crystals at the expense of the water vapor present in a cloud. This process is known as the Bergeron–Findeisen process (e.g., Storelvmo and Tan, 2015). These crystals may fall as snow or melt and fall as rain.

4.2 MEASURING RAINFALL

A standard rain gauge is used for manually recording rainfall by collecting falling rainwater in a funnel-shaped collector connected to a graduated measuring tube (Figure 4.1). The collector's total surface area, which includes both the funnel's top and the larger tube, is ten times that of the smaller tube. As a result, the reading in the small tube is magnified by a factor of 10, enhancing precision. Consequently, rainfall can be measured down to a hundredth of an inch. Rainfall amounts that exceed the small tube's capacity are captured in the larger tube, which can be accurately measured by transferring the water to the emptied small tube.

It's important to note that the accuracy of collection can be affected by wind, which may cause splashing. To obtain precise results, the rain gauge should be positioned close to the ground in an open area, with its opening parallel to the surface of the ground.

Recording-type automatic rain gauges offer several advantages over standard gauges by enabling the plotting of rainfall against time, thus facilitating the extraction

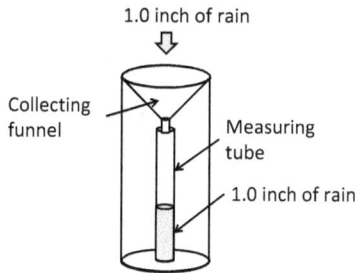

FIGURE 4.1 Standard rain gauge.

of valuable information regarding the intensity and duration of rainfall. Such data play a crucial role in hydrological analysis, specifically in estimating a watershed's response to surface and subsurface flows during storms. Typically, these instruments are installed at fixed locations and provide automatic rainfall records, eliminating the need for manual readings. Commonly used recording rain gauge types include the tipping (or tilting) bucket type, siphon float type, and weighing bucket type (refer to The Constructor, 2023, for further information).

4.3 ESTIMATING AVERAGE RAINFALL

Rainfall can vary significantly across a given area of interest, such as a watershed. Some analyses require estimating the average amount of rainfall over this area, also referred to as the rainfall effective uniform depth (Fetter and Kreamer, 2022). There are three commonly used methods for estimating this average value. The first and simplest method involves calculating the arithmetic mean of rainfall measurements from all rain gauge stations within the watershed. However, this approach may lack accuracy as it disregards the spatial distribution of the stations and does not utilize stations located outside the watershed boundaries.

The second method, known as the Thiessen polygon method, entails creating a series of polygons, each encompassing a single rain gauge station. The precipitation value measured at each station is assigned to the area covered by its corresponding polygon (American Meteorological Society, 2012). It is worth noting that, like all methods, the accuracy of this method tends to improve with an increase in the number of stations.

The third method, known as the isohyets method, involves using available data to establish contour lines of equal rainfall values across the entire watershed. However, this method is considered inferior to the Thiessen method because it requires the creation of new contour lines each time rainfall calculations are needed. In contrast, the Thiessen method's areas around the stations remain constant regardless of changes in rainfall data.

These three methods will be further described through an example. Figure 4.2 provides a schematic representation of a watershed equipped with seven rain gauge stations, three located within the watershed's boundaries and four outside. The average rainfall value can be estimated solely based on the values recorded within the watershed using the arithmetic mean.

$$R_{av} = \frac{6+4.6+3.7}{3} = 4.77 \text{ in}$$

The first step in the Thiessen method is to connect each station to the nearest ones, as illustrated by the dotted lines in Figure 4.3. Stations outside the watershed can also be included in this procedure. Next, bisectors are drawn for each of the dotted lines at right angles (90°), as indicated in Figure 4.4. These lines are then extended in both directions,

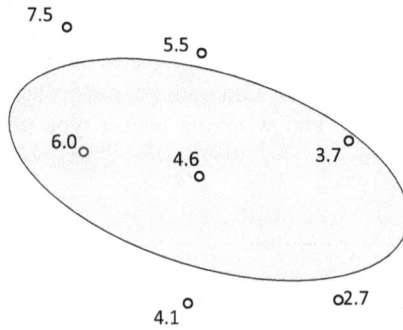

FIGURE 4.2   A watershed with seven rain-gauging stations.

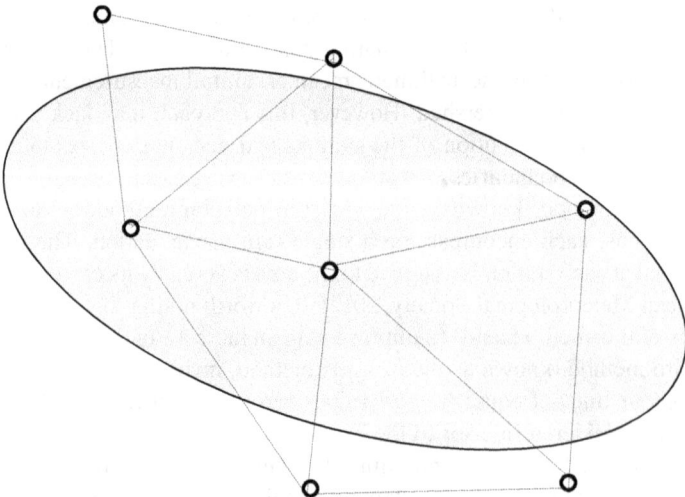

FIGURE 4.3   Connecting the gauges to the nearest neighbors. Lines should not intersect.

creating polygons that are bounded by the watershed boundary (see Figure 4.5). These polygons, which we will refer to as "subareas," represent the segments served by the stations.

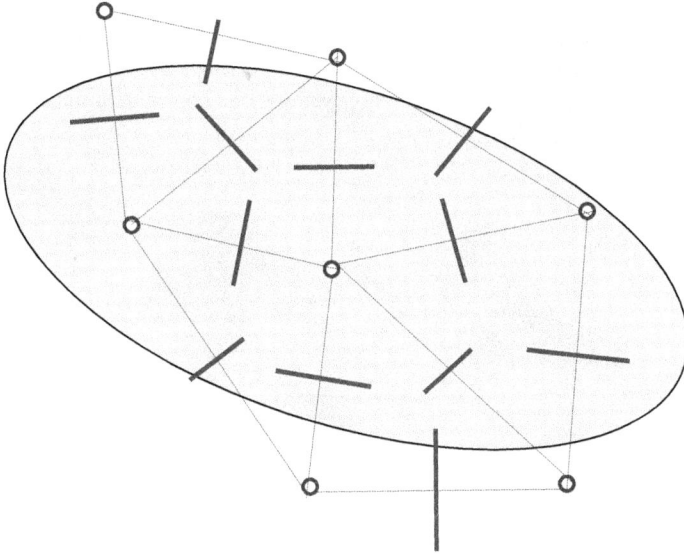

FIGURE 4.4 Bisectors are drawn in the middle of the dotted lines at right angles (90°).

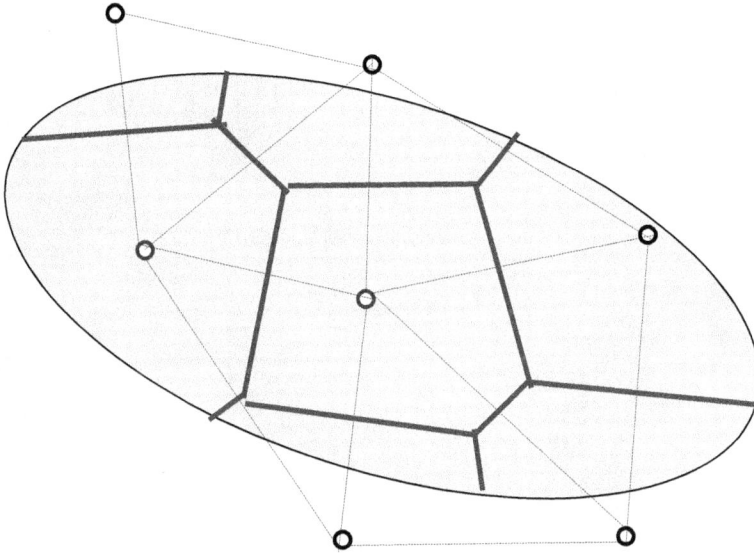

FIGURE 4.5 Bisector lines are extended until they intersect or cut the boundary, forming polygons around the stations.

In the next step, it is necessary to determine the relative size of each subarea in relation to the total area of the watershed for the calculations. In practical applications, a planimeter or more advanced equipment can be used to measure these areas accurately. However, for illustrative purposes, we describe a simple but somewhat tedious approach that involves using square graph paper. This paper is placed over the watershed plot, and the number of squares within each subarea is counted (see Figure 4.6). The sum of the squares in all subareas represents the entire watershed, allowing us to estimate the fraction of each subarea relative to the whole watershed. The calculations for the current example are presented in Table 4.1.

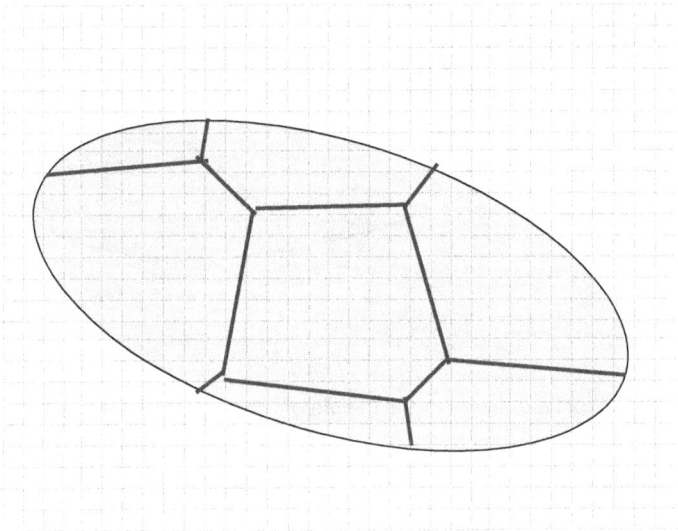

FIGURE 4.6 Square graph paper is used to roughly estimate the fraction of each sub area relative to the total watershed area.

TABLE 4.1   Calculating the average rainfall by the Thiessen method

SUBAREA NUMBER	NUMBER OF SQUARES (A)	FRACTION OF TOTAL AREA (B)	RAINFALL VALUE (C)	WEIGHTED RAINFALL (D)
1	16	0.05	7.5	0.38
2	65	0.21	6.0	1.25
3	39	0.12	5.5	0.69
4	85	0.27	4.6	1.25
5	12	0.04	4.1	0.16
6	60	0.19	3.7	0.71
7	36	0.12	2.7	0.31
Totals	313	1		4.74

The fractions in column B are calculated by dividing the respective values in column A by the total number of squares, which equals 313 in this example. The values in column D are obtained by multiplying the respective values in columns B and C. The average rainfall value is 4.74, determined by summing the values in column D for all subareas. It's important to note that this result is close to the arithmetic mean value, but this is not always the case.

Example 4.1 demonstrates a case with simple geometry that utilizes the Thiessen method. In such cases, there is no need for graph paper.

The isohyets method involves creating contour lines that represent equal rainfall amounts, as depicted in Figure 4.7. This process can be facilitated using readily available computer software. However, for our specific demonstration, we manually draw these lines at 0.5-in increments, as shown in the figures.

Given the limited number of weather stations available, generating precise contours is challenging. Nevertheless, a reasonable starting point is to select the contour labeled as 4.5 in because it closely aligns with the data from more stations compared to others, including stations with readings of 3.7, 4.1, and 4.6 in. Therefore, this contour line should fall between these three stations. The remaining contours are then sketched based on rough interpolation of the labeled values for these stations. In this example, the lowest and highest contour values are 3 and 7.5 in, respectively.

The subsequent step involves calculating the areas of zones located between the contour lines but still within the watershed boundaries by counting the graph paper squares. We then determine the fractions relative to the total watershed area. Finally, we estimate the weighted rainfall values and sum them to find the average rainfall value. The completed calculations can be found in Table 4.2.

To calculate the fractions in column D, divide the respective values in column C by the total number of squares, which is 312 in this example. For column E, multiply the respective values in columns C and D. The resulting average rainfall value is 4.74, calculated by summing the values in column E for all subareas. This value closely aligns with predictions made using other methods.

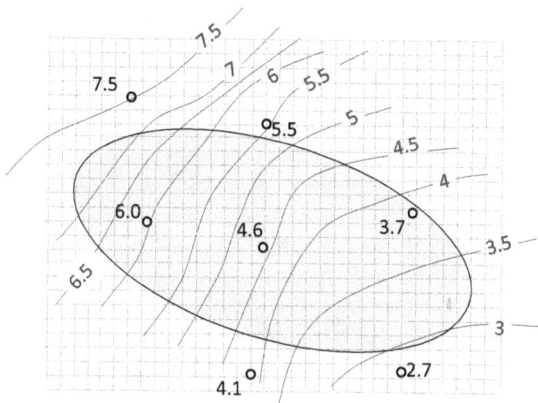

FIGURE 4.7 The isohyets method.

TABLE 4.2   Calculating the average rainfall by the isohyets method

CONTOUR RANGE (A)	AVERAGE VALUE IN THE CONTOUR RANGE (B)	NUMBER OF SQUARES (C)	FRACTION OF TOTAL AREA (D)	WEIGHTED RAINFALL (E)
<3	2.80	1	0.00	0.01
3–3.5	3.25	52	0.17	0.54
3.5– 4	3.75	62	0.20	0.75
4–4.5	4.25	39	0.13	0.53
4.5–5	4.75	38	0.12	0.58
5–5.5	5.25	28	0.09	0.47
5.5–6	5.75	38	0.12	0.70
6–6.5	6.25	25	0.08	0.50
6.5–7	6.75	18	0.06	0.39
7–7.5	7.10	11	0.04	0.25
Totals		312	1	4.72

Example 4.1: Simple Geometry Case with the Thiessen Method

Estimate the average rainfall for the geometry shown in Figure 4.8 using the Thiessen method. The area is a square with sides of 2 km each. Additionally, estimate the total rainfall amount in cubic meters. There are five stations located as shown, with rainfall amounts marked in inches.

The steps of the calculations are illustrated and described below.

Step 1: Connect the stations with dotted lines and draw the bisectors (Figure 4.9):

Step 2: Extend the bisector lines until they meet (Figure 4.10):

Step 3: Define the relevant areas for each station and complete Table 4.3 based on the geometries in Figure 4.11. Areas highlighted in light gray pertain to edge stations, while the darker gray area belongs to the central station. The area calculation is based on triangles, which involves multiplying half of the height by the base. Each edge area equals 0.5 km².

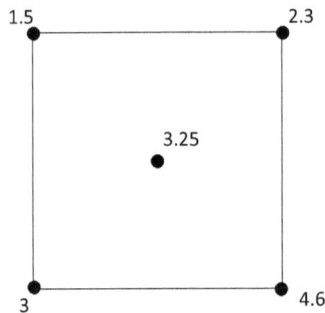

FIGURE 4.8   Geometry used for Example 4.1.

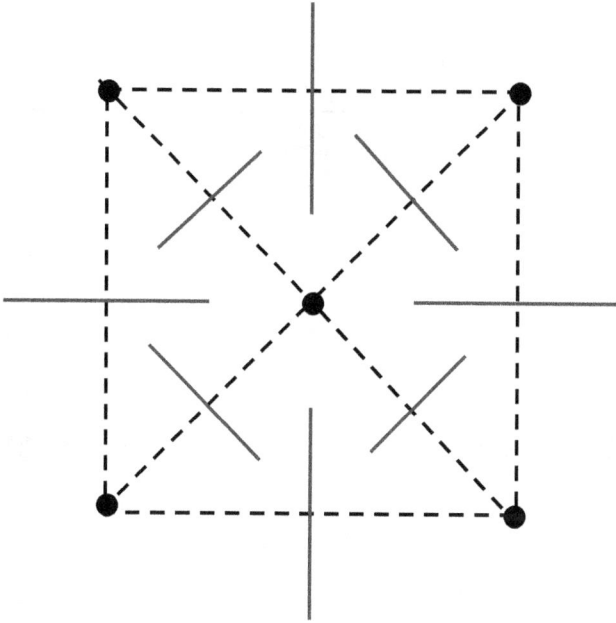

FIGURE 4.9 Step 1 for Example 4.1: Connecting the stations with the dotted lines and drawing the bisectors.

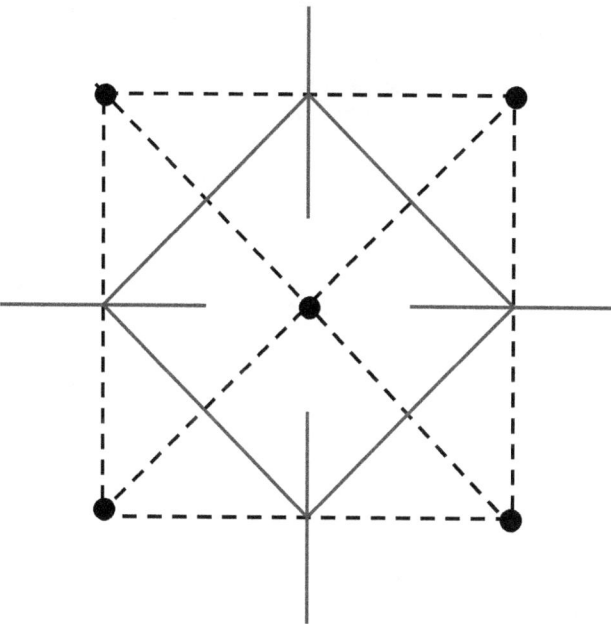

FIGURE 4.10 Step 2 for Example 4.1: Extending the bisector lines until they intersect.

TABLE 4.3 Data for Example 4.1

AREA (KM²)	FRACTION OF TOTAL AREA	RAIN AMOUNT (IN)	WEIGHTED RAINFALL (IN)
0.5	0.125	1.5	0.1875
0.5	0.125	2.3	0.2875
0.5	0.125	3	0.375
0.5	0.125	4.6	0.575
2	0.5	3.25	1.625
Totals 4	1		3.05

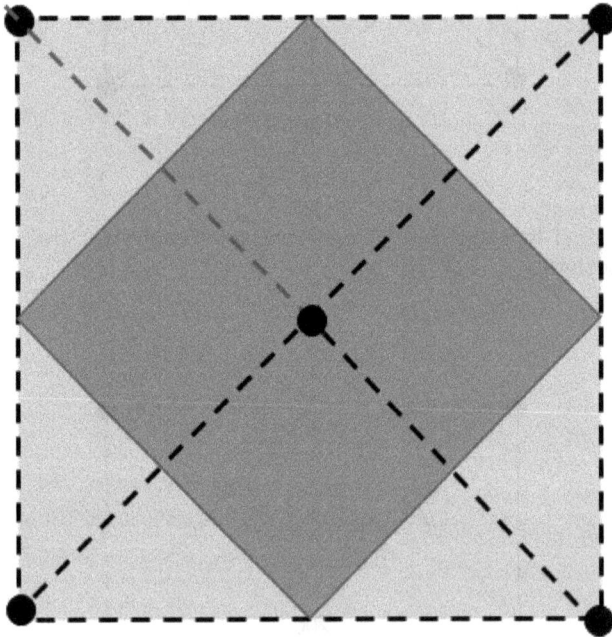

FIGURE 4.11 Step 3 for Example 4.1: Defining the relevant areas for each station.

The central area comprises four triangles, each with an area of 0.5 km², totaling 2 km². To determine the fractions of the total area, divide each area by the overall 4 km².

The answer is 3.05 in (the sum of all values in the last column). Using consistent units of meters, the total rainfall amount is estimated by multiplying this average value by the total area:

$$\frac{3.05 \text{ in} \times 2.54 \dfrac{\text{cm}}{\text{in}}}{100 \dfrac{\text{cm}}{\text{m}}} \times 2 \text{ km} \frac{1,000 \text{ m}}{\text{km}} \times 2 \text{ km} \frac{1,000 \text{ m}}{\text{km}} = 309,880 \text{ m}^3$$

Where 1 in is about 2.54 cm, 1 m is 100 cm, and 1 km is 1,000 m. This value can be converted to gallons by multiplying by 264.2 to yield about 82 million gallons.

As previously mentioned, rainfall exhibits non-uniform spatial distribution. For instance, on the island of Oahu, Hawaii, the average annual rainfall varies significantly, ranging from approximately 500 to around 7,000 mm (refer to Figure 4.12). The availability of surface and groundwater resources depends on this distribution, potentially resulting in water shortages in areas with higher demand. It is important to note, however, that these average values fluctuate seasonally and over the years. For example, in Hilo, Hawaii, the annual average is about 3,500 mm, but January experiences variations ranging from 5 to 1,300 mm.

Figure 4.13 presents the average rainfall data for the continental United States, which also displays substantial variability in precipitation rates. Regions with low rainfall, such as the State of Nevada, heavily rely on groundwater resources due to the lack or unreliability of surface water sources. In some cases, water must be transported within a state, between states, or even across international borders.

Figure 4.14 provides an overview of the global distribution of mean annual rainfall values for the year 2020 (Our World in Data, 2024), highlighting

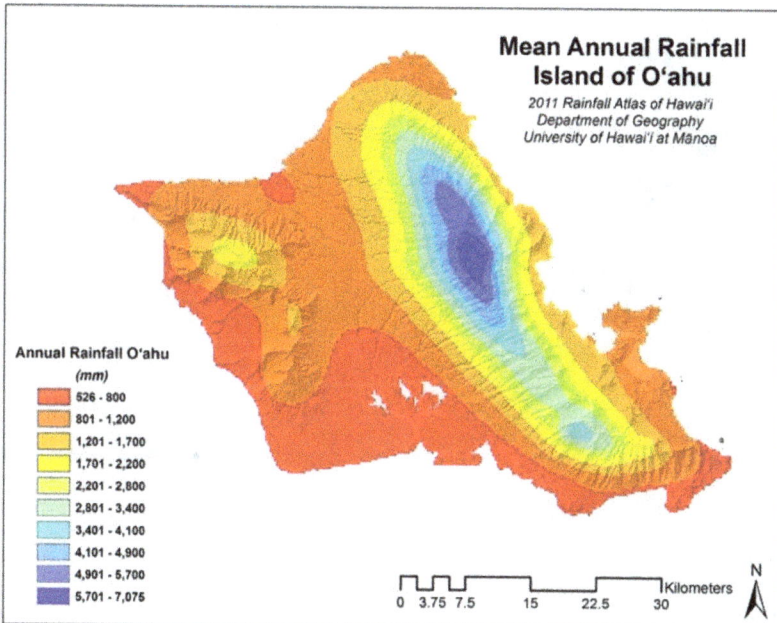

FIGURE 4.12 Average rainfall on the Island of Oahu, Hawaii (Giambelluca et al., 2013).

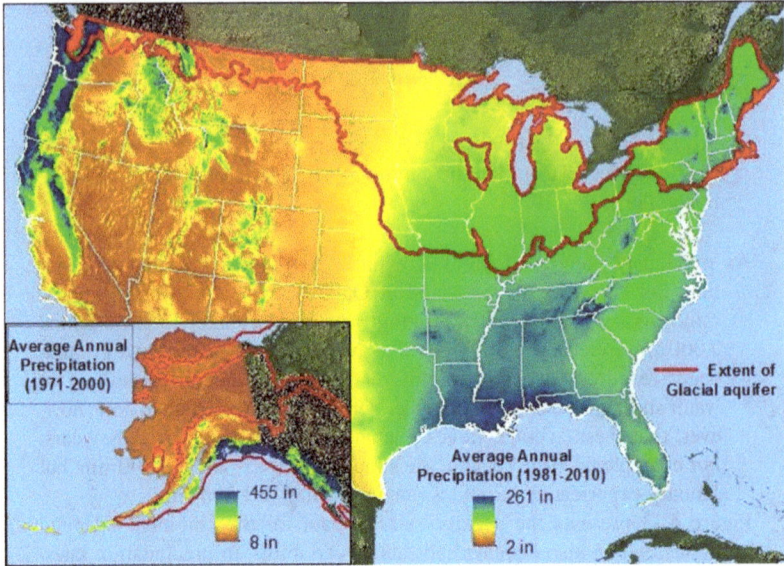

FIGURE 4.13   Average rainfall across the continental United States (USGS, 2023).

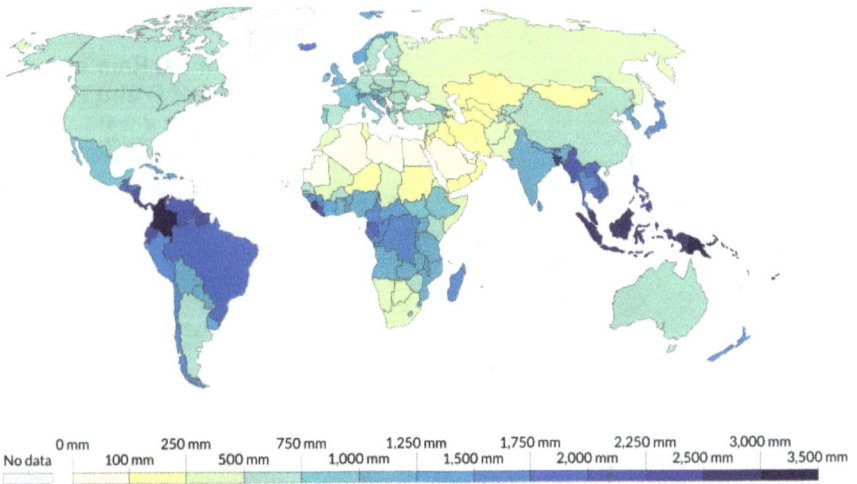

FIGURE 4.14 Average precipitation per year. Adapted from Average precipitation per year (OWID 0083) by Our World in Data, licensed under the Creative Commons Attribution 3.0 Unported License. Available at: https://commons.wikimedia.org/wiki/File:Average-precipitation-per-year_(OWID_0083).png. License details: https://creativecommons.org/licenses/by/3.0/.

significant disparities. Many nations suffer from water scarcity, and prolonged periods of drought can lead to humanitarian crises. As discussed in Section 1.8, climate change has intensified extreme weather conditions, including severe storms and extended droughts. These storms have increased the frequency and severity of floods, while potential conflicts can arise between countries sharing river resources.

4.4 ASSIGNMENTS

1. Redo the calculations in Example 4.1 for a watershed measuring 10 by 10 km, using the same data given. Use the new calculations to gain insights regarding the community water demand. Assume this watershed is located in California, where the average domestic use is 100 gallons per day per person, with a population of 10,000 in the area. Additional demand for agriculture is 50 gallons per day. Groundwater recharge is approximately 40% of rainfall, with the remainder either evaporating or running off.
2. Redo Example 4.1 assuming that the lower right station is not present. Hint: Use the area values calculated in the example to estimate the new areas.
3. Apply the three methods discussed in the text to estimate the average rainfall for the watershed depicted in Figure 4.15.

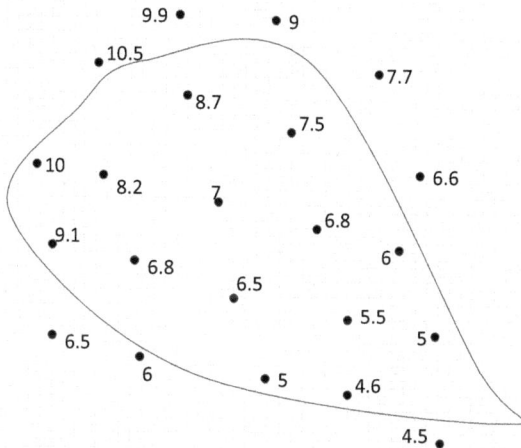

FIGURE 4.15 Watershed boundaries and rainfall stations for assignment 3. Numbers refer to rainfall amounts in centimeters.

REFERENCES

American Meteorological Society. (2012). Thiessen polygon method. Retrieved from https://glossary.ametsoc.org/wiki/Thiessen_polygon_method

Fetter, C. W., & Kreamer, D. (2022). *Applied Hydrogeology* (5th ed.). Long Grove, IL: Waveland Press.

Giambelluca, T. W., Chen, Q., Frazier, A. G., Price, J. P., Chen, Y.-L., Chu, P.-S., Eischeid, J. K., & Delparte, D. M. (2013). Online rainfall atlas of Hawai'i. *Bulletin of the American Meteorological Society, 94*(3), 313–316. https://doi.org/10.1175/BAMS-D-11-00228.1

Licensed under the Creative Commons Attribution 3.0 Unported License. Retrieved from https://creativecommons.org/licenses/by/3.0/

Our World in Data. (2024). Average precipitation per year (OWID 0083) [Image]. Wikimedia Commons. Retrieved from https://commons.wikimedia.org/wiki/File:Average-precipitation-per-year_(OWID_0083).png

Storelvmo, T., & Tan, I. (2015). The Wegener-Bergeron-Findeisen process – Its discovery and vital importance for weather and climate. *Meteorologische Zeitschrift, 24*(4), 455–461. https://doi.org/10.1127/metz/2015/0626

The Constructor. (2023). Types of rain gauges for measuring rainfall data. Retrieved from https://theconstructor.org/water-resources/types-of-rain-gauges/12801/

U.S. Geological Survey. (2023). Map of annual average precipitation in the U.S. from 1981 to 2010. Retrieved from https://www.usgs.gov/media/images/map-annual-average-precipitation-us-1981-2010

Surface Water

5

5.1 LAKES

A lake is a component of Earth's surface water, accumulating water in a depression relative to the surrounding terrain. Lakes are highly valued for their recreational, aesthetic, and water-supply benefits, as well as their critical roles in providing habitats and food resources for fish, aquatic life, and wildlife. Lake water can originate from streams, surface runoff, or groundwater seepage, while water can exit through streams, ground seepage, evaporation, or human withdrawals. The storage of water in a lake, as indicated by water levels, relies on the balance between total inflows and total outflows. For instance, during drought conditions, water levels and lake storage decline due to reduced inflows relative to outflows.

Lake sizes can range from small fishing ponds to major bodies of water. Notable examples of the latter category include Lake Baikal in Siberia and Lake Michigan in the United States, with respective areas of 31,500 and 58,000 km^2 (Safaris African, 2023). While most lakes contain freshwater, some can be classified as saline lakes depending on their water sources and drainage characteristics. For instance, the Great Salt Lake in Utah has higher salinity levels than the oceans (Utah Division of Water Resources, 2025).

The character and functions of a lake's ecosystem depend on various factors, including the lake's physical attributes, climate, geological features of the surrounding drainage areas (referred to as the lake-shed), dimensions, and climate that influence water circulation, temperature, chemical composition, and biological characteristics. Climate and the geological features of the lake-shed predominantly affect elements of the hydrologic cycle. Surface-water runoff, chemical inputs, and sediment loads entering the lake are influenced by overall lake-shed features such as topography, size, and land use/cover.

Unfortunately, the fundamental functions of lakes can be adversely impacted by rapid environmental changes due to their fragile nature. Key sources of environmental problems are primarily human-induced, stemming from land-use practices, excessive water consumption, and illegal waste disposal, with pollutants entering lakes through surface runoff or subsurface flow. For instance, nutrients like phosphorus and nitrogen from agricultural lands can lead to algal overgrowth (eutrophication), while bacterial activities can harm the lake's wildlife. Climate change, manifested through persistent

DOI: 10.1201/9781003587149-5

droughts and rising temperatures, also has detrimental physical, chemical, and biological effects on lakes.

Mono Lake in California provides a clear example of environmental challenges resulting from both natural and human activities, including climate change and water overuse (Mono Lake Committee, 2023; Herbst, 2014). Droughts and diversion of the lake's water for domestic purposes disrupted the natural water cycle that maintains the lake's delicate balance. As documented by Fetter and Kreamer (2022), these environmental issues stem from a decline in the lake's water volume, resulting in increased salinity. Such changes can negatively impact businesses reliant on brine shrimp from the lake, as well as migratory birds and other species that depend on lake-based food sources, including insects. According to Fetter and Kreamer, it would take more than 20 years to reach a target water level set by regulatory agencies that eliminated water diversion from the lake.

5.2 OPEN CHANNELS

As discussed in Chapter 3, precipitation water is divided into subsurface and surface components. The surface component comprises water that moves in open channels under the influence of gravity, flowing from higher to lower elevations. These channels may be referred to as rivers, streams, or creeks, depending on their size, with each term denoting progressively smaller watercourses. From a scientific perspective, these terms are used interchangeably, and the discussion should not be affected by such differences.

Figures 5.1 and 5.2 depict views of water basins or watersheds. A watershed is defined by its topography and is delineated by ridges or mountain crests. Watersheds can be subdivided into smaller units, each served by a single stream branch or tributary. These tributaries may converge to form larger streams or rivers. In the figures, major streams flow through valleys, collecting water from surface runoff as well as contributions from tributaries.

In Figure 5.1, the mountain crest separating the two watersheds serves as a dividing line. Tributaries flow either east or west of this line, directing streams toward their respective outflow points. The streams on the left side flow into a flat area, where most do not directly reach the ocean, represented by the black-shaded area. A large number of tributaries on the steeper right-side topography flow into a few streams that traverse valleys with gentler slopes. These streams eventually enter a neighboring watershed through a single channel at the bottom of the map before reaching the ocean.

As shown in Figure 5.2, various land uses are present, including forests, agricultural areas, and urban zones. Depending on these land-use practices, potential sources of contaminants may exist, which can significantly impact the water quality in the streams and other receiving surface water bodies.

In principle, during its descent down a slope, some of the stream water may evaporate, while some may infiltrate into the ground if the water level in the stream, referred to as the stream stage, exceeds the water table in an underlying aquifer (as illustrated

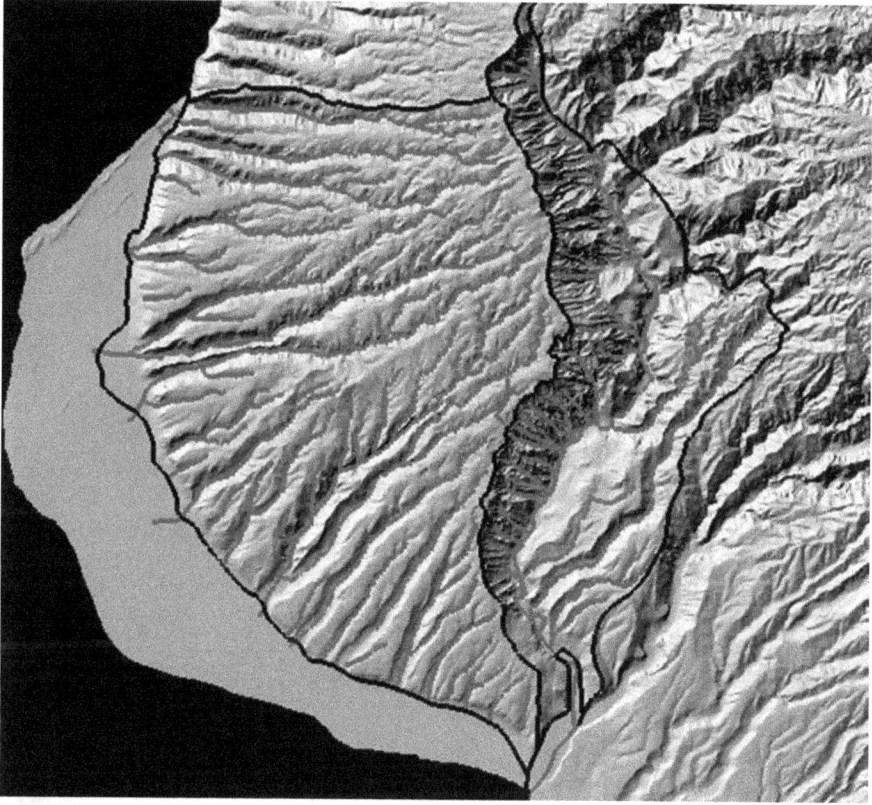

FIGURE 5.1 Example of two neighboring watersheds illustrating the main streams and all respective tributaries. The topography sets the boundaries for each watershed.

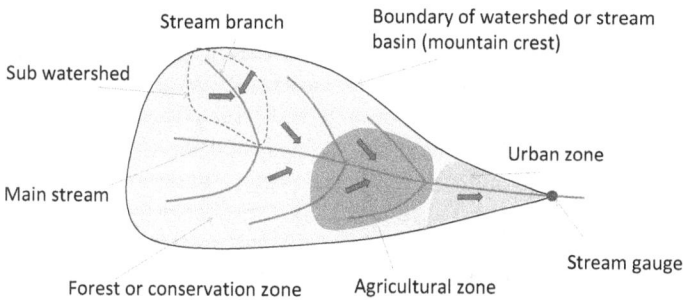

FIGURE 5.2 A schematic representation of a watershed.

in Figure 5.3). Conversely, in some sections of the stream, water may be supplied from a groundwater aquifer if the water table is higher than the stream stage (as depicted in Figure 5.4). These scenarios categorize the stream as either a losing or gaining stream,

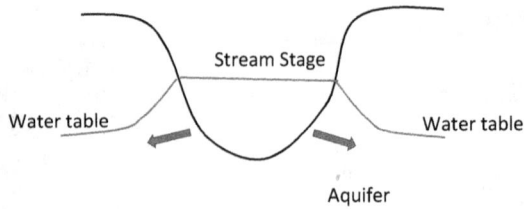

FIGURE 5.3   A schematic representation of a losing stream.

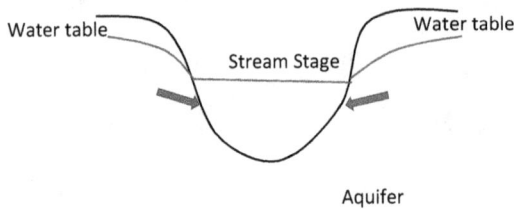

FIGURE 5.4   A schematic representation of a gaining stream.

depending on whether it loses or gains water. It is important to note that these conditions can vary over time and along the length of the stream, contingent upon the specific conditions of the site, which primarily determine the relative elevations of the stream stage and the water table.

As illustrated in Figure 5.2, water can exit at a single point or outlet, connecting a river to the ocean. Constructing a flow gauging station at such a point is ideal for managing stream uses. Changes can occur relative to the schematics in Figures 5.1 and 5.2. For example, a lake will form if water flows to a location surrounded by higher land on all sides. If a dam is built across a stream to hinder its flow, a lake that forms upstream of the dam will be a reservoir.

A water budget for surface water can be evaluated for a watershed by treating it as an independent unit. It's important to note that, in theory, water should not travel between neighboring watersheds. Unfortunately, the same principle does not apply to the groundwater component because watershed boundaries do not necessarily coincide with groundwater aquifer boundaries. Additionally, treating the watershed as an independent unit is not applicable in cases involving stream diversions, as they allow water to be shared between watersheds.

According to the USGS, approximately 70% of the freshwater used in the United States in 2015 came from surface-water sources, including streams and freshwater lakes (USGS, 2023). The remaining 30% came from groundwater. Surface water is a vital natural resource used for drinking and other public purposes, irrigation, and by the thermoelectric power industry to cool electricity-generating equipment.

5.2.1 Bernoulli Equation and Open Channel Flow

The fundamentals and applications of the Bernoulli equation in open channel flow are detailed in many publications, including Chow (1959), Henderson (1966), and Subramanya (2009). Open channel flow refers to flow in which the water surface is at atmospheric pressure, as is the case with streams and rivers. It is also applicable to sub-surface drainage pipes that are not running full. The Bernoulli equation is derived from the principle of conservation of energy, which states that the sum of all forms of energy in the flowing water remains constant at all points along a streamline. The equation can be expressed as follows (see Figure 5.5):

$$\frac{v^2}{2g} + z + \frac{P}{\rho g} = \frac{v^2}{2g} + z + d = \text{constant} \tag{5.1}$$

In equation (5.1), v is the fluid flow speed at a point on a streamline, g is the acceleration due to gravity, z is the elevation of the point above an arbitrary level, p is the pressure at the point, and ρ is the density of water. Since the water surface is at atmospheric pressure, the water depth d can be used in place of the pressure energy term. The three terms in equation (5.1) represent different forms of energy: kinetic, potential, and pressure energy, respectively, and are summed to define the total energy. These terms are expressed in equation (5.1) in terms of energy per unit weight of water, in units of length such as meters or feet.

According to Bernoulli's principle, an increase in water velocity results in an increase in kinetic energy, which must be balanced by a decrease in other forms of energy. Referring to sections 1 and 2 in Figure 5.5, equation (5.1) also implies that:

$$\frac{v_1^2}{2g} + z_1 + d_1 = \frac{v_2^2}{2g} + z_2 + d_2 \tag{5.2}$$

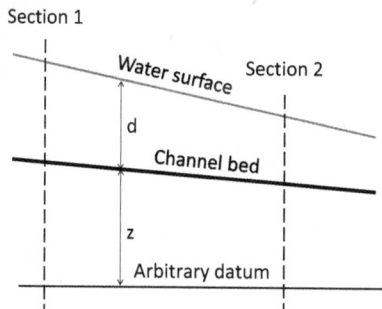

FIGURE 5.5   A channel schematic view along the flow direction.

The Bernoulli equation is valid under the following assumptions:

1. The flow must be in a steady-state condition, meaning the velocity, pressure, and density remain unchanged at any given point.
2. The fluid (water, in this case) must be incompressible, even when the pressure varies.
3. Friction due to viscous forces is neglected.

Considering the principle of conservation, the Bernoulli equation can be used to estimate friction losses between the two sections if all variables in the equation are known. This would be represented by the difference in the total energies at the two sections. Additionally, the equation can be combined with the mass conservation equation to analyze open channel flow, which states that:

$$Q = A_1 v_1 = A_2 v_2 \tag{5.3}$$

where A_1 and A_2 are the areas of the two sections of interest.

Example 5.1: Bernoulli Equation

Consider Figure 5.6, which illustrates the flow through a channel passing beneath a sluice gate, where the water depth is known at the two sections shown. The channel has a rectangular cross section with a width of 18 m. Assuming the Bernoulli equation is valid, estimate the water velocities at the two sections.

With the datum chosen at the channel bed, equation (5.2) yields:

$$\frac{v_1^2}{2 \times 9.81} + 0 + 10 = \frac{v_2^2}{2 \times 9.81} + 0 + 2 \tag{5.4}$$

While equation (5.3) yields:

$$Q = 18 \times 10 \times v_1 = 18 \times 2 \times v_2 \tag{5.5}$$

Equation (5.5) indicates that $v_2 = 5v_1$. Substituting this into equation (5.4) provides a value of $v_1 = 2.56$ m/s, which gives a corresponding value of v_2 as five times that, or 12.79 m/s. These operations, known as elimination and back substitution, are used to solve the two simultaneous equations with the two unknowns v_1 and v_2. Note that the results are independent of the channel width, as it is the same for both

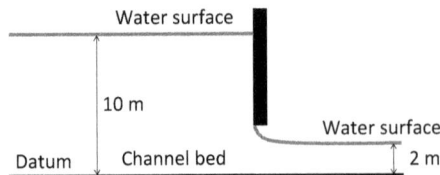

FIGURE 5.6 A channel schematic view along the flow direction (with a sluice gate).

cross sections. Either side of equation (5.4) can be used to estimate the total energy, which would be:

$$\frac{2.56^2}{2 \times 9.81} + 10 = 10.33 \text{ m}$$

if the left side is used.

It should be noted that energy loss can be estimated if the velocities are independently determined, such as by using the techniques discussed in the following sections.

5.2.2 Measuring Flow in Streams

Flow in streams can be measured using both direct and indirect methods (e.g., APPROPEDIA, 2023). Direct techniques involve the use of flow meters to measure flow velocities and weirs, which are structures that can estimate water flow by measuring the water level at the weir site. Indirectly, Manning's equation is utilized to estimate water velocity by measuring the water slope at a particular site and estimating a stream roughness coefficient.

5.2.2.1 Flow Meters

Flow meters are devices used to directly measure stream flow by gauging water velocity and then multiplying it by the stream's cross-sectional area. One common device is the Pygmy meter (see Figure 5.7), which consists of a wheel rotated by water flow. The wheel's rotation rate is calibrated to estimate water velocity.

FIGURE 5.7   The Pygmy meter (USGS, 2018).

The procedure begins by dividing the channel cross section into subsections (as shown in Figure 5.8). The flow meter is employed to measure water velocity at a predetermined point within each subsection, typically along a marked line, suspended cableway, or bridge across the stream. Water depth is measured at the centerline of each subsection. The average velocity within each subsection is estimated by lowering the meter to a point (as depicted in Figure 5.9) positioned at approximately 60% of the water depth.[1] These velocity and depth measurements are used to calculate the discharge for each subsection, as illustrated in Figure 5.8. The total discharge is the sum of discharges from all subsections. Naturally, more accurate estimations are achieved by using a larger number of subsections, but this also requires additional labor.

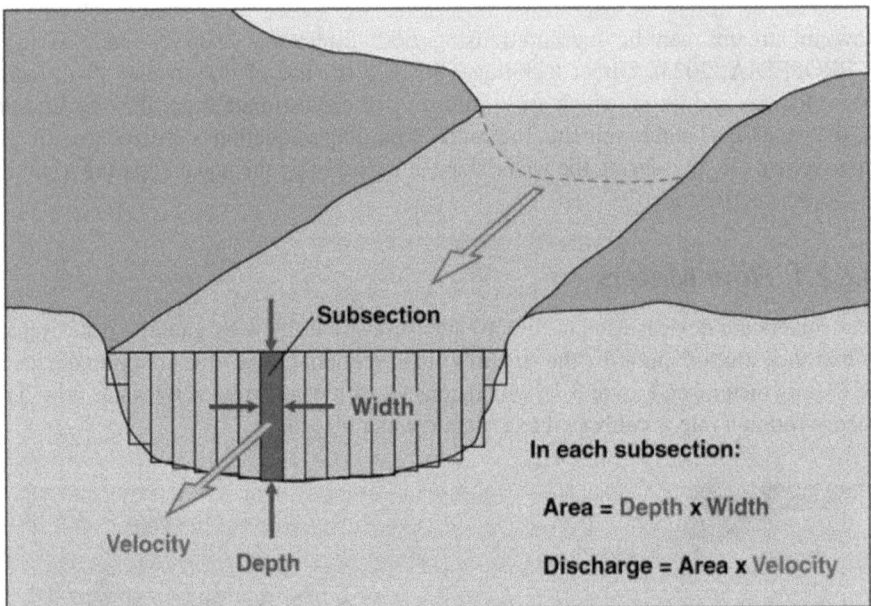

FIGURE 5.8 Stream section calculation scheme (USGS, 2018).

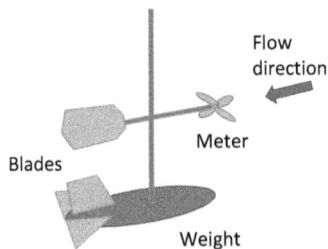

FIGURE 5.9 Sketch of a flow meter setup. The weight is used to hold the meter in place, while the blades keep the meter aligned with the direction of flow.

This calculation procedure is repeated for different river stages, creating a plot known as the stream rating curve, which establishes the relationship between the river stage and discharge. Figure 5.10 provides an example of such a curve. Over time, discharge can be directly estimated from this plot for a given stage, eliminating the need to measure velocity. Stream stage can be manually measured using a measuring stick or through more advanced techniques, such as the use of a water level sensor like the one displayed in Figure 5.11. These measurements can be automatically reported to the appropriate office or agency.

5.2.2.2 Weirs

Weirs are structures constructed across small to medium-sized streams, typically spanning a few meters, for measuring flow (see Figure 5.12). In this setup, water overflows over the top of the weir, as illustrated in the figure. Various types of weirs can be employed, including rectangular and V-notch shapes (refer to Figure 5.13). The water level, denoted as "H," is precisely measured, and an appropriate equation is utilized to estimate the flow rate. To ensure precise measurements, it is crucial to direct all stream water into the weir while also preventing sediment accumulation. Depending on the specific type of weir in use, the following equations are applied to estimate the flow rate within the stream. For a rectangular weir, the equation is as follows (Fetter and Kreamer, 2022):

$$Q = 1.84(L - 0.2H)H^{2/3} \tag{5.6}$$

In equation (5.6), Q is the discharge rate in cubic meters per second, L is the length of the weir crest in meters, and H is the height of the water above the crest, also in meters. For the V-notch weir, the equation reads:

$$Q = 1.379H^{2.5} \tag{5.7}$$

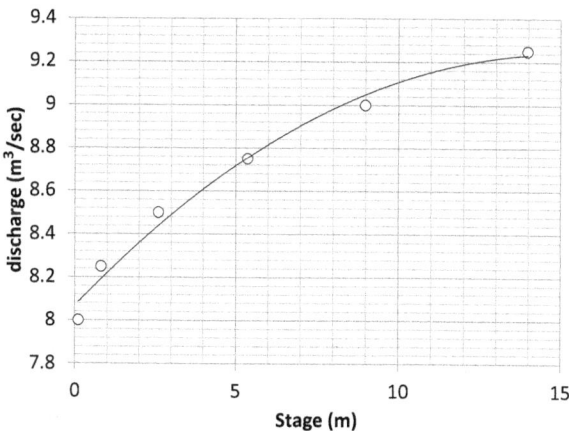

FIGURE 5.10   Example of a stream rating curve.

FIGURE 5.11   A sensor used to measure the stream stage.

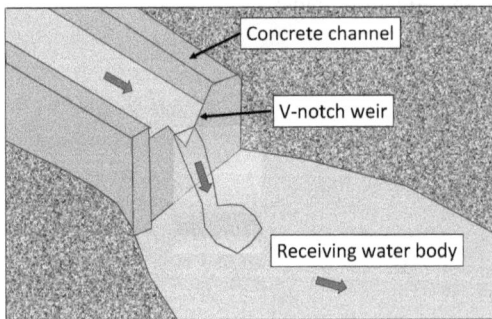

FIGURE 5.12   A sketch of a V-notch weir, which is used in measuring flow in channels or streams.

Equation (5.7) is valid for a notch angle of ninety degrees. It is essential to utilize these two equations with meter units for L and H, as this will result in Q being expressed in cubic meters per second. When using feet units, it requires converting L and H to meters (by multiplying each by 0.305), and subsequently, converting the resulting Q to cubic feet per second by multiplying the value by 34.25.

(a)   (b)

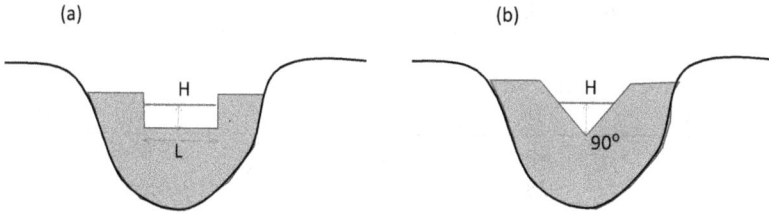

FIGURE 5.13 Schematic plot of (a) rectangular weir, and (b) V-notch weir.

Example 5.2 demonstrates applications to estimate stream discharge via weirs. It should be noted that the equations listed above provide theoretical expressions for water flow. However, measured flow rates are typically lower due to friction and other natural field conditions. The ratio between the actual and theoretical flow rates is termed the coefficient of discharge. Example 5.3 describes the method for estimating a weir's coefficient of discharge.

Example 5.2: Weir Discharge

A rectangular weir of length 3 m across a stream showed a height of water above the crest of 1.5 m. Estimate the discharge.

$$Q = 1.84(L - 0.2H)H^{2/3} = 1.84 \times (3 - 0.2 \times 1.5) \times 1.5^{2/3} = 6.51 \text{ m}^3/\text{s}$$

Assuming a V-notch weir is replacing the rectangular weir, estimate the height of water in this case.
The following equation:

$$Q = 1.379H^{2.5}$$

can be rearranged to estimate H:

$$H = \left(\frac{Q}{1.379}\right)^{1/2.5} = \left(\frac{6.51}{1.379}\right)^{0.4} = 1.86 \text{ m}$$

Example 5.3: Weirs Coefficient of Discharge

A laboratory experiment is used to estimate the coefficient of discharge for a rectangular weir of length $L = 4$ cm (0.04 m) across a laboratory channel. The results are shown in Table 5.1 for a number of runs, specifically the values of the collected volume of water and time of collection, and the height of water above the crest (H).
Measured discharges are first estimated by dividing the collected volumes by the times of collection. Next, the theoretical discharge for each run is estimated by using the following equation:

$$Q = 1.84(L - 0.2H)H^{2/3}$$

TABLE 5.1   Data for Example 5.3

RUN NUMBER	MEASURED VOLUME (M³)	TIME OF COLLECTION (S)	MEASURED DISCHARGE (M³/S)	H (M)	THEORETICAL DISCHARGE (M³/S)	COEFFICIENT OF DISCHARGE
1	0.06	26	2.31E-03	0.0136	3.91E-03	0.590
2	0.06	23	2.61E-03	0.0182	4.63E-03	0.564
3	0.06	20	3.00E-03	0.0209	5.00E-03	0.600
4	0.06	18	3.33E-03	0.027	5.73E-03	0.582
5	0.06	24	2.50E-03	0.0161	4.31E-03	0.579
6	0.06	16	3.75E-03	0.0355	6.54E-03	0.573

Note that L and H should be in meters to estimate Q in m³/s. Finally, the coefficient of discharge for each run is estimated by dividing the measured discharge by the respective theoretical discharge. The completed calculations are shown in the table.

Figure 5.14 illustrates the relationship between H and both measured and theoretical discharges, as well as the value of the coefficient of discharge (CD). The average CD is about 0.61. A linear relationship reasonably approximates the relationships between H and both discharges. The value of the coefficient of discharge is close to 0.6, which roughly agrees with the theoretical value.

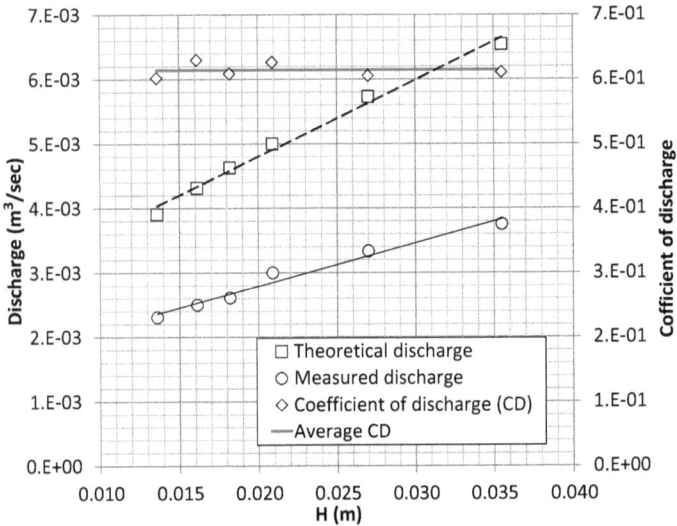

FIGURE 5.14   The relationship between H and both measured and theoretical discharges, as well as the coefficient of discharge for Example 5.2.

5.2.2.3 The Manning Equation

The Manning equation (see, e.g., Fetter and Kreamer, 2022) can be used to estimate the average velocity in a stream, which when multiplied by the area of the stream, would provide the discharge. The equation reads

$$V = k_m \frac{R^{2/3} S^{1/2}}{n} \tag{5.8}$$

In equation (5.8), V is the average velocity, S is the water slope at that location (the drop in water stage between two points divided by the distance separating those points), R is the hydraulic radius, defined by

$$R = \frac{\text{Cross section area } A}{\text{Cross section wetted perimeter } p} \tag{5.9}$$

The cross-sectional area (A) and wetted perimeter (p) are defined in Figure 5.15. The variable n in equation (5.8) is known as the Manning roughness coefficient, which reflects the resistance of water flow in the channel. Therefore, n should be small for smooth channel materials, such as smooth concrete channels ($n = 0.012$), and large for rough materials, such as mountain streams ($n = 0.05$) (Fetter and Kreamer, 2022).

The value of the factor K_m in equation (5.8) depends on the units used. This equation is empirical in nature and is valid for both meter-seconds and foot-seconds units, where the factor K_m equals unity for the former and 1.49 for the latter. Consequently, for V in meters per second, the area A should be in square meters, R and p in meters, the slope S is dimensionless (meter drop per meter distance; no units), and n is also dimensionless. For V in feet per second, the respective units for A, R, and p are square feet, feet, and feet, and again, S and n are dimensionless. After estimating the stream discharge, which is the velocity multiplied by the stream area, a rating curve can be plotted to estimate discharge for a given stream stage.

Example 5.4: Stream Discharge via Manning Equation

For the cross section illustrated in Figure 5.15, various stream stages along with their respective area A and wetted perimeter p are estimated and listed in Table 5.2. The water slope S is about 0.01, and Manning n is 0.04. Estimate the discharge and plot the rating curve (the data in the table is based on DimensionEngine, 2015).

Equation (5.8) is used to calculate V, i.e.,

$$V = k_m \frac{R^{2/3} S^{1/2}}{n}$$

with $k_m = 1$, which is then multiplied by the respective area A to estimate the discharge for each area. A spreadsheet would be ideal for completing the calculations (see Appendix A). For example, for the first row in Table 5.3, $R = A/p = 0.11$, $R^{2/3} = 0.23$, $S^{1/2} = 0.1$, and $n = 0.04$. The resulting $V = 0.57$ m/s and $Q = VA$ is 0.11 m³/s.

Figure 5.16 displays the stage versus discharge (the circles). The line is a best-fit line (called a trendline in Excel©) that can be used to estimate discharge for a given stage without having to measure water velocity and calculate stream area (see Section 2.3.4).

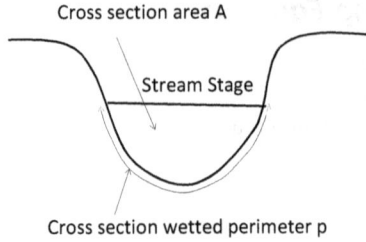

FIGURE 5.15 Sketch of a stream cross section showing the definition of stream stage, cross section A, and wetted perimeter p.

TABLE 5.2 Data for Example 5.4

STAGE M	AREA A M²	WETTED PERIMETER (P) M	HYDRAULIC RADIUS (R) M	$R^{2/3} M^{2/3}$	$S^{1/2}$	VELOCITY (V) M/S	DISCHARGE (Q) M³/S
8.00	0.19	1.75	0.11	0.23	0.10	0.57	0.11
8.25	0.89	3.99	0.22	0.37	0.10	0.92	0.81
8.50	1.97	5.14	0.38	0.53	0.10	1.32	2.60
8.75	3.30	6.30	0.52	0.65	0.10	1.62	5.36
9.00	4.92	7.87	0.63	0.73	0.10	1.83	8.99
9.25	6.96	9.67	0.72	0.80	0.10	2.01	13.97

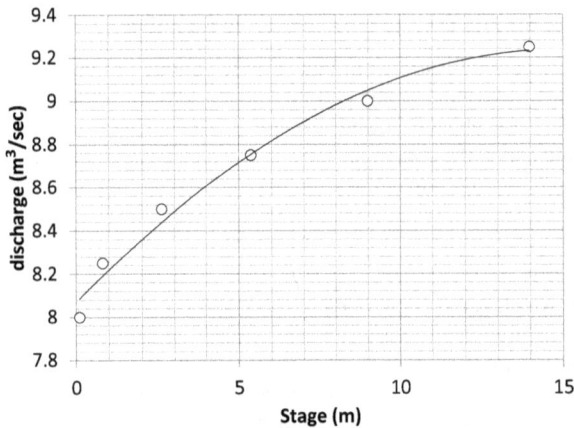

FIGURE 5.16 River stage versus the discharge (the circles) and the best fit line for Example 5.3.

5.2.3 Streamflow Hydrograph

A streamflow hydrograph is a graphical representation of discharge over time at a specific stream station. This time period can range from short durations, such as representing an individual rainstorm over a few hours, to continuous records of flow spanning many decades. Streamflow data serve various critical purposes, including forecasting flow conditions and floods, making water management decisions, assessing water availability, managing water quality, and meeting legal requirements (USGS, 2023).

The United States Geological Survey (USGS) maintains a comprehensive online repository of streamflow data (USGS, 2023). Based on such data, Figures 5.17–5.19 depict daily flow hydrographs for three river sites in the United States, covering the period from January 2020 to February 2023. These figures also compare the discharge values to the median[2] daily discharge values. This comparison aids in assessing whether the discharges are above or below their "normal" values. The inset in the figures lists the number of years used to calculate the medians. Notably, these records can also be employed for predicting future values.

Figure 5.17, representing the Sabine River at Logansport, Louisiana, illustrates a drought period from January to July 2022, during which the discharge values fall relatively below the median values. However, aside from this period, there does not appear to be a persistent drought similar to those affecting other parts of the hydrograph. Similarly, for a site on the Mississippi River (Figure 5.18), discharges closely align with their median values, except for a drought period between July 2022 and January 2023. However, as depicted n Figure 5.19, a severe drought significantly reduced the discharges for available records at a site on the Colorado River.

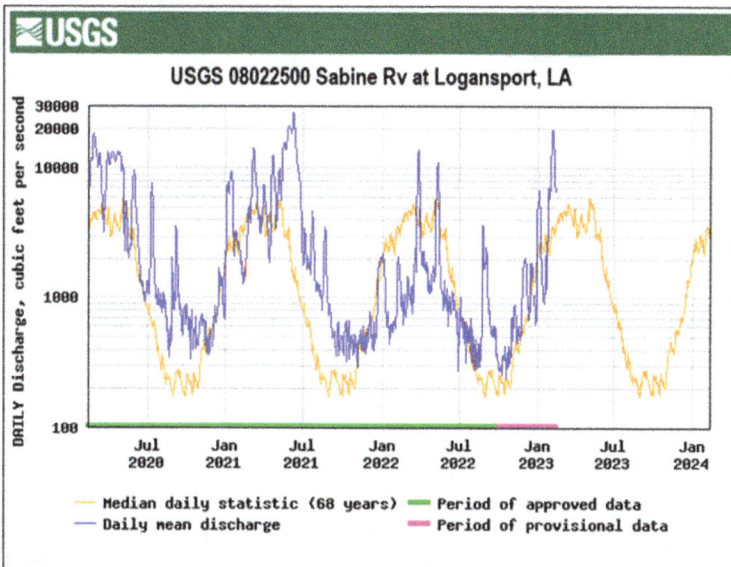

FIGURE 5.17 Flow hydrograph, Station 08022500, Sabine River at Logansport, Louisiana (USGS, 2023).

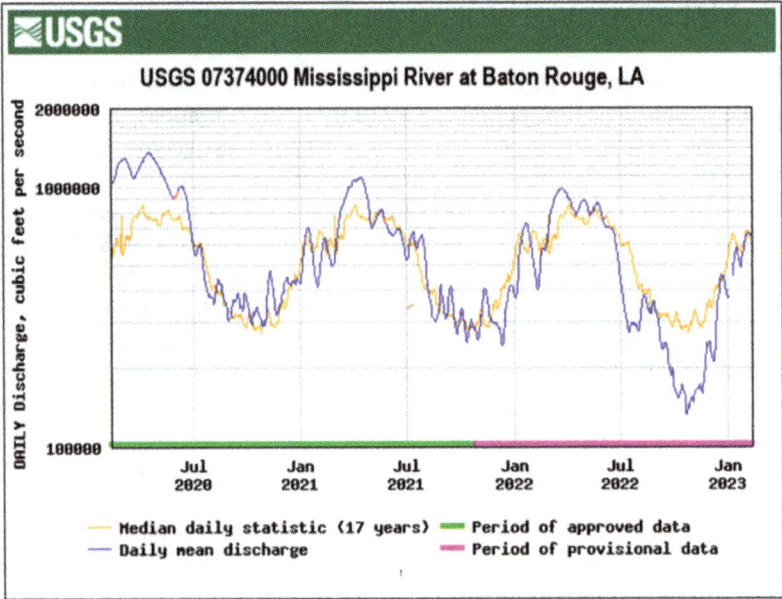

FIGURE 5.18   Flow hydrograph, Station 07381000, Mississippi River at Baton Rouge, Louisiana (USGS, 2023).

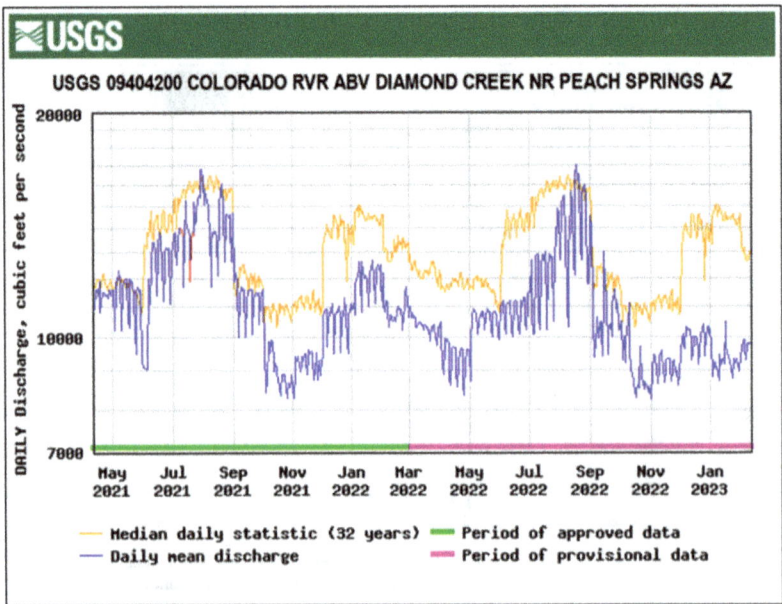

FIGURE 5.19   Flow hydrograph, Station 09404200, Colorado River, Above Diamond Creek Nr Peach Springs Arizona (USGS, 2023).

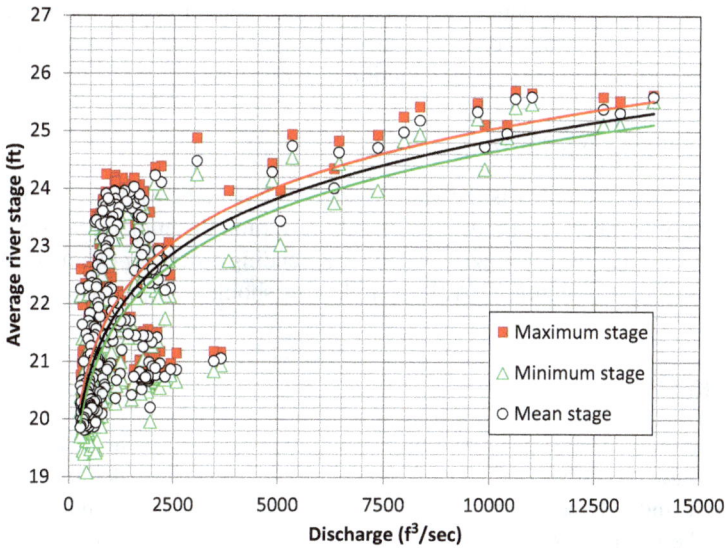

FIGURE 5.20 River rating curve for Station 08022500, Sabine River at Logansport, Louisiana (October 2021–October 2022). The solid lines represent the best fit lines.

Available USGS data also include river stages, which, when combined with discharge data, can be used to create a river rating curve. The data available for the Sabine River in Louisiana were utilized to construct the rating curve shown in Figure 5.20. The stage data include mean, minimum, and maximum values. The figure demonstrates reasonable accuracy for the best-fit lines, particularly for higher discharges exceeding 2,500 ft³/s. Below this threshold, changes in water stage have a limited impact on discharge. For instance, a discharge value of 1,500 ft³/s was recorded for stages ranging from 20 to 24.4 ft. These discrepancies necessitate on-site discharge measurements due to the unsuitability of relying solely on a rating curve.

5.2.4 River Duration Curve

River duration curves serve as a valuable tool for estimating the probability that a specific flow discharge in a river will be equaled or exceeded. These curves play a crucial role in assessing various scenarios, such as the likelihood of the flow being too low to meet water supply needs or too high, leading to flooding or structural damages.

The data utilized in creating a river duration curve consist of daily or annual flow discharge values over specific durations. The process begins by arranging these discharge values in descending order, from largest to smallest, and assigning a rank based on this sequence. The estimated probability, expressed as a percentage, of equaling or surpassing these values is determined as follows:

$$p = \frac{m}{(n+1)} \times 100 \tag{5.10}$$

where m is the rank of each value and n is the total number of values. The rank can also be easily obtained by using the RANK function in Excel, without the need to arrange the values in descending order. This function is defined as

RANK(number, ref, order)

In this function, number is the value whose rank is being evaluated, ref is a reference to a list of numbers, and order specifies how to rank the number. Set order to zero for descending order, or to a nonzero value for ascending order. For example, if the data are located in Excel's column A1:A21, the following should be typed in cell B1 and then copied for B2:B12:

RANK(A1, $A$1:$A$12, 0)

The duration curve is generated by plotting discharge values against their respective probabilities on either linear or semi-logarithmic paper (provided in Appendix B). Figures 5.21 and 5.22 serve as illustrative examples of these representations. In these instances, the semi-logarithmic representation offers enhanced resolution for extreme values at low probabilities, while the linear representation provides better resolution for probabilities exceeding 40%.

The duration curve can also be created by plotting the discharge values against the respective probabilities on normal probability paper, which is also included in Appendix B. Example 5.5 describes the procedure. Such a graph can be useful in examining the distribution of the discharge data, as illustrated in Figure 5.23. When the data points closely align with a straight line, the data set can be assumed to follow a normal distribution. Such a distribution is employed for data analysis when there is an equal likelihood of being above or below the mean of the data. The normal distribution of

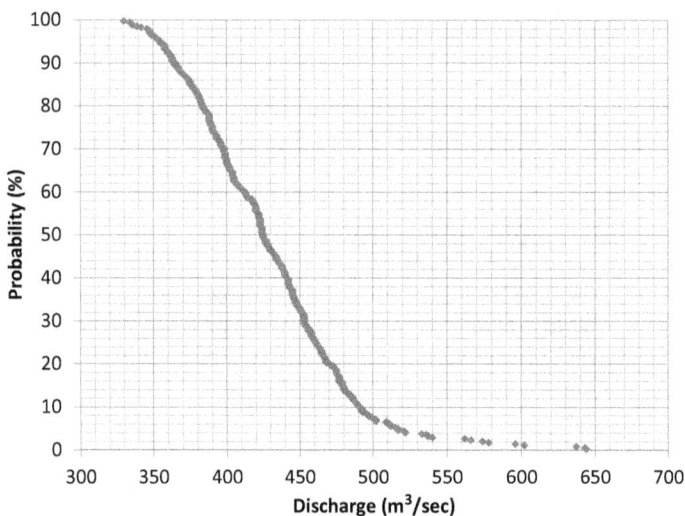

FIGURE 5.21   River duration curve on a linear paper.

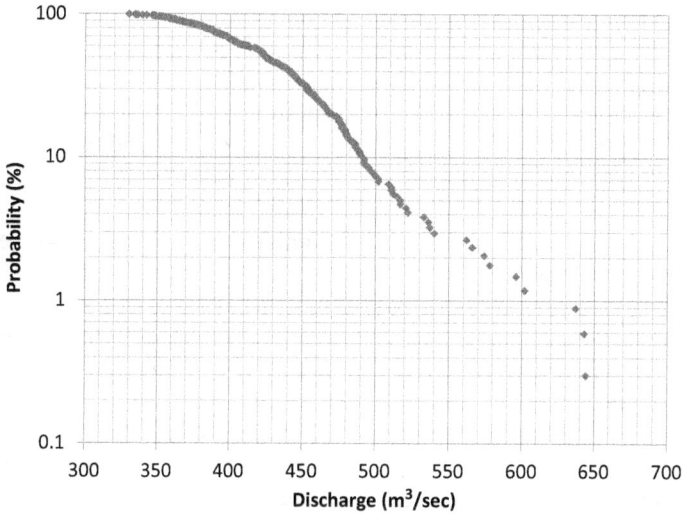

FIGURE 5.22   River duration curve on a semi-log paper.

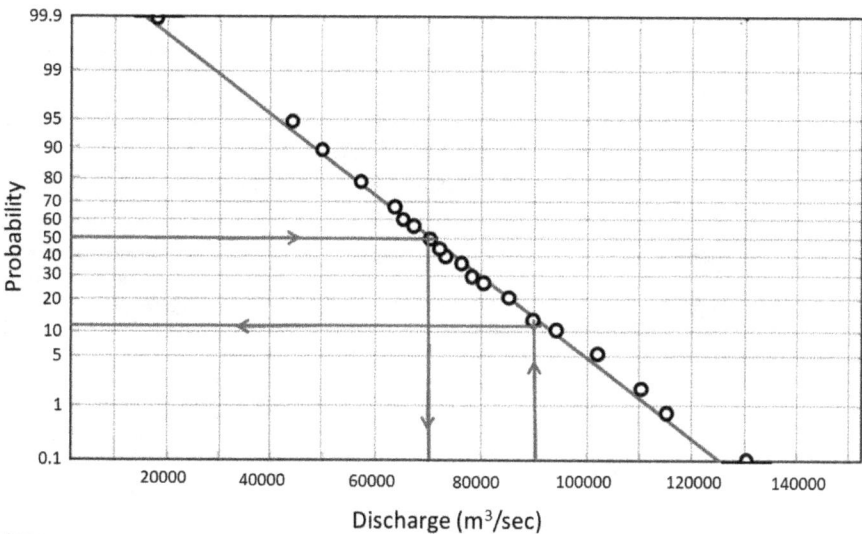

FIGURE 5.23   A duration curve of discharges that can be represented by normal distribution. The lines with arrows are used to estimate probability for a given discharge or vice versa.

a specific variable is appealing as it depends solely on the variable's mean value and standard deviation.

The river duration curve can be employed to estimate the probability of a particular discharge value being equaled or exceeded or conversely to evaluate the discharge for a known probability. In the example shown in Figure 5.23, the median discharge is estimated at a 50% probability, approximately 70,000 m³/s. The probability of discharge equaling or exceeding 90,000 m³/s is approximately 12%.

The river duration curve is also useful in designing a dam to protect against a specified flood, typically estimated to occur at a certain probability, often based on a 100-year flood. However, it is crucial to understand that a "100-year flood" does not mean it happens once every hundred years. Instead, it signifies a probability of occurrence in any given year of 1 in 100 or 1%. Similarly, a "500-year flood" represents a probability of occurrence of 0.5%, and so on. This definition helps eliminate confusion regarding when such a flood might occur in subsequent years. Opting for an extremely low or near-zero probability is impractical because it would result in unnecessary dam construction costs for events that are highly unlikely. For Figure 5.23, the design flood discharge is approximately 110,000 m³/s with a probability of 1%.

River duration curves are also a valuable tool for assessing flow variability among various rivers or within the same river over different time spans, such as year-to-year variations. Figure 5.24 provides an example illustrating data for a specific river over three distinct periods. The plot reveals that low flows, with a probability of 40% or higher, remain nearly identical across these periods. However, higher flow levels exhibit significant variability, measuring approximately 80,000, 14,20,000, and 1,47,000 m³/s for the three time periods at a 0.1% probability level. When designing a dam, it is crucial to consider the worst-case scenario, which involves higher flood levels of around 1,22,000 m³/s at a 1% probability.

Figure 5.25 illustrates a scenario characterized by extreme discharge values occurring at probabilities below 20% and above 80%. Discharges within these ranges experience relatively minor variations. The corresponding ranges of extreme discharge values

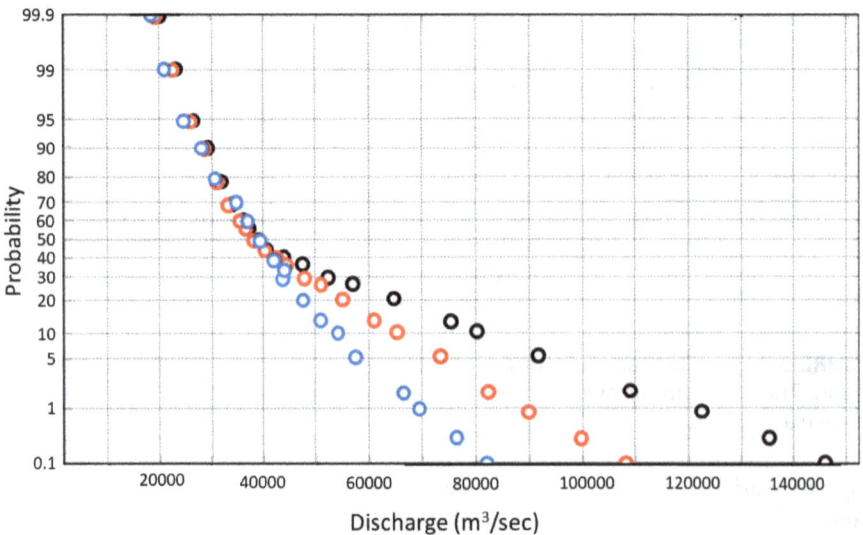

FIGURE 5.24   Duration plots for three time periods for a specific river.

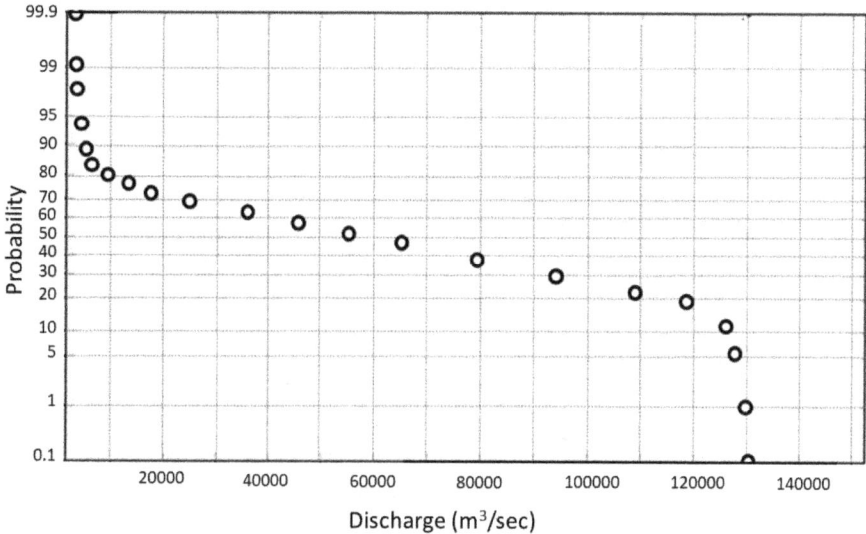

FIGURE 5.25 Linear probability plot for two cases characterized by extreme discharge values occurring at probabilities below 20% and above 80%.

are 1,20,000–1,30,000 m³/s and 200–1,000 m³/s. To address uncertainties within these ranges, it is advisable to employ worst-case scenarios by utilizing conservative discharge values of 200 m³/s and 1,30,000 m³/s, respectively. These values should be considered for fulfilling demands during low-flow periods and ensuring flood protection.

Example 5.5: River Duration Curve

Annual river discharges (in m³/s) are in Table 5.3A. Calculate the probabilities and plot the river duration curve. Estimate the probability of a discharge equaling or exceeding 140 m³/s as well as the discharge for a probability of 90%.

The calculations are shown in Table 5.3B. The first column shows ranks of discharges after sorting the values from largest to smallest (second column). The third column lists the probabilities calculated using equation 5.10 with $n = 10$. (Note that the dates are not included in the calculations or in the duration curve plot.) For example, for rank (m) of 1 (the first row in Table 5.3B):

$$p = \frac{m}{(n+1)} \times 100 = \frac{1}{(10+1)} \times 100 = 9.1\%$$

Figure 5.26 depicts the discharges versus the respective probability on the normal probability paper provided in Appendix B. The data reasonably fit on a straight line, indicating that the discharges can be approximated by a normal distribution. The average value can be estimated at the probability value of 50%, which is about 62 m³/s. The discharge at a 90% probability and the probability for a discharge of 140 m³/s are shown by the black symbols. The respective values are 5 m³/s and 5%.

TABLE 5.3A Data for Example 5.5

YEAR	DISCHARGE
1954	130
1955	13
1956	64
1957	37
1958	76
1959	110
1960	94
1961	54
1962	45
1963	25

TABLE 5.3B Calculations for Example 5.5

RANK M	DISCHARGE	PROBABILITY P
1	130	9.1
2	110	18.2
3	94	27.3
4	76	36.4
5	64	45.5
6	54	54.5
7	45	63.6
8	37	72.7
9	25	81.8
10	13	90.9

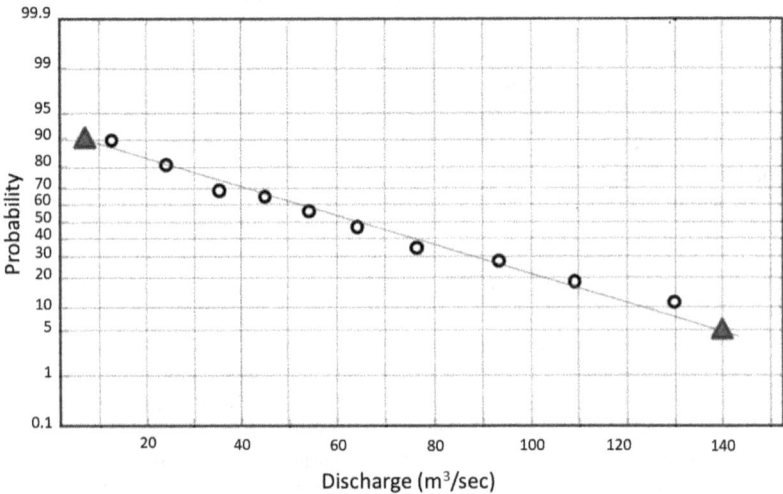

FIGURE 5.26   River duration curve displays discharges versus the respective probability on a normal probability paper.

5.3 ASSIGNMENTS

1. Given the stream channel of a rectangular cross-section (Figure 5.27) lined with concrete, estimate the discharge given the following information: The channel width and water depth are 10 m and 1.5 m, respectively. In the flow direction, perpendicular to this page, the bottom of the stream drops 1 m over a distance of 500 m. The roughness factor for concrete is 0.012.
2. For the same cross-section above, calculate and plot the stream rating curve. (Hint: calculate the discharge for different flow depths, e.g., 0.1, 0.3... 1.5 m.)
3. Estimate the discharge for the following cases:
   a. A V-notch weir with a water height of 1.9 m.
   b. A rectangular weir with a width of 2 m and a water height of 0.2 m.

4. A laboratory experiment is used to estimate the coefficient of discharge for a V-notch weir across a laboratory channel. The results are shown in Table 5.4 for several runs, specifically the values of the collected volume of water, the time of collection, and the height of water above the crest (H). Estimate the coefficient of discharge for the weir and plot both the measured and theoretical discharge against the height of the water above the crest.
5. Figure 5.28 depicts the duration curves for two different streams, nicknamed the Black and Red Rivers. Compare the two cases regarding the median discharge value, the probability of discharge equaling or exceeding 2,000 and 1,000 m³/s, and the discharges for low and high probabilities of 10% and 90%. For dam construction, what is the design discharge for a 1% probability flood? Which of the two streams raises more concerns about extreme conditions, and why?
6. Figures 5.29 and 5.30 display both discharge and river-stage data for the Sabine River in Louisiana. Use the figures to create a river rating curve by using the respective values from the graphs at half month intervals, i.e., January 1 and 15, February 1, etc. Compare your rating curves with those displayed in Figure 5.20. The data in these figures were based on actual measurements during this period.

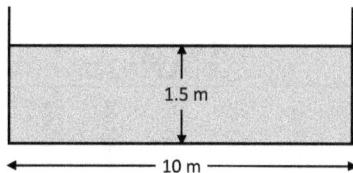

FIGURE 5.27 Cross section for problem 1.

TABLE 5.4   Data for Problem 4

Run Number	Measured Volume (m³)	Time of Collection (s)	Measured Discharge (m³/s)	H (m)
1	0.006	335	1.79E-05	0.0136
2	0.006	160	3.75E-05	0.0182
3	0.006	110	5.45E-05	0.0209
4	0.006	60	1.00E-04	0.027
5	0.006	210	2.86E-05	0.0161
6	0.006	30	2.00E-04	0.0355

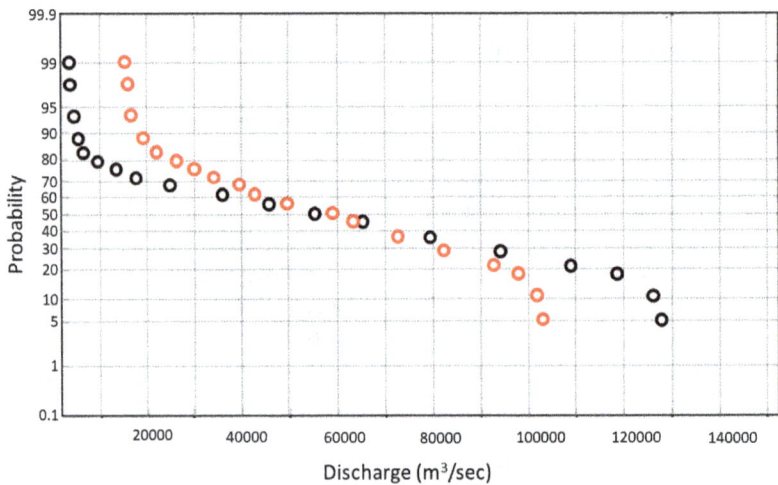

FIGURE 5.28   Duration curves for the Black and Red Rivers for problem 5.

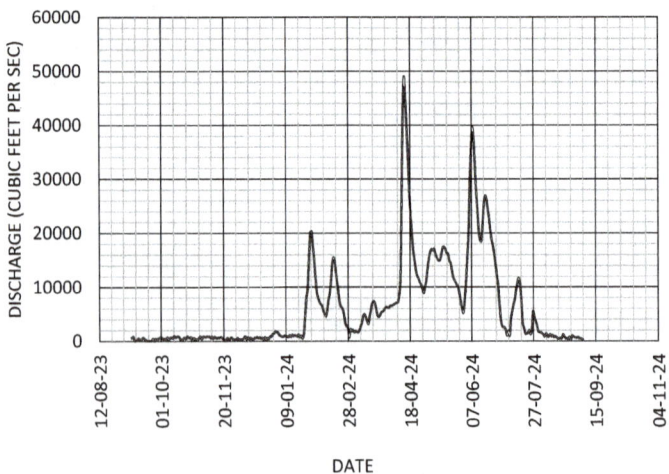

FIGURE 5.29   Stream discharge data for the Sabine River (USGS, 2024).

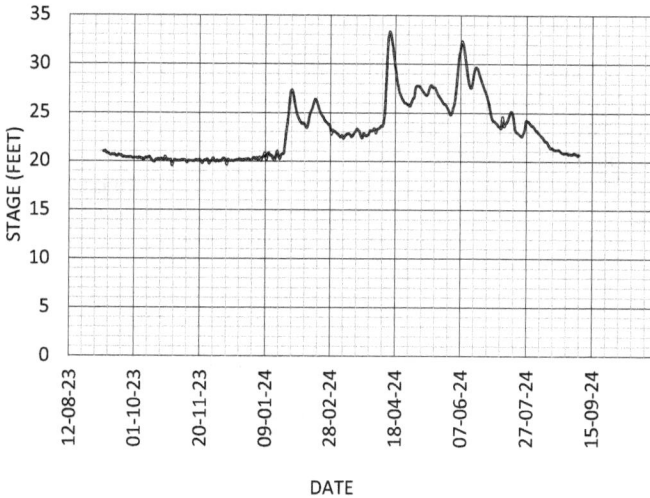

FIGURE 5.30   Stream stage data for the Sabine River (USGS, 2024).

NOTES

1 The velocity values within a subarea are not constant, with a value near zero at the bottom of the stream, and at a maximum value at the water surface. An average value is estimated at about 60% of the water depth.
2 See Section 2.3.3 in Chapter 2 for definition of Median and other statistical terms

REFERENCES

APPROPEDIA. (2023). How to measure stream flow rate. Retrieved from https://www.appropedia.org/How_to_measure_stream_flow_rate

Chow, V. T. (1959). *Open-Channel Hydraulics*. New York, NY: McGraw-Hill.

DimensionEngine. (2015). Developing a Rating Curve. Retrieved from https://www.dimensionengine.com/excel/hydrotools/examples/ratingcurve.html

Fetter, C. W., & Kreamer, D. (2022). *Applied Hydrogeology* (5th ed.). Long Grove, IL: Waveland Press.

Henderson, F. M. (1966). *Open Channel Flow*. New York, NY: Macmillan.

Herbst, D. (2014). Mono Lake: Streams taken and given back, but still waiting. *Lakeline, 34*, 21–24. Retrieved from https://www.researchgate.net/publication/305965303_Mono_Lake_streams_taken_and_given_back_but_still_waiting

Mono Lake Committee. (2023). State of the lake. Retrieved from https://www.monolake.org/learn/stateofthelake/

Safaris Africana. (2023). The 10 largest lakes in the world. Retrieved from https://safarisafricana.com/largest-lakes-in-the-world/

Subramanya, K. (2009). *Flow in Open Channels* (3rd ed.). New Delhi, India: Tata McGraw-Hill.

U.S. Geological Survey. (2018). How streamflow is measured. Retrieved from https://www.usgs.gov/special-topics/water-science-school/science/how-streamflow-measured#:~:text=In%20the%20simplest%20method%2C%20a,also%20measured%20at%20each%20point

U.S. Geological Survey. (2023). Streamflow monitoring. Retrieved from https://www.usgs.gov/centers/pacific-islands-water-science-center/science/science-topics/streamflow-monitoring

U.S. Geological Survey. (2024). USGS current water data for the nation. Retrieved from https://waterdata.usgs.gov/nwis/rt

Utah Division of Water Resources. (2025). Great Salt Lake. Retrieved July 1, 2025, from https://water.utah.gov/great-salt-lake/

Wright, P. (2016). V-notch weir alongside Frank Lane [Photograph]. Wikimedia Commons. Retrieved from https://commons.wikimedia.org/wiki/File:V-notch_weir_alongside_Frank_Lane_-_geograph.org.uk_-_5671760.jpg. Licensed under the Creative Commons Attribution-Share Alike 2.0 Generic License. https://creativecommons.org/licenses/by-sa/2.0/

Groundwater

6

6.1 INTRODUCTION

Groundwater plays a significant role in the water cycle and serves as a primary source for public water supplies and agricultural needs. In the United States, approximately 40% of water used for these purposes is derived from groundwater. According to the American Geosciences Institute (2017), some states heavily rely on groundwater, with the following percentages of their total freshwater withdrawals attributed to groundwater: Kansas (80%), Arkansas (69%), Mississippi (68%), Florida (64%), and Hawaii (63%). Notably, states like Arizona, Florida, Hawaii, and Nevada predominantly source their domestic water supply from groundwater.

6.2 AQUIFERS AND THEIR PROPERTIES

Groundwater occupies and flows through narrow spaces within subsurface materials or geological formations. These materials can be either unconsolidated, such as sediments like sand and gravel (as shown in Figure 6.1), or consolidated, such as fractured rocks (as depicted in Figure 6.2). These spaces manifest as pores, or pore spaces, in unconsolidated or granular materials, and as fractures in consolidated materials. In unconsolidated materials, fluid moves through small, interconnected pores between grains, resulting in slow and relatively uniform flow. In consolidated materials, fluid flows through cracks, with irregular flow highly dependent on fracture networks. Other aquifer types include karst systems, where fluid flows through large, dissolved channels and caves, resembling surface streams. However, the term "porous media" is reserved for cases involving relatively low velocities and is not applicable to such systems.

Water can completely saturate the material at specific depths below the land surface. The upper boundary of this saturated zone is known as the water table, above which drier conditions prevail, characterizing a partially saturated zone. This zone, also referred to as the unsaturated zone, contains both water and air (Figure 6.3). Water development for various purposes primarily focuses on the saturated zone, while the unsaturated zone plays a crucial role in agricultural practices. Additionally, it serves

DOI: 10.1201/9781003587149-6

FIGURE 6.1   Unconsolidated (granular) material. Water is stored in small spaces (pores) between the grains and flows through them. (Photograph by the author.)

*A typical sequence of lava flows contains aa clinker zones (**A**) of relatively high permeability that occur above and below the massive central cores of aa flows (**B**), and many thin pahoehoe flows (**C**). The sequence shown is about 50 feet thick. (photo by Scot K. Izuka, USGS).*

FIGURE 6.2   Examples of different rock-based formations. Water is stored in and flows through fractures and other porous zones (Gingerich and Oki, 2000).

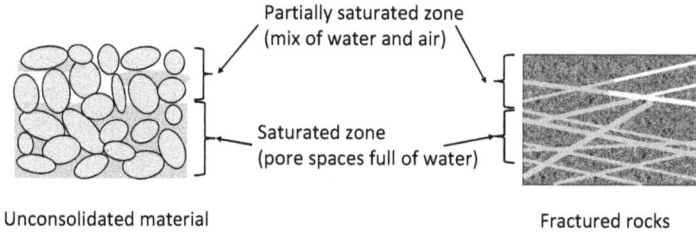

FIGURE 6.3 A schematic representation of unconsolidated and fractured rock materials, illustrating the pore-scale structure. The gray-shaded areas outside the grains or inside the fractures represent water-filled zones, while the white spaces indicate air-filled pores or fractures.

as the initial line of defense for safeguarding groundwater against contamination from human activities and other sources, whether on the surface or below it.

To be suitable for water development, a geological system must possess storage and transmission capabilities for water. Specifically, the term "aquifer" is used to describe a formation that can (1) hold substantial amounts of water and (2) allow it to move at reasonable rates. The absence of either of these features renders the formation an unreliable source of water. For example, a stiff clay formation can store a large amount of water, but the cost of water extraction can be prohibitive due to its limited ability to transmit water. Similarly, a very coarse material system is unsuitable due to its small storage capacity, despite the ease of water transmission.

The storage capabilities of an aquifer are described by several parameters, namely, porosity, effective porosity, specific yield, specific storage, and storage coefficient (also known as storativity). These parameters are defined as follows:

Porosity (sometimes referred to as total porosity) is the ratio of the volume of void spaces in a rock or sediment to the total volume of the rock or sediment. The porosity of unconsolidated material depends on factors such as sorting, packing, and the shape of grains. Sorting is the distribution of grain sizes, while packing is the arrangement of grains within a material. An illustration of pore spaces with grains of different sizes but all spherical in shape is shown in Figure 6.4. Porosity can be reduced when a portion of the domain is blocked, as depicted in the bottom illustration of the figure. In rocks, apart from packing, porosity will depend on the same factors.

Porosity can be easily estimated in the laboratory as the ratio between the sample's pore volume and its total volume (refer to Section 10.4 for specific experimental procedures). A procedure can begin with a dry sample, and the volume of water needed to fully saturate the sample is measured. Alternatively, a fully saturated sample is oven-dried, and the volume of water is estimated based on the difference in the weights of the wet and dried samples. In either case, porosity can be calculated using equation (6.1):

$$n = \frac{V_w}{V_t} \times 100 \tag{6.1}$$

in which n is porosity as a percentage and V_w refers to the volume of water occupying the pore space, while the total volume V_t is given by

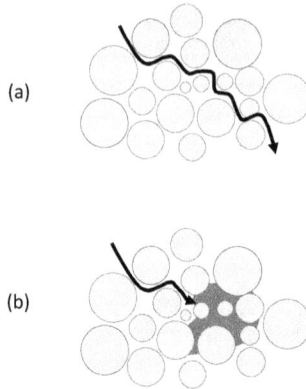

FIGURE 6.4   Permeability is reduced if a portion of the domain is blocked. The black line with the arrow represents the water flow path.

$$V_t = V_{dry} + V_w \tag{6.2}$$

where V_{dry} is the solids' volume.

The method, that starts with a dry sample is straightforward and involves measuring the volume of the sample and the volume of water required to fully saturate it. In the second method, which begins with the sample already saturated with water, the sample is initially weighed (W_{ps}), and its volume is then measured (V_t). After oven drying, the sample is weighed again (W_d). Porosity is estimated using equation (6.1), in which the volume of water (V_w) is determined as follows:

$$V_w = \text{weight of water divided by water density} = \left(W_t - W_{ps}\right)/\rho_w$$

For a gram–centimeter system, the calculation for V_w is simplified by assuming a density of water of 1 g/cm³, which is valid at 20°C. However, V_w can be confirmed by re-saturating the now dry sample.

Porosity can also be estimated as a function of the sample's bulk density ρ_b and particle (or solid) density ρ_d via equation (6.3):

$$n = \left(1 - \frac{\rho_b}{\rho_d}\right) \times 100 \tag{6.3}$$

These densities are estimated by dividing the oven-dry weight of a sample by either its oven-dry volume to estimate ρ_d or by its total volume to estimate ρ_b. The dry volume equals the total volume minus the volume of water:

$$V_{dry} = V_t - V_w$$

Finally, the bulk and particle densities ρ_b and ρ_d are respectively estimated from

$$\rho_b = W_d/V_t$$

$$\rho_d = W_d/V_{dry}$$

Example set 6.1 contains a number of cases related to estimating sample porosity and dry and bulk densities.

The same steps as outlined above can be applied to a partially saturated sample. The key distinction lies in the volume of water, which will be less than the total volume of the void space. In this scenario, you can also estimate a variable known as water or moisture content using equation (6.1), and this value will be less than the porosity estimated by equation (6.3) because the sample is partially dry. Such a scenario is often more representative of field cases, as fully saturated samples are seldom encountered.

Examples 6.1: Porosity and Bulk and Particle Densities

1. Calculate the porosity of a $250\,cm^3$ sample that required $140\,cm^3$ of water to fully saturate the sample from a fully dried condition.

$n = V_w/V_{total} \times 100 = 140\ cm^3/250\ cm^3 \times 100 = 56\%$

2. Calculate the particle density of a soil sample that has a bulk density of $1.6\ g/cm^3$ and a porosity of 40%.

$n = \left(1 - \rho_b/\rho_d\right) \times 100$

$40 = \left(1 - 1.6/\rho_d\right) \times 100$

$1.6 / \rho_d = 0.6$

$\rho_d = 2.67\ g/cm^3$

3. Calculate the bulk density of a sample with a volume of $288\,cm^3$, 15% saturation, and weighs $320\,g$.
   Oven dry weight $= 320 \times (100 - 15/100) = 272$ g (the dry weight is estimated by reducing the total weight by 15%, which reflects the weight of water).

$\rho_b = 272/288 = 0.97\ g/cm^3$

4. Calculate the porosity of a $255\,g$ sample that contains $70\,g$ of water when 50% of the pores are full of water. Particle density is $2.65\ g/cm^3$.
   Oven dry weight $= 255\,g - 70\,g = 185\,g$ soil
   Volume of solids $= 185\ g/2.65\ g/cm^3 = 69.8\,cm^3$ soil

Volume of water for full saturation $= 70\,\text{cm}^3/0.50 = 140\,\text{cm}^3$ water
Total volume of soil $= 140\,\text{cm}^3 + 69.8\,\text{cm}^3 = 209.8\,\text{cm}^3$

$$n = \left(V_{air} + V_w\right)/V_{total} \times 100 = 140\ \text{cm}^3/209.8\ \text{cm}^3 \times 100 = 66.7\%$$

Effective porosity is a portion of the total porosity, defined as the ratio of the total volume of voids available for fluid transmission to the total volume of the porous medium. The volume of effective voids is represented by the amount of interconnected pore space and fracture openings available for fluid transmission.

Specific yield is also a subset of the total porosity, representing the ratio of the volume of voids within a soil or rock mass that can be drained by gravity to the total volume of the mass. Estimating specific yield in the laboratory is straightforward and involves draining a fully saturated sample of known volume and observing the change in the volume of gravity-drained water over time. The specific yield is then calculated as the final collected amount divided by the sample volume. Section 10.5 outlines the laboratory procedure for estimating specific yield.

Specific storage is defined as the volume of water that a unit volume of the aquifer releases from storage when there is a unit decline in hydraulic head.[1] The storage coefficient, also referred to as storativity, represents the volume release per unit area, in contrast to specific storage where the release is per unit volume.

The ability of an aquifer to transmit water is characterized by permeability, hydraulic conductivity, and transmissivity. Permeability measures the ability of water to flow through porous media. For water to move through the formation, the pore spaces between grains or fractures in the rock must be interconnected, as illustrated in Figure 6.4. Consequently, permeability can be reduced if a portion of the domain is blocked, as shown in the bottom illustration of the figure. Permeability is an intrinsic property of the porous material and is independent of the type of fluid or liquid being transmitted, whether it be oil, water, or air.

The term hydraulic conductivity is also used to assess an aquifer's transmission ability but takes into account the characteristics of the liquid present in the aquifer. The viscosity and density of the fluid are controlling factors in hydraulic conductivity, and these properties can change with temperature for the same fluid. Hydraulic conductivity and permeability are related by equation (6.4):

$$K = \frac{k\rho g}{\mu} \tag{6.4}$$

in which K is hydraulic conductivity, k is permeability, ρ and μ are density and viscosity of the fluid, and g is gravity acceleration.

In some instances, popular media cause confusion between permeability and hydraulic conductivity. To clarify this confusion, it's essential to recognize the units associated with each parameter. Hydraulic conductivity is measured in units of velocity, which represents the rate of flow over distance per time, and it can be expressed in units like centimeters per day. Permeability, on the other hand, is measured in units of square length, such as square centimeters.

6.3 AQUIFER WATER FLOW AND DARCY'S LAW

Aquifer types include unconfined and confined aquifers (see Figure 6.5). In the simple case of a confined aquifer, water rises in the wells due to the pressure exerted by the confining layer, known as the aquitard. This confining layer significantly restricts or reduces water flow. The dotted line in the figure represents the potentiometric level, which is the level the water would reach if allowed to rise. This differs from the water table in unconfined aquifers, as it indicates the potential for water to reach that level.

In contrast, unconfined aquifers lack a confining layer and feature a distinct water table that separates the saturated zone below from the unsaturated zone above. In unconfined aquifers, the water level in the wells coincides with the water table (assuming no significant vertical flow).

Water movement within porous media aquifers is influenced by two primary driving forces: water pressure and gravity. Water flows from high energy to low energy. Obviously, water flows from areas of higher elevation to lower ones due to gravity, such as from mountains to the ocean. However, increased water pressure can override gravity, enabling water to move in any direction, including against gravity. To create the necessary pressure differential for water to flow toward a specific point, pumping can be employed. Therefore, it is essential to consider both gravity and pressure mechanisms when accurately assessing groundwater flow.

Figure 6.6 presents a schematic representation of an unconfined aquifer utilized for water extraction through pumping, aimed at illustrating these two processes that govern water flow. The pre-pumping water table is represented by the dotted line in the figure and slopes from right to left, indicating water movement from the mountain side towards the ocean, specifically from Point B to Point A and then to Point C. The pre-pumping level in Well A is denoted as h_0. In the diagram, water is pumped at Point A, where the water level can be monitored. Water levels can also be observed at monitoring wells at Points B and C. If vertical flow is negligible, the top of the water in each

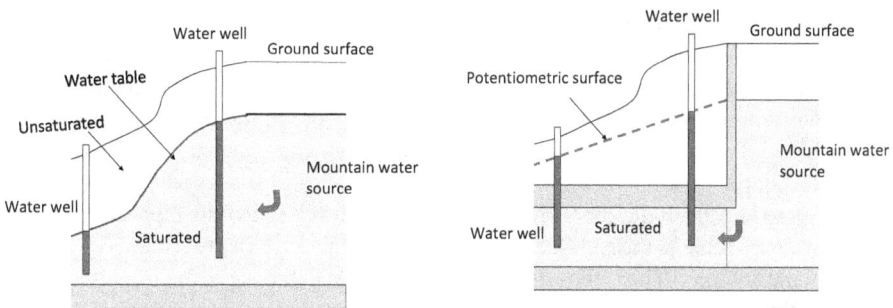

FIGURE 6.5   Left: unconfined aquifer; right: confined aquifer (the darker gray areas outside the wells represent aquitards).

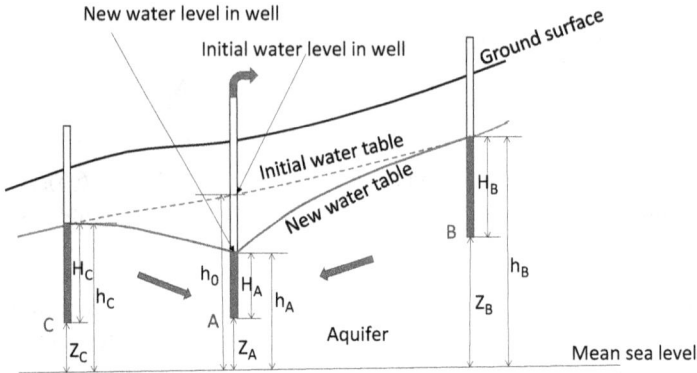

FIGURE 6.6   Water flow in an unconfined aquifer toward a pumping well.

well aligns with the water table; otherwise, variations in the well's water level reflect either downward or upward flow.

The "new" water-table line represents the water table after some time has elapsed since pumping began. The pressure heads at the three points, namely H_A, H_B, and H_C, indicate the water level above each point, while the elevations of these points above mean sea level, Z_A, Z_B, and Z_C, represent the gravity head. Note that an arbitrary datum can be selected, but typically the mean sea level is chosen. The hydraulic head values (h_A, h_B, and h_C) are obtained by summing the pressure head and the elevation head at each respective point.

Under pumping conditions, the hydraulic head at Point A decreases due to a reduction in the pressure head from its initial value of h_0 to h_A. As a result, the water flow reverses, moving against gravity in the lower side of the aquifer towards the pumping well. As is always the case, the flow direction is determined by the sum of the pressure head and the elevation head, irrespective of the relative values of pressure heads or gravity heads. In this scenario, water moves towards the pumping well because both hydraulic heads h_B and h_C are greater than h_A.

Figure 6.7 illustrates a general scenario that combines unconfined and confined aquifers, along with a third type known as a perched aquifer. The perched aquifer forms due to the presence of a low-permeability aquitard located above the water table in specific locations. An unconfined aquifer develops along with a corresponding water table, which may result in a spring if it intersects with the ground surface. Perched aquifers, due to their relatively small size, are generally not considered significant water sources.

In Figure 6.7, the confined aquifer is separated from the unconfined aquifer by an aquitard. The water within the confined aquifer is under pressure, creating a potentiometric level (depicted as the dotted line). This potentiometric level represents the water's level if allowed to rise, such as through the digging of wells like well 1 where the level exceeds the top of the aquitard. This contrasts with the water table in the unconfined aquifer (represented by the solid line), which signifies the actual water level rather than a potential one. However, both levels fluctuate in response to natural and manmade inflows and outflows of water. For well 1, the pressure head is denoted as H_1 due to the confined condition.

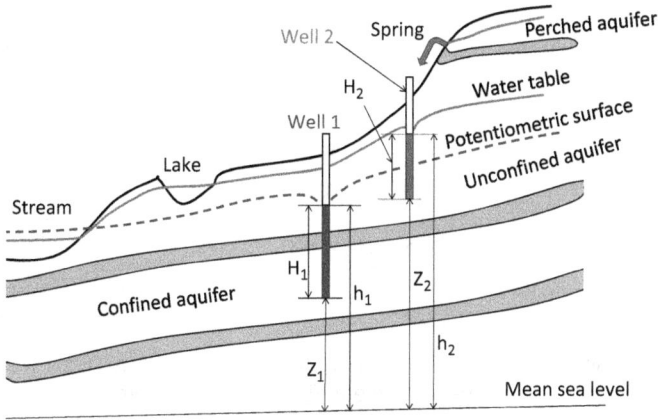

FIGURE 6.7   Types of aquifers. The gray areas represent aquitards of low permeability that drastically or fully restrict water flow.

In both aquifers, the hydraulic head adheres to the same definition, namely, $h_1 = H_1 + Z_1$ and $h_2 = H_2 + Z_2$, for well 1 and well 2, respectively. Example 6.2 provides calculation methods to determine various heads and the corresponding direction of water flow.

Groundwater moves slowly, typically at a maximum rate of about one meter per day. Consequently, water can reside in an aquifer for extended periods, often spanning hundreds or even thousands of years. This long retention time significantly influences aquifer storage and the occurrence of contamination, as well as efforts to clean it up. The rate of groundwater flow is estimated by using Darcy's law (Darcy, 1856). According to this law, the discharge (volume of flow through the aquifer over a specific time period) in a given area of the aquifer is directly proportional to several factors: the cross-sectional area (perpendicular to the flow direction), the hydraulic conductivity of the porous media, and the gradient of the hydraulic head at that location. The discharge can be calculated by multiplying these three quantities together, as follows:

$$Q = -KAS \tag{6.5}$$

In equation (6.5), Q is the discharge (with units of volume per unit time, e.g., cubic centimeters per second), A is the flow area (in units of length squared, e.g., square centimeters), and S is the hydraulic head slope [dimensionless (no units)]. The negative sign is included because water flows in the direction of decreasing head (from high to low energy), resulting in a negative value for S. It is important to emphasize that Q, K, and A are all positive quantities, and the negative sign only accounts for the negative value of S.

In the schematic plot of Darcy's experiment in Figure 6.8, the area is that for a circle, and the slope S is given by equation 6.6:

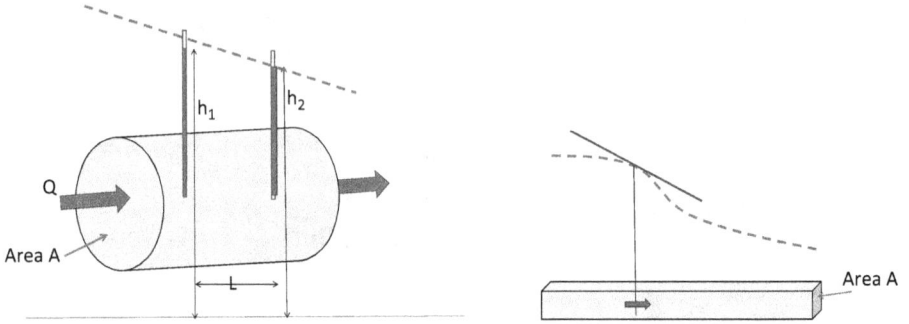

FIGURE 6.8 Schematic representation of factors involved in Darcy's law: (a) Circular domain with fixed hydraulic head gradient, (b) Rectangular domain with variable hydraulic head gradient.

$$S = \frac{h_2 - h_1}{L_2 - L_1} \qquad\qquad (6.6)$$

in which $L = L_2 - L_1$ is the distance between the two points where the small tubes are inserted (units of length, e.g., meters).

Example 6.2: Hydraulic Head

To assess potential vertical flow, three wells, A, B, and C, were drilled close to each other, inside the same casing, at points 1, 2, and 3 in Figure 6.9. The water levels in the wells were 80, 60, and 100 m below the ground surface. The ground surface is 250 m above mean sea level, which is the chosen datum. Using Figure 6.9, calculate the elevation, pressure, and hydraulic heads for all wells, and identify the direction of flow relative to points 1, 2, and 3.

Elevation head (Z) is defined as the distance from the datum to the bottom of a well, with values of 20, 110, and 30 m for wells A, B, and C, respectively. The total depth of well A is 250 m, calculated as 250 m minus 20 m, resulting in a depth of 230 m. Similarly, the total depths of wells B and C are 140 m and 220 m, respectively.

The pressure head is defined as the height of water in the well above its bottom. It can be calculated by subtracting the depth of water in the well from the total depth of the well. The values for the pressure head are 150, 80, and 120 m for wells A, B, and C, respectively.

Therefore, the respective hydraulic head values, represented as Z+H, are 170 m for well A, 190 m for well B, and 150 m for well C.

It is evident that water is flowing downward from point 2 to points 1 and 3, moving from higher to lower hydraulic heads. This conclusion can be drawn from the figure, as the water level in Well B is higher than that in the other wells, even without performing any calculations.

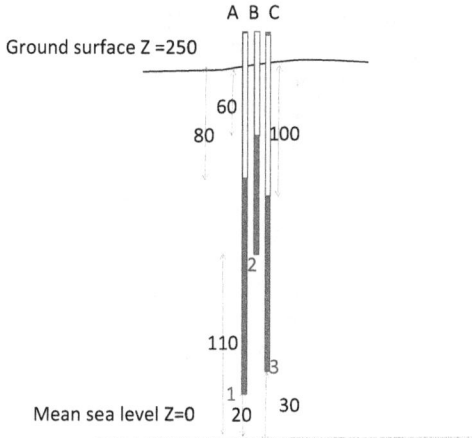

FIGURE 6.9 Sketch for Example 6.2 for three wells within the same casing.

	WELL A	WELL B	WELL C
Z	20	110	30
H	150	80	120
h=Z + H	170	190	150

In real aquifers, the slope is estimated using the hydraulic headlines depicted in Figure 6.7. These lines are typically not perfectly straight, so an average slope between two points is employed. At any specific point, the slope is determined as the tangent of the hydraulic head line at that location (refer to the schematic plot on the right-hand side of Figure 6.8). In this figure, S represents the estimated slope of the tangent indicated by the dotted line. For a given value of hydraulic conductivity, the discharge within the aquifer varies due to the variable slope of the hydraulic head.

The discharge (Q) in equation (6.5) can be used to define q, which represents the discharge per unit area and is commonly referred to as the specific discharge or Darcy's flux:

$$q = \frac{Q}{A} = -KS \tag{6.7}$$

The variable q in equation (6.7) has units of velocity (length per unit time, e.g., meter per day), but it is not true velocity, because the area A contains both solids and void spaces. The true area available for flow is a fraction of the area, which is the area A multiplied by porosity n. Thus, the velocity is estimated by equation (6.8):

$$v = \frac{Q}{An} = \frac{q}{n} = -\frac{KS}{n} \tag{6.8}$$

In the literature, v is termed average linear velocity because the pore sizes are not uniform and the actual velocity distribution is complex. As will be explained later, v, rather than q, is used in assessing contamination fate and transport in porous media.

Example 6.3: Darcy's Law

A. Utilizing Figure 6.8b, estimate the discharge per unit width from an aquifer to the ocean, assuming that the aquifer is confined with a thickness of 100 m and a hydraulic conductivity of 650 m/day. The average potentiometric (hydraulic head) decline is 1.0 meter per kilometer.

   The flow area is estimated by multiplying the aquifer width, which is one meter, by the aquifer thickness, providing the discharge per unit width. (Note that area in Darcy's law is that perpendicular to the flow direction as shown in the figure.) Therefore, the area A will be 1.0 m multiplied by 100 m, which is 100 m². The average slope is −1.0 meter per kilometer, which would be −1.0 meter per 1,000 m or −0.001. The discharge is estimated from equation (6.5) to give

$$Q = -KAS = -650 \frac{m}{day} \times 100 \ m^2 \times -0.001 = 65 \ m^3/day$$

B. Estimate the discharge also per unit width for the aquifer depicted in Figure 6.10, where three wells exist with the shown water level values above the mean sea level. The same values in example A are assumed for K and aquifer thickness.

   The average hydraulic head slopes can be estimated using the water levels in the wells and the separating distances. So, the slope between the two first wells is estimated by the following expression (note the units have to be consistent by using meters in this case):

$$S = \frac{(142 - 150 \ m)}{2,000 \ m} = -0.004$$

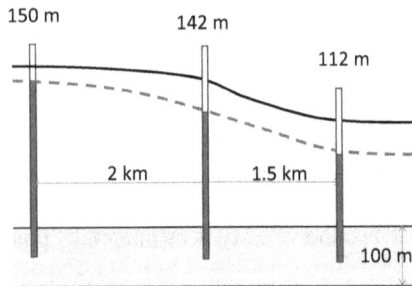

FIGURE 6.10   Sketch for case B in Examples 6.3 for an aquifer with water levels known at three wells.

Similarly, the slope between the second and third wells is −0.02. Hence, by using equation (6.5), the average discharges will be 260 and 1,300 m³/day, for the first and second sections of the aquifer, respectively.

C. Estimate specific discharge (Darcy's flux) and average linear velocity for the aquifer in example A, which has a porosity of 0.3. Compare the resulting values and explain the differences.

Darcy's flux is estimated from equation (6.7) to give

$$q = -KS = -650\frac{m}{day} \times -0.001 = 0.65 \text{ m/day}$$

The average linear velocity is estimated from equation (6.8):

$$v = -\frac{KS}{n} = -\frac{650 \text{ m/day} \times -0.001}{0.3} = 2.17 \text{ m/day}$$

Note that v is larger than q because the area available for flow is smaller than the total area. So, for the same value of discharge, v must be larger. This is also clear because v is estimated by dividing q by a fraction (porosity is always less than 1.0). Just as an illustration, the time of travel of a contaminant for 1.0 km will be estimated by dividing this distance by v, or 1,000 m by 2.17 m/day, which is roughly 460 days.

6.4 SIMPLE ANALYTICAL FLOW SOLUTIONS UTILIZING DARCY'S LAW

As discussed earlier, flow in groundwater aquifers can be represented by Darcy's law (equation 6.5). For a one-dimensional case in a confined aquifer, the equation can be written as follows:

$$Q = -Kb\frac{dh}{dx} \tag{6.9}$$

where Q, K, b, and dh/dx represent the discharge per unit width, hydraulic conductivity, aquifer thickness, and slope of the potentiometric surface, or hydraulic head h, respectively. Here, the slope S in equation (6.5) is substituted by dh/dx where dh is the change in head and x is the direction. Equation (6.9) can be easily integrated to yield

$$Q = -Kb \cdot \frac{h_2 - h_1}{L} \tag{6.10}$$

Equation (6.10) can be used to estimate Q for given K and b, as well as h_1 and h_2, the head values at two locations separated by a distance L. This equation can be rearranged to yield

$$h = h_1 - \frac{Q}{Kb} \cdot x \tag{6.11}$$

where h_1 is the head value at $x=0$. Equation (6.11) can be used to estimate the head h at any distance x in the direction of flow.

Flow in a one-dimensional unconfined aquifer can be described by Darcy's law in the following form:

$$Q = -Kh\frac{dh}{dx} \tag{6.12}$$

Here, the variable h also represents aquifer thickness, in contrast to the variable b in equation (6.9), which was applicable to a confined aquifer. This equation can be integrated to yield

$$Q = \frac{K}{2} \cdot \frac{h_2^2 - h_1^2}{L} \tag{6.13}$$

Equation (6.13) can be used to estimate Q for a given K, as well as h_1 and h_2, the head values at two locations separated by a distance L. This equation can be rearranged to yield

$$h = \sqrt{h_1^2 - \frac{2Q}{K} \cdot x} \tag{6.14}$$

in which h_1 is the head value at $x=0$. The expression in equation (6.14) gives the water level at any distance x.

6.5 LABORATORY EXPERIMENTS TO ESTIMATE HYDRAULIC CONDUCTIVITY

Darcy's law can be employed in laboratory settings to estimate the hydraulic conductivity (K) of porous media, provided that the values of Q, A, and S in equation (6.5) are known. While it can be applied to consolidated materials such as rock formations, it is better suited for granular materials. Within certain limitations, primarily related to water velocity, the law remains valid regardless of the direction of flow or tube orientation. However, the law is not valid for cases of high velocities resulting from large openings, such as significant fractures, or unrealistically steep head gradients.

In a laboratory setup, two approaches are used to estimate hydraulic conductivity: the constant head method and the variable head method. The apparatus for this purpose is called a permeameter, as shown in Figure 6.11, which is available from commercial vendors (e.g., Gilson, 2024). The procedure begins by filling the chamber with the sample, with porous plates placed above and below to prevent material particles from being

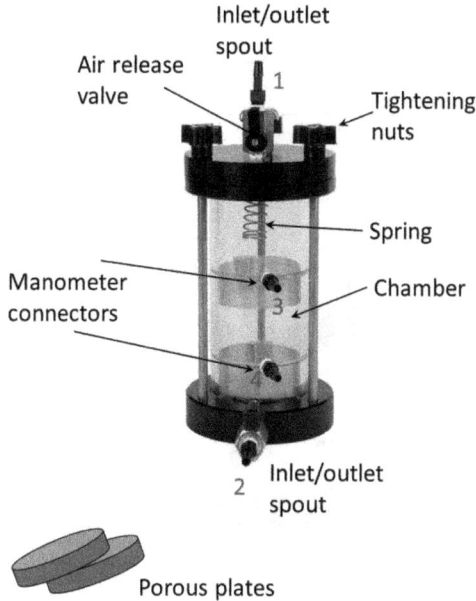

FIGURE 6.11 The permeameter: An instrument for estimating hydraulic conductivity. (After Gilson, 2024.)

carried away by the flowing water. The water flow can be directed either downward or upward through the inlet/outlet ports marked as 1 and 2 in the figure. The driving head is measured as the difference between the water levels in the manometers connected at points marked 3 and 4. The air release valve is used to remove any trapped air from the sample.

These two approaches for measuring hydraulic conductivity are described here, with more details provided in Chapter 10, Sections 10.7 and 10.8.

6.5.1 Constant Head Method

The schematic plot of this procedure is shown in Figure 6.12. In this approach, two small tubes, referred to as piezometer tubes or manometers, are placed at points 3 and 4 within the permeameter. These tubes indicate the water levels at these respective points. After sealing both the top and bottom components, water is allowed to flow through point 1, exiting at point 2, and ultimately draining into a sink or glassware for collecting volumes. Once the water levels in the manometers stabilize, a specific volume of water is collected in a measuring cup, and the collection time is meticulously recorded using a stopwatch. The discharge rate is then estimated by dividing the collected volume (typically measured in cubic centimeters or milliliters) by the recorded time (typically in seconds), resulting in a discharge rate expressed in cubic centimeters (or milliliters) per second. This entire

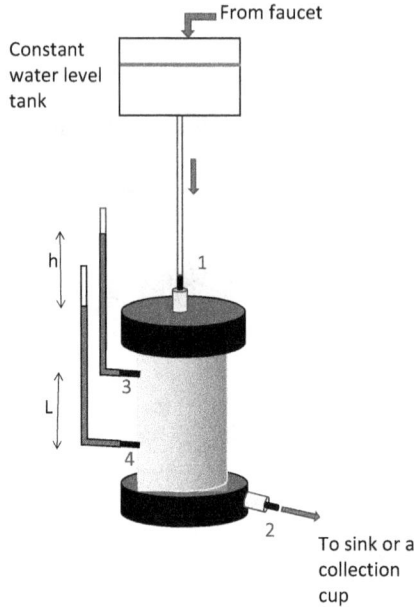

FIGURE 6.12   Schematic illustration of the constant head experiment setup.

process is repeated several times, and the average discharge rate is subsequently calculated. Finally, equation (6.15) is employed to estimate the hydraulic conductivity:

$$K = \frac{QL}{Ah} \tag{6.15}$$

In equation (6.15), the slope S is replaced by h/L, where h is the water-level difference in the manometers, and L is the vertical distance between points 3 and 4. Q is the average discharge, and A is the area of the chamber (a circle). To account for human errors, the whole procedure is repeated several times for different positions of the constant-head tank, and the results of all values of K are used to estimate an average value. Most errors are related to the difficulties in maintaining a constant head, in addition to accurately measuring water volumes and collection times.

6.5.2 Variable Head Method

The setup for the variable head method is illustrated in Figure 6.13. The permeameter is used, but a tube is attached at point 1, replacing the constant water-level tank. The experiment starts by establishing a steady flow into the system through the tube before shutting off the source of water. At that time, the water level (h_o) is recorded as well as the time t to reach level h. Equation (6.16) is used to calculate the hydraulic conductivity:

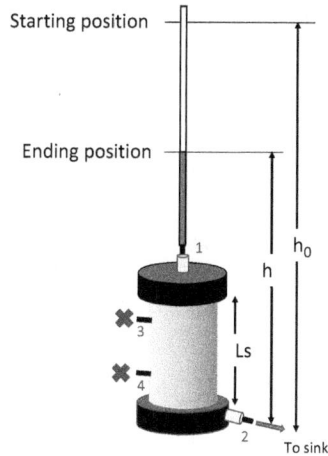

FIGURE 6.13 Schematic illustration of the variable head permeability test.

$$K = \frac{2.3 d_t^2 L}{d_c^2 t} \log\left(\frac{h_o}{h}\right) \tag{6.16}$$

where d_t and d_c are the diameters of the tube and chamber, respectively, L is the length of the sand column, t is time, h_o is the initial head, and h is the head after the water level has been lowered.

Section 10.8 describes in details the procedure for the variable head test, including an example of measured experimental data.

Example 6.4: Constant Head Experiment

The area of the chamber A is 31.67 cm², and the distance L is 6.35 cm. Ten runs were completed. For each run, the outflow volume was collected three times, and the respective head value h was recorded. The average discharge is estimated by dividing each collected volume V by the respective time of collection t and then averaging the three values for each run:

$$Q = \frac{1}{3}\left(\frac{V_1}{t_1} + \frac{V_2}{t_2} + \frac{V_3}{t_3}\right)$$

Equation (6.15) was used to estimate the hydraulic conductivity K for each run. For the ten runs, the average K value is 0.214 cm/s. Again, spreadsheets (e.g., Excel©) are ideal for completing the repeated calculations (see Appendix A). The results are shown in Table 6.1.

The last two columns in the table are used to create Figure 6.14 (the circles). Although there is some scatter due to human errors, the relationship is close to a linear type, which validates Darcy's law. In this case, for given values for A and L, Q is directly proportional to h. The slope of the trendline (0.214 m/s) represents the value of hydraulic conductivity, which matches the average value calculated from the table above. The line should pass through the origin because Q=0 when h=0.

TABLE 6.1 Results for Example 6.4

RUN NUMBER	t_1 S	V_1 cm³	t_2 S	V_2 cm³	t_3 s	V_3 cm³	Q cm³/s	h cm	K cm/s	h/L	Q/A cm/s
1	10.47	109	10.25	111	10.35	118	10.88	10.0	0.218	1.575	0.344
2	10.27	121	5.30	67	5.65	63	11.86	11.0	0.216	1.732	0.374
3	5.23	59	5.50	64	5.30	62	11.54	10.9	0.212	1.717	0.364
4	5.32	65	5.48	63	5.28	60	11.69	10.8	0.217	1.701	0.369
5	5.23	61	5.28	63	5.50	64	11.74	10.8	0.218	1.701	0.371
6	5.30	61	5.03	61	5.08	63	12.01	11.2	0.215	1.764	0.379
7	5.15	62	5.28	64	5.25	64	12.12	11.4	0.213	1.795	0.383
8	5.13	64	5.23	64	5.22	66	12.45	11.7	0.213	1.843	0.393
9	5.25	64	5.25	64	5.27	64	12.18	11.5	0.212	1.811	0.384
10	5.13	66	5.17	66	5.25	63	12.54	12.0	0.210	1.890	0.396

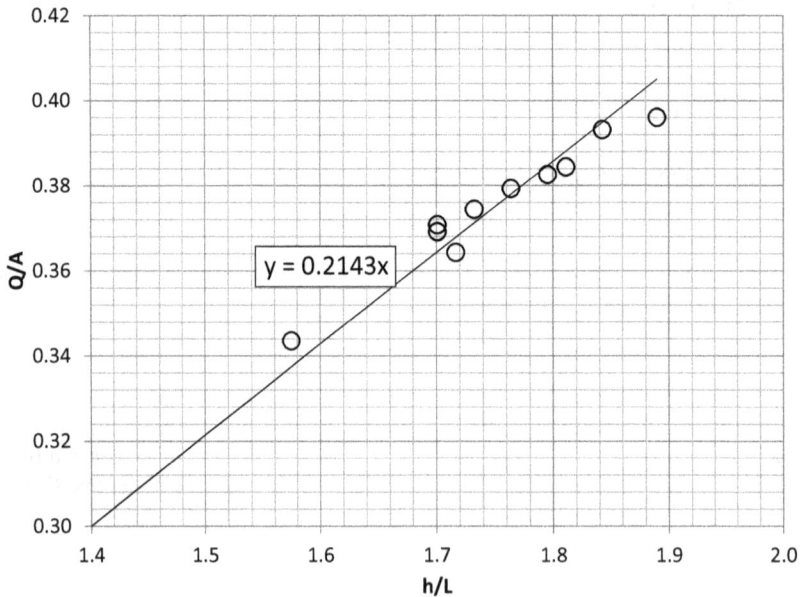

FIGURE 6.14 Relationship between Q/A and h/L for Example 6.4.

6.6 AQUIFER HOMOGENEITY AND ISOTROPY

An aquifer can be classified as either homogeneous or heterogeneous, a categorization that arises from various geological processes involved in the formation of the aquifer. Heterogeneous aquifers are characterized by spatially variable features, such as hydraulic conductivity or porosity. In contrast, homogeneous aquifers exhibit spatially invariant conditions, although such conditions are rarely encountered in practice. The heterogeneous nature of aquifers is evident in Figure 6.2, which illustrates a volcanic aquifer system.

While some aquifer features, like porosity, may vary spatially, they remain consistent in different directions. However, hydraulic conductivity exhibits directional variability, with values changing depending on the direction considered. This directional characteristic, known as anisotropy, results from the geological processes contributing to aquifer formation. Figure 6.2 also highlights this feature, where the relative resistance to flow is influenced by the nature of fractures, including their frequency and orientation. In the figure, the vertical hydraulic conductivity for formation B surpasses the horizontal value, reflecting the predominant orientation of fractures in the vertical direction.

6.7 AQUIFER EFFECTIVE HYDRAULIC CONDUCTIVITY FOR A LAYERED SYSTEM

An illustrative example is presented here to expound upon the concepts elucidated in Section 6.6. In Figure 6.15a, a schematic representation of an aquifer system is depicted, comprising three layers characterized by distinct conductivities and thicknesses.

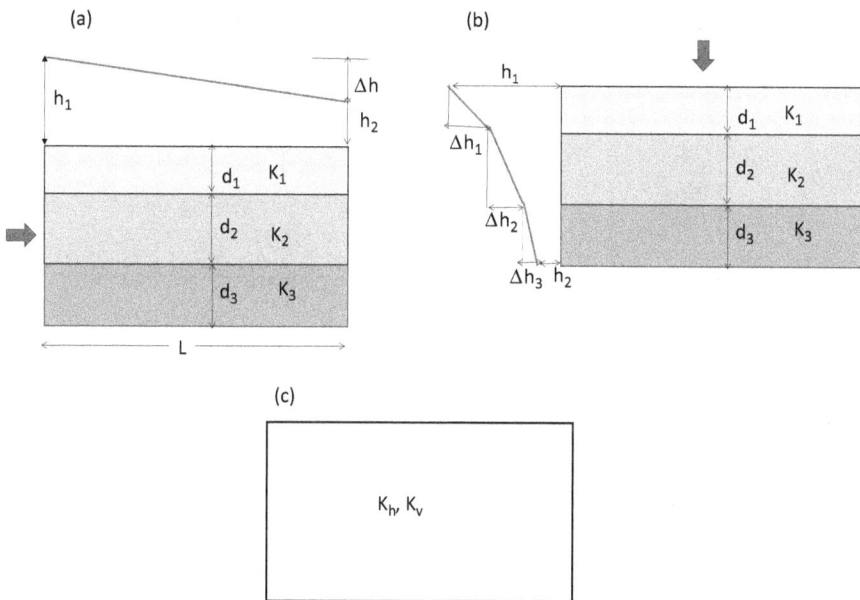

FIGURE 6.15 A layered aquifer system: (a) horizontal flow, (b) vertical flow, (c) the system with effective conductivities.

Each of these layers is homogeneous and isotropic. It is conceivable to replace this system with an alternative one that preserves the flow characteristics of the original system. In the case of horizontal flow illustrated in Figure 6.15a, the total discharge equals the sum of the individual discharges for the three layers. For layer 1, the discharge per unit width is determined by Darcy's law:

$$Q_1 = K_1 d_1 \frac{\Delta h}{L} \tag{6.17}$$

in which d_1 is the area of flow per unit width and Δh is the drop in head over a distance L. A similar expression can be written for discharges for the other layers with the same values for Δh and L. The total discharge is thus

$$Q = Q_1 + Q_2 + Q_3 \tag{6.18}$$

Expressing discharges in terms of Darcy's law in equation (6.18), and considering that the head drop is the same for all the layers, the following expression for effective horizontal hydraulic conductivity will result via equation (6.19):

$$K_h = \frac{(K_1 d_1 + K_2 d_2 + K_3 d_3)}{d} \tag{6.19}$$

in which d is the total thickness $d = d_1 + d_2 + d_3$.

For vertical flow, as shown in Figure 6.15b, the same value of discharge will pass through the three layers. Darcy's law can be used to estimate the incremental drops in head, as shown in equations (6.20) and (6.21):

$$\Delta h = \Delta h_1 + \Delta h_2 + \Delta h_3 \tag{6.20}$$

$$\Delta h = \frac{Q d_1}{K_1} + \frac{Q d_2}{K_2} + \frac{Q d_3}{K_3} = \frac{Q d}{K_v} \tag{6.21}$$

Equation (6.21) can be rearranged to provide the following expression for vertical effective hydraulic conductivity.

$$K_v = \frac{d}{\left(\dfrac{d_1}{K_1} + \dfrac{d_2}{K_2} + \dfrac{d_3}{K_3} \right)} \tag{6.22}$$

By defining effective conductivities, the system illustrated in Figure 6.15a and b is transformed into the configuration shown in Figure 6.15c, where conductivities K_h and K_v represent the effective conductivities in the horizontal and vertical directions, respectively. These effective conductivities are determined by equations (6.19) and (6.22). It is important to note that the new system is homogeneous but anisotropic. In this context, "homogeneous" means that the hydraulic conductivity values remain consistent throughout the system, regardless of location. However, the system is "anisotropic" because it exhibits differing horizontal and vertical hydraulic conductivity values.

As an illustrative example, consider a three-layer aquifer system with respective thicknesses of 1, 3, and 2 m, and hydraulic conductivity values of 7, 6, and 10 m per day. In this case, K_h and K_v can be calculated as follows:

$$K_h = \frac{(7 \times 1 + 6 \times 3 + 10 \times 2)}{6} = 7.5 \text{ m/day}$$

$$K_v = \frac{6}{\left(\dfrac{1}{7} + \dfrac{3}{6} + \dfrac{2}{10}\right)} = 7.11 \text{ m/day}$$

The expressions for K_h and K_v are valid for any number of layers by simply adding new terms to the equations.

Of interest, it is also important to note that the discharge in the horizontal direction, as depicted in Figure 6.15a, will be directly proportional to the hydraulic conductivity of each layer multiplied by its respective thickness. In other words, for layer 1, Q_1 will be proportional to K_1 multiplied by d_1, and this relationship holds true for all subsequent layers. This conclusion is derived from the fact that the head gradient is the same across all layers in the direction of flow. Consequently, the discharge for layer one can be expressed as follows:

$$Q_1 = Q \frac{K_1 d_1}{K_1 d_1 + K_2 d_2 + K_3 d_3} \tag{6.23}$$

In equation (6.23), Q is the total discharge. Similar expressions can be written for the other two layers. For the example described above, if Q is 20 m³/day, the values of individual discharges will be 8.9, 8, and 3.1 m³/day, which will add up to the total of 20 m³/day.

6.8 ASSIGNMENTS: AQUIFER PROPERTIES

1. Calculate the bulk density of a 420 cm³ soil sample with an oven dry weight of 575 g. Estimate the porosity if the particle density is 2.65 g/cm³.
2. Calculate the bulk density of a 450 cm³ soil sample that weighs 615 g and is 10% saturated.
3. Calculate the volume of a soil sample that is 12% saturated, weighs 650 g and has a bulk density of 1.35 g/cm³.
4. A soil sample has a volume of 215 cm³ and a weight of 514.7 g. After saturating the sample, it is weighed at 594.2 g. The sample is then drained by gravity until it reaches a constant weight of 483.4 g. Finally, the sample is dried in

TABLE 6.2   Data for Problem 5

	A	B	C
Ground elevation	400	420	445
Total well depth			80
Elevation head z		320	
Depth to water	30		60
Pressure head H	50		
Hydraulic head h		385	

the oven for 10 h, resulting in a weight of 452.1 g. Assuming the density of water is 1 g/cm³, compute the following: (a) water content of the sample, (b) volumetric water content of the sample, (c) degree of saturation, (d) porosity, (e) specific yield, and (f) dry bulk density.

5. Given three observation wells A, B, and C that are spaced in a row 750 ft apart, complete the data in Table 6.2. Estimate the hydraulic head gradients between A and B and between B and C.

6. Estimate the effective horizontal and vertical conductivities for a five-layer aquifer system. The layers are of equal thickness 6 m each and have conductivities of 13, 24, 11, 8, and 6 m/s. Estimate the portion of discharge for each layer based on a total discharge of 750 m³/s.

7. Estimate the head gradient and the drop in head if the length of the aquifer system is 100 m.

8. Estimate the incremental and total drop in head for vertical flow when Q is 24 m³/s.

9. Estimate the hydraulic conductivity for the middle section of the confined aquifer illustrated in Figure 6.16 based on the provided information. The dimensions are in meters, and conductivities are in m/s. The total drop in head is 9 m for a discharge of 0.1 m³/s per unit aquifer width. Estimate the effective aquifer hydraulic conductivity.

6.9 RESPONSE OF AQUIFERS TO PUMPING

Aquifer pumping is crucial for developing groundwater resources, especially since most aquifers do not meet the criteria for free-flowing wells. The decline in the water table or potentiometric level is influenced by various factors, including the number and spacing of wells and their pumping rates. It's important to note that drawdowns will inevitably increase in cases with numerous closely situated wells operating at high discharge rates. In some instances, individual wells operate in close proximity as part of a specialized

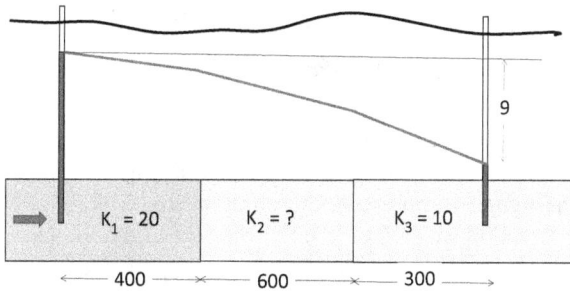

FIGURE 6.16 Sketch for problem 8 to estimate conductivity for the middle section of an aquifer consisting of three sections.

management scheme designed to meet varying demands. Additionally, the depth to which a well penetrates the aquifer affects the aquifer's response, leading to changes in drawdowns.

The type, size, and properties of the aquifer also play a pivotal role in determining drawdowns. These factors encompass different aquifer types, such as confined, leaky confined, or unconfined aquifers, each exhibiting distinct characteristics in response to pumping. For instance, leaky confined aquifers tend to produce lower drawdowns compared to confined aquifers under similar conditions, owing to the additional water that leaks into the aquifer.

The flow within the aquifer is further influenced by its properties and the spatial distribution of parameters, particularly in cases characterized by heterogeneous and anisotropic conditions. Hydraulic conductivity and storage parameters are critical in determining the expected drawdowns. Aquifers with high hydraulic conductivity are expected to experience lower drawdowns, as indicated by Darcy's law, where a large hydraulic conductivity results in a relatively small slope of the hydraulic head for a given discharge.

The response of extensively extended aquifers to pumping differs from those affected by nearby boundaries. For example, a large water source at an aquifer boundary, similar to a leaky confined aquifer, can provide inflow, thereby reducing expected drawdowns. Conversely, limited extent aquifers bounded by lower hydraulic conductivity zones experience reduced inflow to the well area, resulting in higher drawdowns.

Groundwater recharge plays a vital role in replenishing aquifers and enhancing the sustainability of water sources by reducing drawdowns. As should be expected, extended periods of drought coupled with excessive pumping can lead to substantial and concerning drawdowns.

Drawdown can occur in two main conditions: steady-state and unsteady (transient). A steady-state condition exists when the inflow to the well, through recharge and inter-aquifer flow, equals the pumping discharge rate. In contrast, the transient case, where drawdown changes with time, is more common due to the variability in well pumping rates, which depend on factors like demand, time of day, season, weather, and other variables.

6.9.1 Equation for Steady-State Pumping in a Confined Aquifer

An equation for estimating the response of pumping under steady-state conditions in a confined aquifer can be derived by making simplifying assumptions, as shown in Figure 6.17. In the figure, P represents the pumping well, while M_1 and M_2 are two monitoring or observation wells. The symbol b refers to the thickness of the aquifer, while h_w, h_1, and h_2 denote the water levels in the pumping and monitoring wells, and h_0 represents the initial pre-pumping water level. It is important to note that monitoring wells do not impact the flow; their sole purpose is to observe water levels. The dotted lines represent the potentiometric level, indicating where the water would reach if allowed, e.g., through the installation of wells.

To derive an equation for drawdown in the aquifer, several simplifying assumptions are made:

1. **Aquifer Flow**: The assumption is that aquifer flow is strictly horizontal, with no vertical flow components, such as those that might occur through a leaky aquitard.
2. **Aquifer Characteristics**: The aquifer is assumed to be large, horizontal, perfectly confined, homogeneous, isotropic, and of constant thickness.
3. **No Interference**: There is no interference from nearby water streams or geological zones that could influence water flow to the well.
4. **Uniform Hydraulic Head**: Prior to pumping, the hydraulic head is assumed to be uniform throughout the aquifer.
5. **Single Well**: Pumping is carried out through a single, fully penetrating well that operates at a fixed rate.
6. **Infinitesimally Small Well Diameter**: Finally, it is assumed that the well diameter is infinitesimally small, which is generally valid due to the relatively small radius compared to the size of the aquifer.

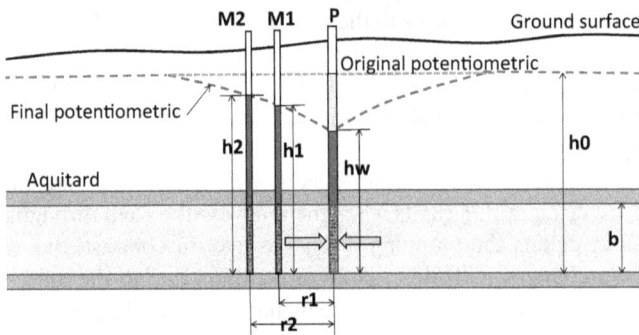

FIGURE 6.17 Flow to a well in a confined aquifer.

The equation can be expressed in the following form:

$$Q = 2\pi K b \frac{(h_2 - h_1)}{\ln\left(\dfrac{r_2}{r_1}\right)} \tag{6.24}$$

In equation (6.24), commonly referred to as the Thiem (1906) equation, Q represents the rate of well discharge (pumping), K denotes the aquifer hydraulic conductivity, b signifies the aquifer thickness, while h_2 and h_1 stand for the respective hydraulic head values at monitoring wells located at distances r_1 and r_2 from the pumping well. The constant π is approximated as 3.14, and ln represents the natural logarithm of the ratio r_2/r_1, which can be calculated using a scientific calculator. The well discharge rate, Q, can be determined using this equation provided that the values of all the variables on the right-hand side are known. The equation can also be rearranged to estimate the hydraulic head at any distance from the pumping well using the following equation:

$$h_2 = \frac{Q}{2\pi b K} \ln\left(\frac{r_2}{r_1}\right) + h_1 \tag{6.25}$$

Here, h_2 is estimated at distance r_2 when all variables are known, specifically the well pumping rate Q and the head value h_1 at distance r_1. If needed, the water level in the pumping well can be estimated by replacing h_2 and r_2 with h_w and r_w, which represent the water level value and radius related to the pumping well. However, it should be noted that the actual (measured) water level in the well will be smaller than the theoretical value estimated based on equation (6.25) due to mechanical head losses in the pumping well.

Equation (6.24) can also be rearranged to estimate the hydraulic conductivity when all variable values on the right-hand side of equation (6.26) are known:

$$K = \frac{Q}{2\pi b (h_2 - h_1)} \ln\left(\frac{r_2}{r_1}\right) \tag{6.26}$$

6.9.2 Equation for Steady-State Pumping in an Unconfined Aquifer

Figure 6.18 displays an unconfined aquifer with P representing the pumping well, while M_1 and M_2 are two monitoring or observation wells. The symbol h_0 refers to the original location of the water table of the aquifer, while h_w, h_1, and h_2 refer to the water table heights in the pumping and monitoring wells, respectively. It is important again to note that monitoring wells do not affect the flow, and their function is limited to observing the water levels. The area above the water table is unsaturated, containing both water and air, but for the purposes of assessing an idealized aquifer–well system, it will be assumed to be fully dry.

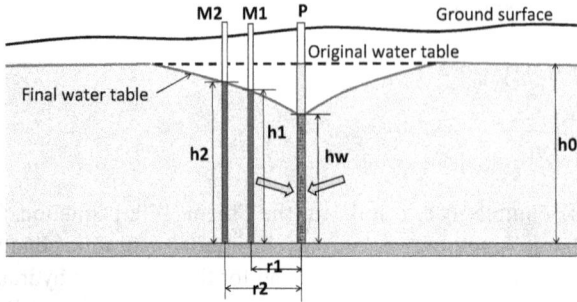

FIGURE 6.18   Flow to a well in an unconfined aquifer.

Parallel expressions for discharge, drawdowns, and hydraulic conductivity can be developed for the unconfined aquifer depicted in Figure 6.18. These equations are listed in equations (6.27) through (6.29).

$$Q = \pi K \frac{\left(h_2^2 - h_1^2\right)}{\ln\left(\dfrac{r_2}{r_1}\right)} \tag{6.27}$$

$$h_2^2 = \frac{Q}{\pi K} \ln\left(\frac{r_2}{r_1}\right) + h_1^2 \tag{6.28}$$

$$K = \frac{Q}{\pi\left(h_2^2 - h_1^2\right)} \ln\left(\frac{r_2}{r_1}\right) \tag{6.29}$$

In these expressions, and others discussed above, it is important to use consistent units. For example, distances, head values, and aquifer thickness would be in meter (m), hydraulic conductivity in meter per hour (m/h), and discharge in cubic meter per hour (m³/h).

Example 6.5: Steady-State Aquifer Response

1. Estimate the hydraulic conductivity of a confined aquifer with an active well at a discharge rate Q=400 m³/h. The aquifer thickness b is 40 m. There are two observation wells at distances r_1=25 m and r_2=75 m, with respective hydraulic heads h_1=85.3 and h_2=89.6 m.

$$K = \frac{Q}{2\pi b\left(h_2 - h_1\right)} \ln\left(\frac{r_2}{r_1}\right) = \frac{400\,\dfrac{m^3}{h}}{2 \times 3.14 \times 40 \times (89.6 - 85.3\ m)} \ln\left(\frac{75\ m}{25\ m}\right)$$

$$= 0.41\,\frac{m}{h}$$

The value of $\ln(75/25)=\ln(3)$ is calculated using a scientific calculator as explained in Chapter 2. The result is about 1.1.

2. Estimate the water level in a pumping well (h_w) operating in a confined aquifer given the following information: $Q=113\,m^3/h$, $K=0.42$ m/h, $b=40\,m$, and $h=39.5\,m$ at $r=50\,m$. The well radius (r_w) is $0.5\,m$.

$$h_w = \frac{Q}{2\pi bK}\ln\left(\frac{r_w}{r}\right)+h = \frac{113\dfrac{m^3}{h}}{2\times 3.14 \times 40\ m \times 0.42\ \dfrac{m}{h}}\ln\left(\frac{0.5}{50}\right)+39.5$$

$$= -4.94 + 39.5 = 34.56\ m$$

The value of $\ln (0.5/50)=\ln (0.01)=-4.61$ (a negative value). The head at the pumping well should be lower than that of the monitoring well.

6.9.3 Equation for Unsteady-State (Transient) Pumping in a Confined Aquifer

6.9.3.1 Single pumping well

Figure 6.17 is applicable here, but the hydraulic head (or potentiometric) level shown in the figure represents a snapshot at a given time. As time changes, the level can fluctuate under pumping or injection. Under the same assumptions discussed above, the drawdown or change in hydraulic head is estimated using the Theis (1935) solution:

$$h = h_o - \frac{Q}{4\pi T}W \tag{6.30}$$

In equation (6.30), h_0 is the original hydraulic head before the onset of pumping and h is the hydraulic head at a distance r from the well at time t. The variable W is called the well function (Appendix C) and is estimated based on the value of u. The variable u is given by the expression:

$$u = \frac{Sr^2}{4Tt} \tag{6.31}$$

in which S is the aquifer storage coefficient, r is the distance between the pumping and monitoring wells, T is transmissivity, and t is the time since the start of pumping. The calculations start by using equation (6.31) to estimate u, then extracting the corresponding value of W from the table in Appendix C. Finally, equation (6.30) is used to estimate the drawdown (see Example 6.6a).

This procedure can be used to estimate values for h for a given r at various time points (see Example 6.6b). Conversely, h can be estimated at multiple locations at a given time. Spreadsheets can be ideal for such repeated calculations (see Appendix A).

For a small value of u (less than 0.01), rather than using the table in Appendix C, the value of W can be approximated by

$$W = -0.5772 - \ln(u) \tag{6.32}$$

In equation (6.32), ln(u) is the logarithm of u and can be estimated by using a scientific calculator. The small value of u occurs closer to the well and/or at larger times, as can be inferred from equation (6.31).

Example 6.6a: Drawdowns for One Well

Given the following information:

Well pumping rate $Q=3,000\,m^3$/day; transmissivity $T=600\,m^2$/day; storage coefficient $S=4\times10^{-4}$; initial hydraulic head $h=12\,m$, calculate the drawdown at a distance of 1 km (1,000 m) from the well after 1 year (365 days).

First, we will estimate u (remembering to use consistent units):

$$u = \frac{Sr^2}{4Tt} = \frac{0.0001 \times 1,000^2\,m^2}{4 \times 600\,\dfrac{m^2}{day} \times 365\,day} = 0.000114 = 1.14 \times 10^{-4}$$

The corresponding value for W from the table in Appendix C is between 7.94 and 8.63, or about 8.5. Because u is less than 0.01, equation 6.32 can also be used:

$$W = -0.5772 - \ln(u) = -0.5772 - \ln(0.000114) = -0.5772 + 9.078 = 8.5$$

which agrees with the value obtained from the table.

Finally, the water level is estimated from:

$$h = h_o - \frac{Q}{4\pi T}W = 12 - \frac{3000\,\dfrac{m^3}{day}}{4 \times 3.14 \times 600\,\dfrac{m^2}{day}} \times 8.5 = 12 - 3.32 = 8.68\ m$$

The value 3.32 m represents the drawdown, which is the decline from the original level of 12 m.

Example 6.6b: Drawdowns at Multiple Times

For the same information in Example 6.6a, estimate the monthly drawdowns over a year from the start of pumping.

The calculations are repeated as shown in Table 6.3, and the results are depicted in the plot in Figure 6.19 with a decreasing trend for hydraulic head over time, as expected.

TABLE 6.3   Results for Example 6.6b

MONTH	NUMBER OF DAYS	DAYS SINCE PUMPING START	u	W	h
1	31	31	0.001344	6.04	9.60
2	28	59	0.000706	6.68	9.34
3	31	90	0.000463	7.10	9.17
4	30	120	0.000347	7.39	9.06
5	31	151	0.000276	7.62	8.97
6	30	181	0.00023	7.80	8.90
7	31	212	0.000197	7.96	8.83
8	31	243	0.000171	8.09	8.78
9	30	273	0.000153	8.21	8.73
10	31	304	0.000137	8.32	8.69
11	30	334	0.000125	8.41	8.65
12	31	365	0.000114	8.50	8.62

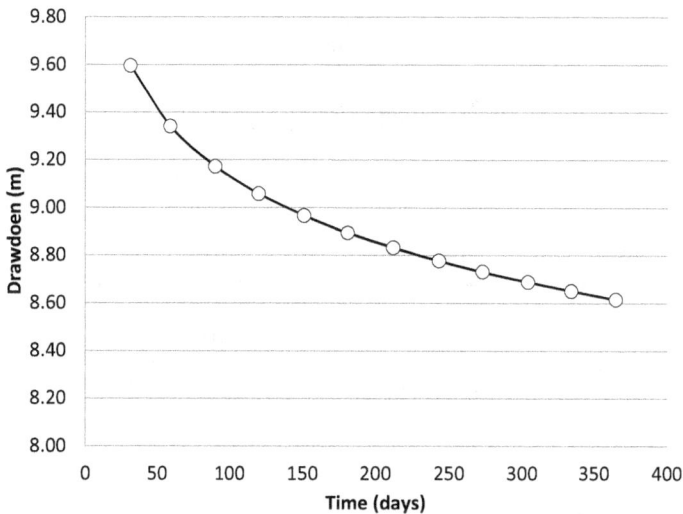

FIGURE 6.19   Results for drawdown vs. time for Example 6.6b.

6.9.3.2 Multiple wells with pumping and injection

The expression in equation (6.30) can be generalized to give

$$s = h_0 - h = \pm \frac{Q}{4\pi T} W \tag{6.33}$$

in which s is the change in head value caused by pumping, and is customarily called the drawdown. The positive sign for s is due to the fact that h is less than h_0 in this case. However, equation (6.33) is also applicable for well injection, where the change represents an increase in head value (h is greater than h_0). The value of s will be negative in this case and is labeled build up.

Equation (6.33) can be applied to cases with multiple wells. In such cases, the resulting drawdown will be the algebraic sum of all those caused by the individual wells. Example 6.7 demonstrates the case of three wells, but the method can be generalized for any number of wells, each starting at different times.

Example 6.7: Drawdowns for Multiple Wells

Estimate the hydraulic head value in the monitoring well due to the combined pumping and injection from three operating wells (Figure 6.20) using the information provided in the table below regarding Q, r, and t. The wells operate at different rates and started at different times. The aquifer transmissivity and storage coefficient are $600\,m^2/day$ and 0.0001, respectively. The initial hydraulic head $h_o = 12\,m$.

The values of change in hydraulic head caused by each well are calculated by first estimating the respective u and W. The values of W are estimated using the table in Appendix C. Finally, the change in water levels (s) for each well is estimated. The following equations are used for calculating u and s:

$$u = \frac{Sr^2}{4Tt}$$

$$s = h_0 - h = \frac{Q}{4\pi T}W$$

In these equations, the respective values for Q, r, and t for each well are used. The results are shown in Table 6.4. The final water level equals the initial value (i.e., 12 m) minus the decreases due to pumping (i.e., 3.95 and 1.03 m) plus the increase due to injection (i.e., 3.66 m).

The final head value will then be: $h = 12 - 3.95 - 1.03 + 3.66 = 10.68$ m

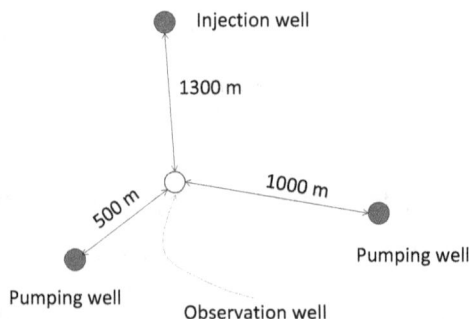

FIGURE 6.20   Sketch for Example 6.7 for pumping and injection wells.

TABLE 6.4 Results for Example 6.7

	Q (m³/ DAY)	r (m)	t (DAY)	u	W	s (m)
Pumping well 1	3,500	1,000	365	0.000114	8.50	3.95
Pumping well 2	700	500	1,200	8.68E-06	11.08	1.03
Injection well	2,850	1,300	2,000	3.52E-05	9.678	3.66

6.9.3.3 Well recovery

Equation (6.33) can also be applied in the case of well recovery, which occurs when a well is shut off after operating for a certain period. Such a test is a field experiment typically conducted at the conclusion of aquifer pumping tests, following the cessation of well operation. During this test, the water-level response, referred to as residual drawdown, is measured in one or more surrounding observation wells.

Figure 6.21 provides a visual representation of the changes in drawdown during both the pumping and recovery phases of the test. Initially, the drawdown increases (or the water level drops, as shown in the figure) until pumping ceases after a duration of t_s. Subsequently, the drawdown will gradually decrease (or the water level will rise) as it progresses towards full recovery. The drawdown value, denoted as s', at a specific time t', defines the residual drawdown, which ultimately reaches a zero value at full recovery.

The residual drawdown (i.e., the value of drawdown after shutting off the well) is estimated from:

$$s' = \frac{Q}{4\pi T}(W_1 - W_2) \tag{6.34}$$

in which W_1 and W_2 are the well functions estimated from the table in Appendix B.1C by using the respective u values given below:

$$u_1 = \frac{Sr^2}{4Tt} \tag{6.35}$$

$$u_2 = \frac{Sr^2}{4Tt'} \tag{6.36}$$

in which t and t' are the total time since the start of the pumping process and the time since the well shut off, respectively (Figure 6.21).

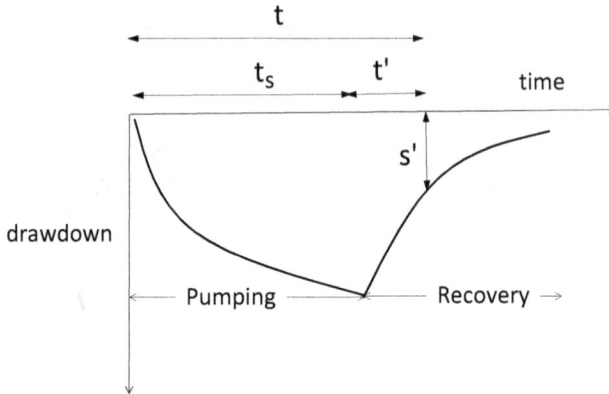

FIGURE 6.21   Schematic representation of the drawdown during the pumping and recovery phases.

Example 6.8: Well Recovery

Calculate and plot the drawdown vs. time for a pumping recovery test given the following information.

Well discharge $Q=250\,m^3/min$; transmissivity $T=50\,m^2/min$; storage coefficient $S=0.00156$; well distance from the pumping well $r=200\,m$; end of the pumping period $=1,000\,min$.

A portion of the calculation data is presented in Table 6.5. In most steps, time increments were set at 20-min intervals, resulting in a total calculation time of 2,000 min, equally divided between pumping and recovery phases. To achieve better resolution, a few additional time steps of less than 20 minutes each were introduced at the first part of the calculations. It is important to note that the drawdown is zero at time equals zero, rendering the Theis solution inapplicable in this scenario, and no calculations are needed.

The "—" sign in Table 6.5 signifies that the information is not relevant, particularly during the pumping phase. For this phase, drawdown is calculated using equations (6.35) and (6.30) for u_1 and s, respectively. Conversely, during the recovery phase, calculations require the use of u_1 and u_2, as determined by equations (6.35) and (6.36), while the residual drawdown s' is estimated by using equation (6.34). Values for W_1 and W_2 can be obtained from the table in Appendix C or by applying equation (6.32) for values of u_1 and u_2 that are less than 0.01. At the end of the pumping period, the maximum drawdown reaches 9.37 m.

The full results are displayed in Figure 6.22. The vertical axis points downward to reflect the change in water level, which declines until the end of the pumping period and increases thereafter, signifying the recovery phase. Note that a longer time is needed for full recovery (or a zero residual drawdown).

TABLE 6.5   Results for example 6.8

t	$t'=t-t_s$	u_1	W_1	u_2	W_2	s OR s'
0	–	–	–	–	–	0
1	–	0.312500	0.88	–	–	1.10
3	–	0.104167	1.79	–	–	2.23
.	.	.	.	.	.	.
.	.	.	.	.	.	.
.	.	.	.	.	.	.
20	–	0.015625	3.60	–	–	4.50
40	–	0.007813	4.28	–	–	5.35
60	–	0.005208	4.69	–	–	5.86
80	–	0.003906	4.97	–	–	6.21
100	–	0.003125	5.19	–	–	6.49
.	.	.	.	.	.	.
.	.	.	.	.	.	.
1,000	–	0.000313	7.49	–	–	9.37
1,020	20	0.000306	7.51	0.015625	3.60	4.90
1,040	40	0.000300	7.53	0.007812	4.28	4.06
1,060	60	0.000295	7.55	0.005208	4.69	3.58
1,080	80	0.000289	7.57	0.003906	4.97	3.25
.	.	.	.	.	.	.
.	.	.	.	.	.	.
2,000	1,000	0.000156	8.19	0.000313	7.49	0.87

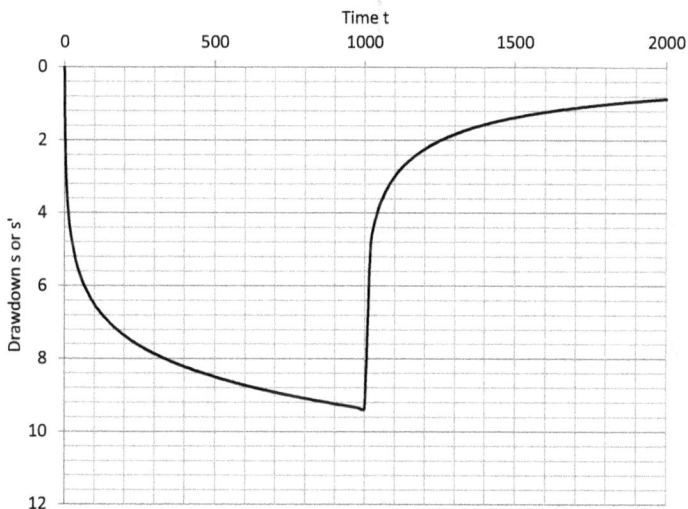

FIGURE 6.22   Results for Example 6.8 for drawdown vs. time.

6.9.3.4 Boundary effects

Among other assumptions, the Theis solution is defined based on the premise that the aquifer is extensive in size and unaffected by any physical formations that might influence the water level response to well pumping or injection. However, in real field scenarios, certain geological formations, such as faults, can act as barriers, leading to conditions where no flow occurs. This situation can be addressed by utilizing image wells, allowing the application of the Theis approach.

This procedure is illustrated in Figure 6.23, where a barrier exists at a distance R_w from the pumping well. The method commences by introducing an image pumping well on the opposite side of the barrier, at the same distance R_w, and operating at the same pumping rate as the actual well. The Theis solution is then considered valid after eliminating the barrier, resulting in a new hypothetical aquifer with two pumping wells. Subsequently, the drawdown at any point can be estimated using the Theis equation by combining the drawdowns caused by the two wells, which can be expressed as follows:

$$s = s_1 + s_2 = \frac{Q}{4\pi T}(W_1 + W_2) \tag{6.37}$$

with W_1 and W_2 estimated based on the respective values of u, which are calculated by utilizing the pertinent distances from the wells r_1 and r_2:

$$u_1 = \frac{Sr_1^2}{4Tt}$$

$$u_2 = \frac{Sr_2^2}{4Tt}$$

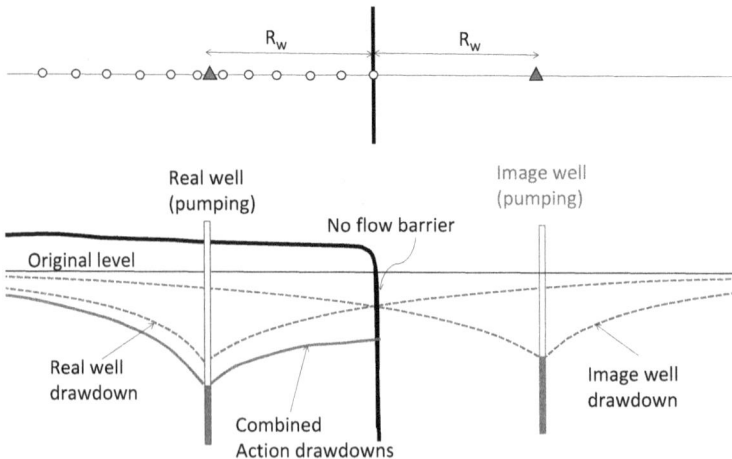

FIGURE 6.23 The case of a flow barrier existing at a distance Rw from a pumping well. The circles in the top figure schematically represent locations where the drawdowns are estimated. The drawdown $s = s_1 + s_2$.

In this context, S and T denote the storage coefficient and transmissivity, respectively, while u_2 and W_2 represent the image well. Figure 6.23 provides a schematic representation of individual drawdowns (shown as dotted lines) for the two wells at a specific time, along the line connecting the two wells (Figure 6.23a). The resulting drawdowns are combined to determine the collective value at any given point (represented by the heavier solid line). When compared to a scenario without a barrier, as one would expect both physically and mathematically, the water level should be lower, indicated by larger drawdowns. This is because the barrier reduces the available aquifer volume for water extraction.

Another situation that violates the Theis assumptions concerns the presence of a nearby large water body, such as a river, capable of continuously supplying water to the aquifer. Under this condition, the water level in the aquifer at the river's location will remain unchanged (i.e., it will experience zero drawdown) despite pumping activities. Similar to the scenario with a barrier, an image well becomes necessary to apply the Theis solution, but in this case, it must be an injection well to represent the flow from the river (as depicted in Figure 6.24). The figure illustrates schematic representations of drawdown and buildup using dotted lines. The combined effects can be estimated using equation (6.37) by summing the contributions of both wells, but with a negative value for the image term, as follows:

$$s = s_1 - s_2 = \frac{Q}{4\pi T}(W_1 - W_2) \tag{6.38}$$

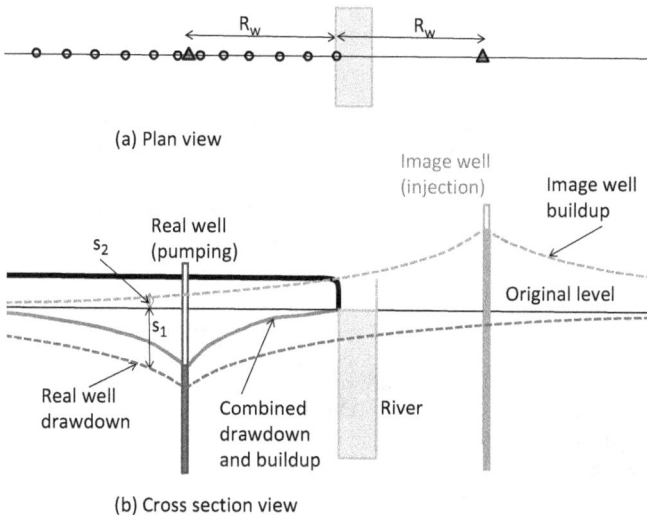

(a) Plan view

(b) Cross section view

FIGURE 6.24 The case of a large river existing at a distance Rw from a pumping well. The circles in the top figure schematically represent locations where the drawdowns are estimated. The drawdown $s = s_1 - s_2$.

which would produce the (final) drawdown, incorporating the effect of river flow. Once again, W_1 and W_2 are estimated using the appropriate u, with W_2 denoting the injection image well. The heavier solid line in the figure illustrates this representation. As anticipated, there will be a reduced drawdown due to the additional water supplied by the river. Consequently, in this scenario, the condition of zero drawdown at the aquifer–river interface is met.

Example 6.9: Aquifers with Boundaries

Calculate and plot the drawdowns vs. distance from the boundary (also known as the water level profile) for cases with a barrier and a river given the following information: Well pumping $Q=3,000\,\text{ft}^3/\text{min}$; Transmissivity $T=800\,\text{ft}^2/\text{min}$; Time $t=1,000\,\text{min}$; and distance to the barrier or river $R_w=550\,\text{ft}$.

The calculation details are presented in Table 6.6, and utilizing a spreadsheet can greatly simplify the process. It is important to consider only the points within the (real) aquifer's vicinity, which are visually represented as circles in Figures 6.15a and 6.16a. To cover a distance of 1,100 ft, equivalent to twice the value of R_w, sixteen points were employed. It is advisable to increase the density of points closer to the well to enhance resolution. Additionally, exercise caution when placing a point at $X=R_w$ (or $r_1=0$), as these locations do not yield valid solutions. The calculations become straightforward once one meticulously evaluates the distances between the wells and the calculation points (i.e., r_1 and r_2 for each point), taking into account their positions relative to both the wells and the boundary. Figures 6.23a and 24a can be used to illustrate the relationship between r_1, r_2, the point's location X, and the distance R_w for a specific point. Values for r_1 and r_2 can be computed as follows:

$$r_1 = ABS\left(X - R_w\right)$$

$$r_2 = X + R_w$$

where r_1 is calculated as the absolute value of $(X - R_w)$, which is a positive value. Next, drawdowns are calculated by using equations (6.37) and (6.38) for the barrier and river cases, respectively. Utilizing the same data, the calculated buildup values should match those for the barrier case, except they should be negative for buildup (values are not listed in the table). The respective results are illustrated in Figures 6.25 and 6.26 for the cases with the barrier and river.

The plots exhibit the expected features, including the symmetrical nature of drawdown values around the pumping well when the image well is not considered. In this case, the drawdown is 3.28 ft at equal opposite distances from the well of 550 ft. With the barrier, the combined drawdown is generally larger, and the profile is not symmetrical, with increased values closer to the barrier's side. At the barrier location ($X=0$), an equal drawdown value of 3.28 ft is calculated for both the real and image wells, resulting in a combined drawdown of 6.56 ft. The corresponding value at the same distance of 550 ft on the other side is lower, at 5.91 ft.

For the river case, the combined drawdown is zero at $X=0$ where the river is located, resulting in an overall lower aquifer drawdown. The profile is also not symmetrical, with a larger combined drawdown of 0.66 ft at the same distance of 550 ft on the other side of the well (Figure 6.26).

TABLE 6.6   Results for Example 6.9

POINT NUMBER	X ft	r_1 ft	u_1	W_1	s_1 ft	r_2 ft	u_2	W_2	s_2 ft	s FOR THE BARRIER CASE (ft)	s FOR THE RIVER CASE (ft)
1	0	550	9.45E-06	10.99	3.28	550	9.45E-06	10.99	3.28	6.56	0.00
2	100	450	6.33E-06	11.39	3.40	650	1.32E-05	10.66	3.18	6.58	0.22
3	200	350	3.83E-06	11.90	3.55	750	1.76E-05	10.37	3.10	6.65	0.46
4	300	250	1.95E-06	12.57	3.75	850	2.26E-05	10.12	3.02	6.77	0.73
5	400	150	7.03E-07	13.59	4.06	950	2.82E-05	9.90	2.96	7.01	1.10
6	500	50	7.81E-08	15.79	4.71	1,050	3.45E-05	9.70	2.90	7.61	1.82
7	540	10	3.13E-09	19.01	5.67	1,090	3.71E-05	9.62	2.87	8.55	2.80
8	545	5	7.81E-10	20.39	6.09	1,095	3.75E-05	9.61	2.87	8.96	3.22
9	555	5	7.81E-10	20.39	6.09	1,105	3.82E-05	9.60	2.87	8.96	3.22
10	560	10	3.13E-09	19.01	5.67	1,110	3.85E-05	9.59	2.86	8.54	2.81
11	600	50	7.81E-08	15.79	4.71	1,150	4.13E-05	9.52	2.84	7.56	1.87
12	700	150	7.03E-07	13.59	4.06	1,250	4.88E-05	9.35	2.79	6.85	1.27
13	800	250	1.95E-06	12.57	3.75	1,350	5.70E-05	9.20	2.75	6.50	1.01
14	900	350	3.83E-06	11.90	3.55	1,450	6.57E-05	9.05	2.70	6.25	0.85
15	1,000	450	6.33E-06	11.39	3.40	1,550	7.51E-05	8.92	2.66	6.06	0.74
16	1,100	550	9.45E-06	10.99	3.28	1,650	8.51E-05	8.79	2.63	5.91	0.66

Distance from boundary

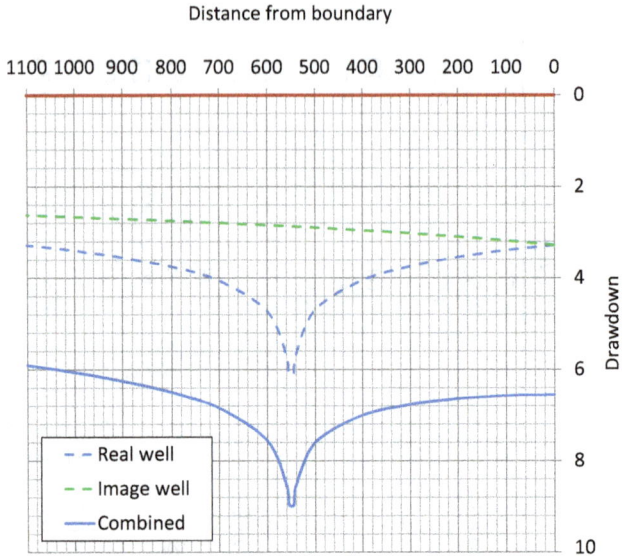

FIGURE 6.25   The water level profiles for the barrier case. The red line represents the pre-pumping water level.

Distance from boundary

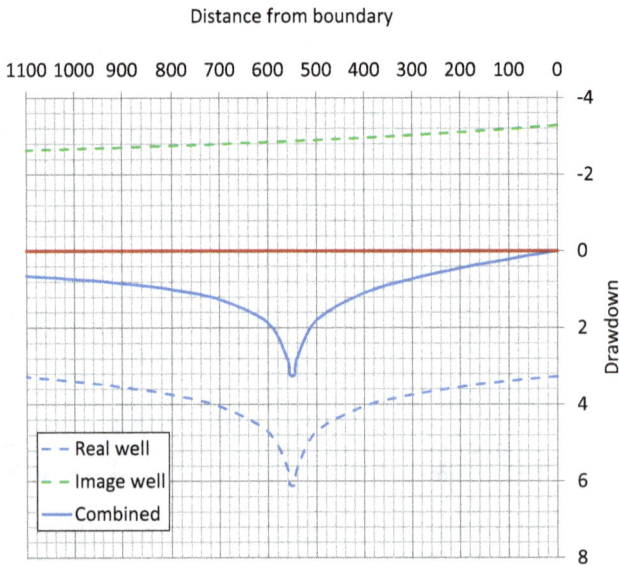

FIGURE 6.26   The water level profiles for the river case. The red line represents the pre-pumping water level.

6.9.3.5 Other general cases

Applications to cases with variable pumping rates are feasible. Furthermore, the Theis equation can be employed in relatively shallow unconfined aquifers by substituting the storage coefficient with the specific yield and utilizing an average transmissivity value. More accurate solutions for unconfined aquifers, which consider the influence of the unsaturated zone, are also available. Additional solutions can be found in existing literature, such as those for leaky confined aquifers. For comprehensive coverage of these topics, readers are referred to the classic book by Fetter and Kreamer (2022).

6.10 ASSIGNMENTS: AQUIFER PUMPING RESPONSE

1. Calculate the drawdowns at distances of 15, 50, 100, 300, 1,000, 1,500, and 3,000 m from a pumping well for a confined aquifer after 30 days. Transmissivity and storativity are 4,500 m²/day and 0.0006, respectively. The pumping rate is 5,000 m³/day.
2. For the same data in problem 1, estimate the drawdowns at a distance of 100 m from the pumping well at times of 0.1, 1, 10, 500, and 1,000 days. Estimate the steady-state drawdown at that location.
3. Calculate the steady-state water levels at distances of 45, 150, 300, 900, 3,000, 4,500, and 9,000 ft from a pumping well for a confined aquifer with a transmissivity of 45,000 m²/day. The pumping rate is 150,000 ft³/day from a well with a radius of 1 ft. The water level in the pumping well is 16 ft.
4. Estimate the hydraulic conductivity of an unconfined aquifer with an active well at a discharge rate $Q=400$ m³/h. There are two observation wells at distances 25 and 75 m, with respective water levels of 85.3 and $h_2=89.6$ m
5. Use the data in Example 6.7 to estimate the water level after an additional 120 days.
6. Use the data in Example 6.7 to estimate the water level if pumping rates are increased by 20%.
7. Use the data in Example 6.8 to complete Table 6.7. Validate your answers for drawdowns against the Figure 6.22 shown in the same example.

TABLE 6.7 Data for Assignment 7

T	$t' = t - t_s$	u_1	W_1	u_2	W_2	s OR s'
250						
500						
750						
1,250						
1,500						
1,750						

6.11 AQUIFER TEST ANALYSIS

There are many aquifer test procedures that are essential for hydrogeologic investigations as a means of estimating aquifer hydrogeologic properties, such as hydraulic conductivity and storage coefficient. Three methods will be covered here, and the interested reader can consult other references, including Fetter and Kreamer (2022) and the commercial software AQTESOLV (2023).

6.11.1 The Log–Log Curve Matching

In this test, groundwater is extracted from a pumping well, and water level measurements are observed in both the pumping well and one or more observation wells. Data collection involves recording drawdowns in both the pumping well and the observation wells at various time intervals. It is also essential to know the distances between the observation wells and the pumping well. The log–log type curve matching is based on the similarity between the well function (W vs. 1/u) and the drawdown vs. time measurements, provided they are both plotted on logarithmic paper. Figure 6.27 illustrates the plot for W vs. 1/u.

The following steps are taken to calculate transmissivity and storage coefficient. Hydraulic conductivity can be estimated by dividing transmissivity by the aquifer thickness. Specific storage is estimated by dividing the storage coefficient by the aquifer thickness.

1. The drawdown values, i.e., the changes in hydraulic head from the original values, are plotted against their respective times on log–log graph paper with a scale similar to that in Figure 6.27. Appendix C contains a replica of this figure as well as a table of the pertinent data. A blank plot of the same scale is also included in Appendix B.
2. The two sheets are then superimposed and carefully aligned, ensuring that the directions of the axes match, and they are adjusted until a reasonable match is achieved (see Figure 6.28).

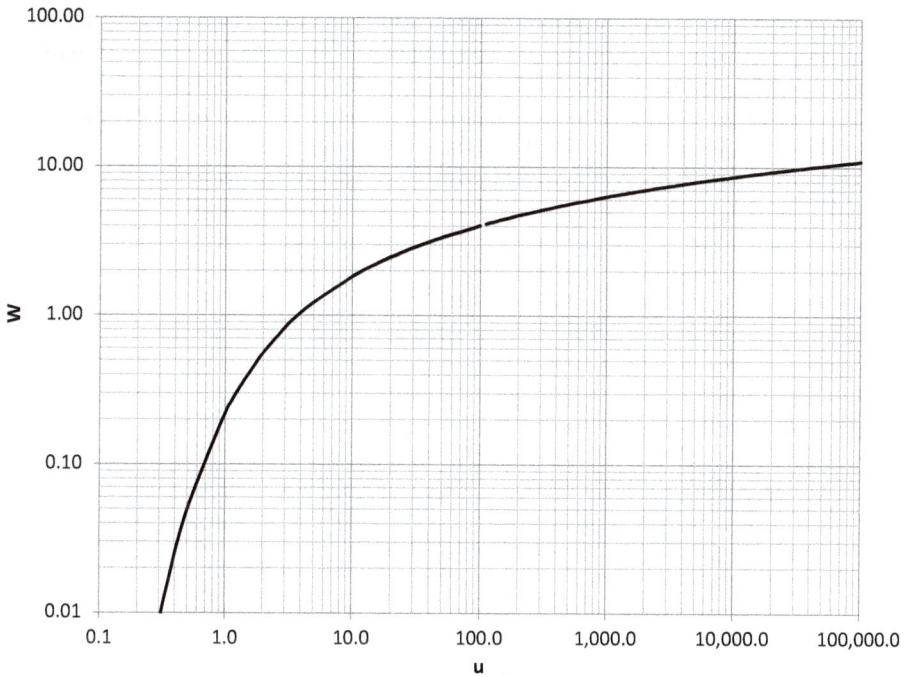

FIGURE 6.27 The well function plot.

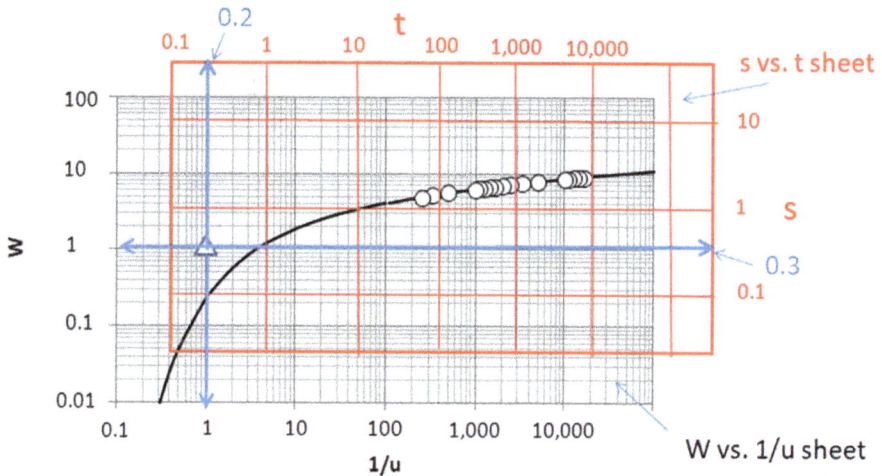

FIGURE 6.28 The matching process between the s vs. t data (the sheet with red lines) and the W vs. 1/u curve (the sheet with black lines). The triangle indicates the selected matching point. The blue lines with arrows are used to read the values s, t, W, and 1/u from the appropriate sheet. Here, the matching point is selected for both W=1 and 1/u =1 just for convenience.

3. An arbitrary matching point anywhere within the two sheets is selected, and the respective values of s, t, W, and 1/u are read from the two sheets, as indicated by the blue lines with arrows. The selected matching point does not need to fall precisely on the curves; any point is acceptable. In this case, a point where W=1 and 1/u=1 is chosen for convenience.

4. The values of s, t, W, and u obtained from the selected matching point are used to estimate transmissivity (T) and storage coefficient (S) using the following equations:

$$T = \frac{QW}{4\pi s} \tag{6.39}$$

$$S = \frac{4uTt}{r^2} \tag{6.40}$$

In equations (6.39) and (6.40), Q and r are the well pumping rate and the distance between the pumping and monitoring well, respectively, and W, u, s, and t are the values obtained at the matching point.

Example 6.10 describes the calculation scheme.

Example 6.10: Log–Log Matching Method

The objective of this example is to use the matching in Figure 6.28 to estimate the hydraulic conductivity and specific storage of the aquifer. The test pumping rate was 1,225 m³/day, and the aquifer thickness is 75 m. The monitoring well is located at a distance of 500 m from the pumping well.

From Figure 6.28, the respective values of W, u, s, and t are 1, 1, 0.3 m, and 0.2 days.

Aquifer parameters are estimated as follows. It is important to use consistent units, meter-day in this case.

$$T = \frac{QW}{4\pi s} = \frac{1{,}225\,\dfrac{\text{m}^3}{\text{day}} \times 1}{4 \times 3.14 \times 0.3\ \text{m}} = 325.1\ \text{m}^2/\text{day}$$

Hydraulic conductivity would be

$$K = \frac{T}{b} = \frac{325.1\,\dfrac{\text{m}^2}{\text{day}}}{75\ \text{m}} = 4.33\ \text{m/day}$$

for the aquifer thickness of 75 m. The storage coefficient is estimated as

$$S = \frac{4uTt}{r^2} = \frac{4 \times 1 \times 325.1\,\dfrac{\text{m}^2}{\text{day}} \times 0.2\ \text{day}}{500^2\ \text{m}^2} = 0.001$$

S is dimensionless (no units). The specific storage is estimated from

$$S_s = \frac{S}{b} = \frac{0.001}{75 \text{ m}} = 1.4 \times 10^{-5} \text{ 1/m}$$

The match in Figure 6.28 is based on the assumptions utilized in the Theis solution and the analytical expressions introduced earlier. Figure 6.29a illustrates an ideal case, while deviations from this ideal scenario can be observed in the remaining figure panels, primarily due to deviations from the assumptions of the Theis solution.

In panel b, the deviation is attributed to aquifer leakage from another aquifer system, indicating that the aquifer is not perfectly confined. Panel c depicts the case of an unconfined aquifer characterized by significant variable vertical flow activity in the unsaturated zone. This dynamic flow behavior is reflected in varying rates of drawdown decline over time.

Panel d demonstrates the influence of nearby boundary conditions. A faster recession of the hydraulic head suggests the presence of a no-flow condition, possibly caused by geological faults, which violates the assumption in the Theis solution requiring a large extent of the aquifer. Conversely, if a large water body is nearby, it can result in additional water flowing into the aquifer, similar to the leaky case depicted in panel b.

To better understand these deviations, it is advisable to conduct field investigations into both the surface and subsurface characteristics of the area. There are established approaches for analyzing non-ideal cases, as described in AQTESOLV (2023).

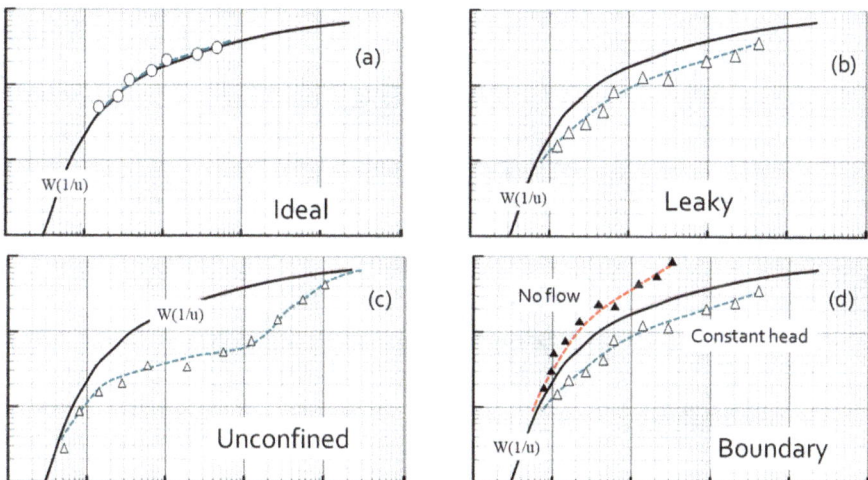

FIGURE 6.29 (a) Plot representing an ideal case, while plots (b) through (d) show the absence of reasonable match signifying the invalidity of the Theis solution.

6.11.2 The Semi-Log Curve Matching (the Jacob Solution)

The semi-log curve matching method is applicable for small values of u, resulting in an algebraic expression for W in equation (6.32). Under this condition, the relationship between drawdown s and time t closely approximates linearity when represented on a semi-log scale, as depicted in Figure 6.30. However, at early times, as shown in the figure, there is some deviation due to the expected relatively high value of u. To account for this, a straight line can be fitted to the data, but this should exclude the early time points. The transmissivity and storage coefficient can be estimated using the following expressions:

$$T = \frac{2.3\, Q}{4\pi \Delta s} \tag{6.41}$$

$$S = \frac{2.25 T t_o}{r^2} \tag{6.42}$$

In equations (6.41) and (6.42), Q is the well pumping rate, Δs is the change in drawdown for a log cycle, which is 5.2 m in the example in Figure 6.30, t_o is the line intercept with the time axis (about 5 days in the figure), and r is the separation distance between the observation and pumping wells. Appendix B contains a blank semi-log graph paper that can be used for plotting purposes.

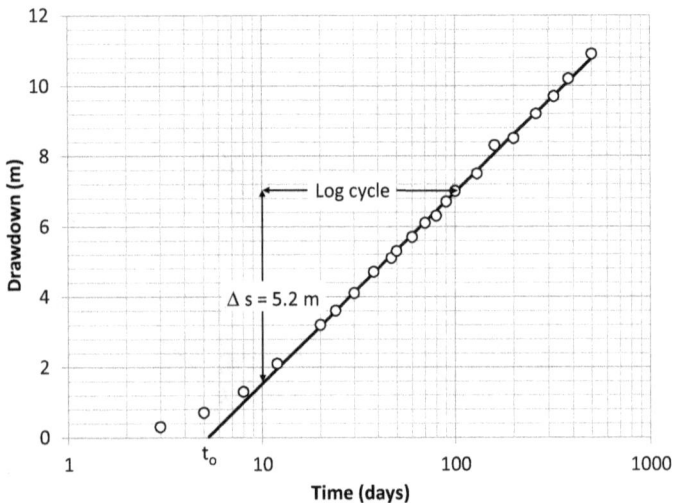

FIGURE 6.30 Semi-log representation of the relationship between drawdown and time under the Jacob approximation.

Example 6.11: Semi-log Matching Method

Utilizing the plot in Figure 6.30, it is required to estimate transmissivity and storage coefficient if the pumping rate $Q = 1,000 \, m^3/day$, and r is $1,000 \, m$.

From Figure 6.30, t_o is 5 days, and Δs is 5.2 m. Therefore, the values of T and S can be estimated as follows.

$$T = \frac{2.3 \, Q}{4\pi\Delta s} = \frac{2.3 \times 1,000 \dfrac{m^3}{day}}{4 \times 3.14 \times 5.2 \, m} = 35.2 \, m^2/day$$

$$S = \frac{2.25 T t_o}{r^2} = \frac{2.25 \times 35.2 \dfrac{m^2}{day} \times 5.2 \, day}{1,000^2 \, m^2} = 0.0004$$

Example 6.12: Least Squares Method for a Pumping Test Case

The least squares method is applied in this case using the data from Example 6.11, which addresses the semi-log matching method. This method postulates a linear relationship between drawdown and the logarithm of time. The relevant data are presented in Table 6.8, where time and drawdown values are labeled as x, and y, respectively. However, the linear regression analysis should be conducted between the logarithm of x, versus y, for which the values are provided in the second and third columns, respectively. The remaining calculated values are also presented in the table, following the same steps described in Example 2.1.

TABLE 6.8 Data and calculations for the pumping test case

x'	X = LOG (x')	y	x²	y²	xy	ESTIMATED y
3	0.30	–	–	–	–	–
5	0.70	–	–	–	–	–
8	1.30	–	–	–	–	–
12	1.08	2.1	1.16	4.41	2.27	1.95
20	1.30	3.2	1.69	10.24	4.16	3.16
24	1.38	3.6	1.90	12.96	4.97	3.59
30	1.48	4.1	2.18	16.81	6.06	4.12
38	1.58	4.7	2.50	22.09	7.42	4.68
47	1.67	5.1	2.80	26.01	8.53	5.18
50	1.70	5.3	2.89	28.09	9.00	5.33
60	1.78	5.7	3.16	32.49	10.14	5.76
70	1.85	6.1	3.40	37.21	11.26	6.13
80	1.90	6.3	3.62	39.69	11.99	6.44
90	1.95	6.7	3.82	44.89	13.09	6.72

(Continued)

TABLE 6.8 (Continued) Data and calculations for the pumping test case

x′	X = LOG (x′)	y	x²	y²	xy	ESTIMATED y
100	2.00	7	4.00	49.00	14.00	6.97
130	2.11	7.5	4.47	56.25	15.85	7.59
160	2.20	8.3	4.86	68.89	18.29	8.08
200	2.30	8.5	5.29	72.25	19.56	8.61
260	2.41	9.2	5.83	84.64	22.22	9.23
320	2.51	9.7	6.28	94.09	24.30	9.72
380	2.58	10.2	6.66	104.04	26.31	10.13
500	2.70	10.9	7.28	118.81	29.42	10.78

$$\sum x = 38.78 \quad \sum y = 124.2 \quad \sum x^2 = 73.79 \quad \sum y^2 = 922.86 \quad \sum xy = 258.84$$

The values of m and b calculated using equations (2.15) and (2.16) from Chapter 2 are 5.45 and –3.9, respectively. The regression equation is as follows:

$$y = 5.45 \cdot Log(x') - 3.9 \tag{6.43}$$

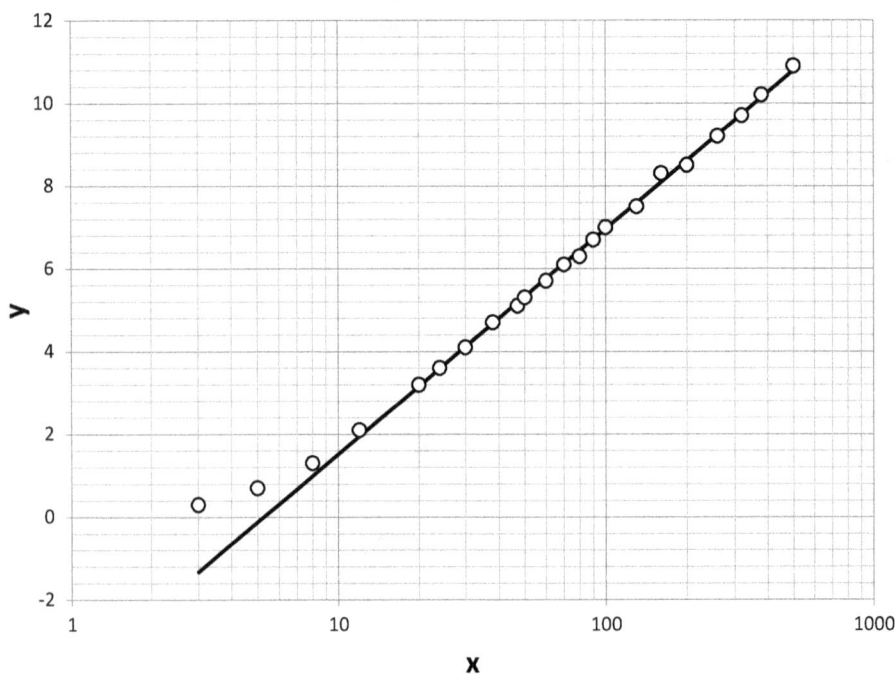

FIGURE 6.31 Comparison between the data and the regression line for Example 6.12.

Figure 6.31 illustrates the data and the linear regression line. The y-intercept, crucial for aquifer pump test calculations, is determined by setting y=0 in equation (6.43). This yields a logarithmic value of x' equal to 0.71. Consequently, the value of x' can be approximated as 5.0, obtained by calculating 10 to the power of 0.71. This value is also visually found from Figure 6.31 at y=0. The slope of the line required for the application of the Jacob formula is represented by the value 5.45 in equation (6.43). Both the slope and intercept values align with those obtained through manual fitting in Example 6.11.

6.11.3 Well recovery

As discussed in Section 6.9.3.3, a recovery test is a field procedure typically conducted at the conclusion of aquifer pumping tests, following the shutdown of the well. Figure 6.21 provides a schematic representation of drawdown behavior during both the pumping and recovery phases. The procedure for estimating transmissivity involves fitting a straight line to a plot of residual drawdown vs. the logarithm of t/t' (Theis, 1935). It is important that the fitted line emphasizes data points closer to the origin of the graph as t/t' approaches unity towards the end of the recovery period. Transmissivity T can be estimated using the following equation:

$$T = \frac{Q}{4\pi\Delta s'} \tag{6.44}$$

In equation (6.44), Q is the pre-recovery pumping rate and $\Delta s'$ is the change in residual drawdown over a log cycle. Example 6.12 describes the calculation scheme for a recovery test.

Example 6.13: Well Recovery

The objective is to estimate the transmissivity of the aquifer following the recovery from pumping 14 m³/h for 30 h. The recorded time values after the end of pumping t' and the respective residuals s' are shown in Table 6.9.

The first step is to calculate t/t', where t is the total time, equaling t_s plus t', with $t_s = 30$ h. Next, s' vs. t/t' are graphed on a semi-log sheet as shown in Figure 6.32. A similar blank sheet is included in Appendix B for future use.

Next, a straight trendline is drawn, emphasizing the data near the end where t/t' approaches unity. The change in residual drawdown is then estimated from the graph, as shown in Figure 6.32, which is 1.6 m. Finally, transmissivity is estimated from the following equation:

$$T = \frac{Q}{4\pi\Delta s'} = \frac{14\frac{m^3}{h}}{4 \times 3.14 \times 1.6\ m} = 0.7\ m^2/h$$

TABLE 6.9   Data and results for Example 6.13

t′ (h)	t (h)	t/t′	s′ (m)
0.05	30.05	601	5.0
0.1	30.1	301	4.5
0.2	30.2	151	3.7
0.5	30.5	61	3.0
1	31	31	2.5
2.5	32.5	13	2.0
10	40	4	1.2
20	50	2.5	0.9
50	80	1.6	0.5
100	130	1.3	0.3
900	930	1.03	0

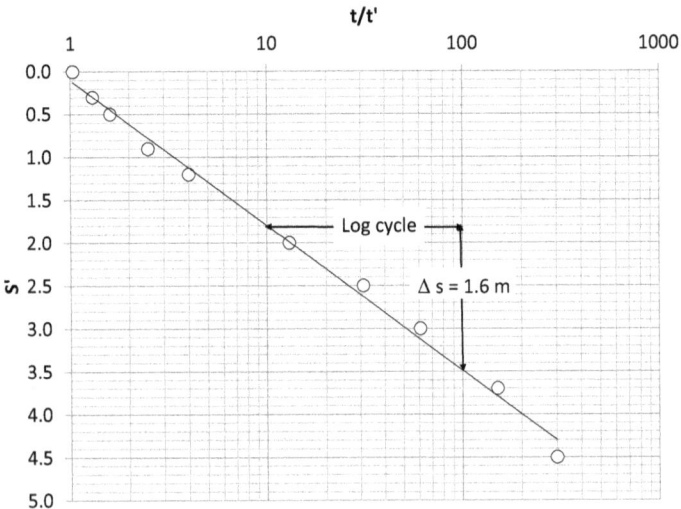

FIGURE 6.32   Plot of t/t′ vs. s for Example 6.13.

6.12 ASSIGNMENTS: AQUIFER TEST

1. A well in a confined aquifer with a thickness of 49 ft was pumped at a rate of 29.4 ft³/min. Drawdown was measured in a fully penetrating observation well located 824 ft away. Use the data in Table 6.10 to estimate aquifer

TABLE 6.10 Data for problem 1

TIME (min)	DRAWDOWN (ft)	TIME (min)	DRAWDOWN (ft)	TIME (min)	DRAWDOWN (ft)	TIME (min)	DRAWDOWN (ft)
3	0.3	30	4.1	80	6.3	260	9.2
5	0.7	38	4.7	90	6.7	320	9.7
8	1.3	47	5.1	100	7	380	10.2
12	2.1	50	5.3	130	7.5	500	10.9
20	3.2	60	5.7	160	8.3		
24	3.6	70	6.1	200	8.5		

TABLE 6.11 Data for Problem 2

TIME (min)	DRAWDOWN (ft)	TIME (min)	DRAWDOWN (ft)	TIME (min)	DRAWDOWN (ft)
6.9	0.01	70.6	0.03	694	0.15
10.2	0.01	106	0.05	1,010	0.19
15.4	0.02	157	0.06	1,490	0.22
22	0.02	227	0.08	2,210	0.25
32.6	0.02	324	0.1	3,290	0.28
50.3	0.03	475	0.13	4,340	0.3

TABLE 6.12 Data for Problem 3

t' (min)	s' (ft)	t' (min)	s' (ft)
1	25.73	48	13.47
2	24.76	100	10.70
5	22.84	200	8.39
10	19.90	400	6.49
20	17.01	500	5.98
30	15.35	1,000	4.74
40	14.19	1,100	4.73

transmissivity, storage coefficient, hydraulic conductivity, and specific storage. Use both the Theis and the Jacob solutions (solution $T \sim 1,400 \, \text{ft}^2/\text{min}$, $S \sim 0.00002$).

2. Estimate transmissivity and specific yield for an unconfined aquifer with a thickness of 170 ft. The discharge value was 42.8 ft³/min. The observations are listed in Table 6.11, taken from a distance of 225.7 ft. Use the Jacob solution and discuss the limitations.

3. Table 6.12 lists data for residual drawdowns from a well recovery test. The well's pumping rate of 100 f³/min was shut off after 500 min. Estimate the aquifer transmissivity (solution $T \sim 0.8 \, \text{ft}^2/\text{min}$).

NOTE

1 Hydraulic head is the driving force for water movement and it combines pressure and gravity forces. Section 6.3 will elaborate on this and related concepts.

REFERENCES

American Geosciences Institute. (2017). Water use in the United States. Retrieved from https://www.americangeosciences.org/critical-issues/water-use

AQTESOLV. (2023). AQTESOLV. Retrieved from https://www.aqtesolv.com/

Darcy, H. (1856). *Les fontaines publiques de la ville de Dijon: exposition et application des principes à suivre et des formules à employer dans les questions de distribution d'eau* [Public fountains of the city of Dijon: Exposition and application of the principles to follow and the formulas to use in questions of water distribution]. Paris, France: Victor Dalmont.

Fetter, C. W., & Kreamer, D. (2022). *Applied Hydrogeology* (5th ed.). Long Grove, IL: Waveland Press.

Gingerich, S. B., & Oki, D. S. (2000). *Ground water in Hawaii* (U.S. Geological Survey Fact Sheet 126-00, 6 p.).

Gilson. (2024). *Permeability of Soil Testing Equipment*. Global Gilson. Retrieved from https://www.globalgilson.com/soil-permeability

Theis, C. V. (1935). The relation between the lowering of the piezometric surface and the rate and duration of discharge of a well using groundwater storage. *Transactions, American Geophysical Union, 16*, 519–524. https://doi.org/10.1029/TR016i002p00519

Thiem, G. (1906). *Hydrologische Methoden*. Leipzig: J. M. Gebhardt. Retrieved from https://pubs.usgs.gov/wsp/wsp1536-E/pdf/wsp_1536-E_d.pdf

Flow Nets

7

7.1 BASICS OF THE FLOW NET DESIGN

A groundwater flow net is a graphical technique used to determine the flow pattern within an aquifer and to estimate key information, such as groundwater flow rate, water velocity, and travel times. Theoretical and practical aspects of flow nets are discussed in classic works, including Bear (1972), Freeze and Cherry (1979), and Fetter (2001). Applications and examples can also be found in Poeter and Hsieh (2020) and The Groundwater Project (2024a, 2024b). While this graphical method relies on strict assumptions, it remains a valuable tool for initial analyses.

The flow net consists of two intersecting families of lines: equipotential lines and flow lines (see Figure 7.1). These lines intersect at right angles (90°), with flow lines defining flow channels or flow tubes. The gaps between equipotential lines represent the distance over which a gradual change in hydraulic head occurs. For a more in-depth reference, consult the work by Poeter and Hsieh (2020).

The assumptions inherent in the flow net design primarily pertain to the aquifer's characteristics and flow conditions. It is assumed that the aquifer is composed of porous material, where Darcy's law holds true. Additionally, the aquifer should exhibit homogeneity (consistent properties throughout space), isotropy (uniform properties in all directions), and full saturation. While not covered here, the method can be extended

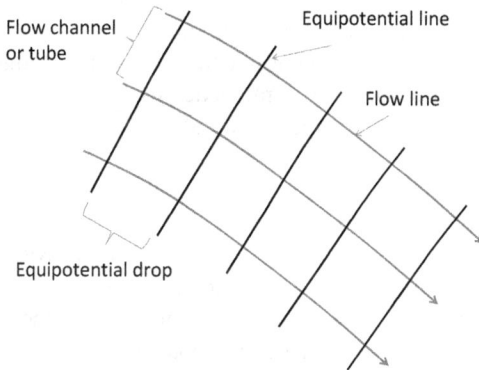

FIGURE 7.1 Elements of the flow net, including families of flow and equipotential lines.

DOI: 10.1201/9781003587149-7

to more complex scenarios, such as layered systems and anisotropic media. The solution applies specifically to two-dimensional steady-state flow (with no time-dependent changes), requiring well-defined aquifer boundary conditions.

The construction of a groundwater flow net begins with defining the flow domain and specifying boundary conditions along the domain perimeter (Poeter and Hsieh, 2020). These boundary conditions can include fixed hydraulic head or no-flow situations. Constant head boundaries are typically associated with water bodies like lakes or rivers with stable water levels. In such cases, the equipotential (head) values correspond to the water levels in these bodies, indicating that water flows from higher equipotential values to lower ones. No-flow conditions can arise from impermeable geological formations or faults surrounding the aquifer. Boundary conditions may also involve the water table in unconfined aquifers or the presence of water wells within the domain.

Once the boundary conditions are defined, the flow net is constructed iteratively to satisfy the conditions outlined above. If the hydraulic conductivity of the material is known, it becomes possible to calculate volumetric flow rates, flow velocities, and travel times within the flow domain.

7.1.1 Rules for Constructing Flow Nets

1. Constant or fixed-head boundaries represent the starting or ending points of equipotential lines, indicating where water enters or exits the aquifer. Water always flows from areas of higher equipotential values to those with lower values.
2. Impermeable boundaries serve as flow lines, typically marking the initial and final flow paths. To maintain clarity, it is advisable to use a limited number of flow lines and equipotential lines to prevent overly complex plots. If necessary, the flow net can be refined later by adding lines between the initially drawn ones.
3. The equipotential and flow lines should intersect at right angles (90°) to ensure accuracy in your representation.
4. A trial-and-error approach, as shown in Figure 7.2, is employed to adjust the lines until the flow net consists of square cells, ensuring that the lines intersect at right angles. It is advisable to start from one edge and progress as illustrated in Figure 7.3. It is recommended to design the net cells as squares (of equal sides) to simplify water flow calculations in the aquifer, as will be explained later.
5. Realistic conditions will include curved lines, and thus a square cannot be drawn in the conventional way. Drawing a circle within the cell will ensure that the lines intersect at right angles and that the cell is approximated as a "square." The diagram in Figure 7.3 involves sketching a circle within the cell, with the flow and equipotential lines serving as tangents. The distances d and S represent the average spacing in this scenario and should be equal for a perfect circle.

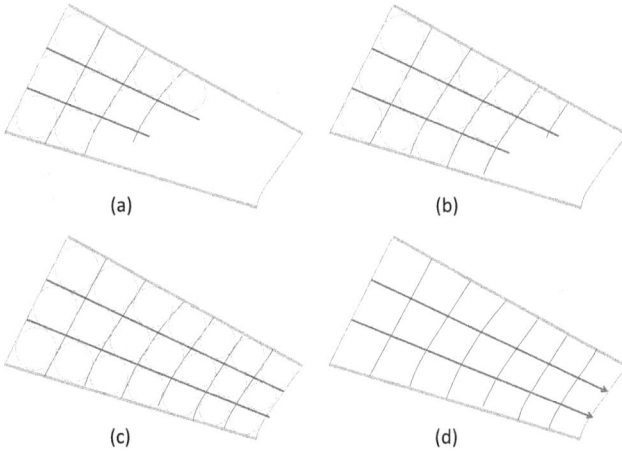

FIGURE 7.2 Steps (a) through (d) for drawing a flow net starting from one edge. Circles are used to help in designing cells as squares and should be removed in the final step.

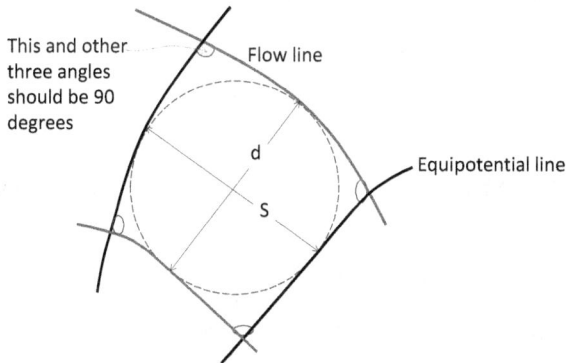

FIGURE 7.3 The flow and equipotential lines intersect at right angles (90°).

6. If necessary, it is acceptable to have partial tubes or partial equipotential lines separating increments within the network.

7. It is important to note that a flow line should never intersect with another flow line, including impermeable boundaries (which are treated as flow lines). The same rule applies to equipotential lines, which should not intersect with one another.

8. It is not necessary to consider hydraulic conductivity when drawing the network, as the aquifer is assumed to be homogeneous and isotropic. However, hydraulic conductivity is essential for calculating flow rates and travel times.

Example 7.1: Simple Aquifer Case

We will begin with a straightforward example involving a rectangular aquifer in the configuration shown in Figure 7.4. The boundary conditions are as follows: a specified head of 75 m on the left side and 15 m on the right side. These head values are due to the presence of two rivers that maintain constant water levels. The gray top and bottom lines on the diagram indicate no-flow conditions, where the movement of water is restricted. In this scenario, the resulting direction of water flow is from the higher equipotential on the left side to the lower value on the right side.

In this particular example, there is no need to follow the procedure outlined in Figure 7.2. This is because the shape of the aquifer naturally results in straight flow and equipotential lines marked with arrows, allowing us to create conventional squares. The top and bottom boundaries, where no flow occurs, correspond to the first and last flow lines. We have divided the aquifer into four tubes, with five flow lines represented by the unlabeled lines marked with arrows. These lines carry the flow between the fixed-head lines, moving from higher to lower elevations. Subsequently, we draw the equipotential lines, ensuring that the square shapes are maintained. In this case, the division resulted in eight sections, which is a whole number. Depending on the size of the aquifer, fractional divisions may be necessary. For instance, the aquifer size might require eight and a half divisions instead of exactly eight, which is acceptable. The change in equipotential increment can be estimated by dividing the head difference by the number of increments, denoted as N, or

$$\Delta h = \frac{h_1 - h_2}{N} \tag{7.1}$$

In equation (7.1), Δh is the equipotential increment, and h_1 and h_2 are the upstream and downstream specified head values, respectively. For this example, Δh is 7.5 m, with $N=8$, and h_1 and h_2 equaling 75 and 15 m, respectively. The labels on the equipotential lines indicate the head values, they incrementing by a fixed amount of 7.5 m.

The flow rate can be calculated within a single tube by using Darcy's law, as expressed in equation (7.2):

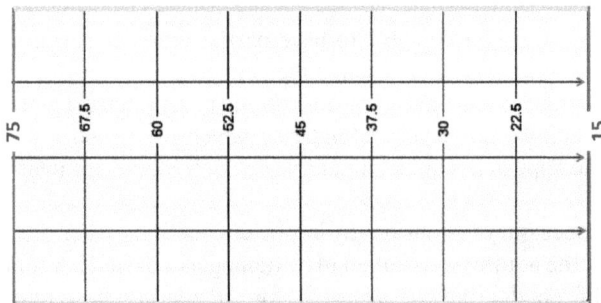

FIGURE 7.4 A flow net for a simple rectangular aquifer (plan view). The same graphing scheme in this plot is used for all subsequent flow-net figures.

$$\Delta Q = Kd \frac{\Delta h}{S} \qquad (7.2)$$

In this equation, ΔQ represents the flow rate within the tube, where K is the hydraulic conductivity, d denotes the width of the tube, and S represents the incremental distance between two successive equipotential lines. It's important to note that this equation is formulated for a two-dimensional aquifer. However, since aquifers are inherently three-dimensional, the calculations in the current treatment should aim to estimate the flow rate per unit width of the aquifer.

If the cells are designed as squares, then d will be equal to S, and the equation simplifies to

$$\Delta Q = K\Delta h \qquad (7.3)$$

The flow rates within all the tubes are equal, considering that water does not cross between tubes. The total flow rate is thus estimated by multiplying the single tube value given by equation (7.3) by the number of tubes N_t:

$$Q = N_t \Delta Q \qquad (7.4)$$

For this example, N_t is four, and if K is 10 m/day, the total flow rate Q via equation (7.4) would be

$$Q = 4 \times 10 \times 7.5 = 300 \frac{m^3}{day} \text{ per meter of aquifer thickness.}$$

Example 7.2: Irregularly Shaped Aquifer

In this example, an irregularly shaped aquifer is presented. To proceed, follow the steps outlined in Figure 7.2, resulting in Figure 7.5. While the hand-drawn representation may not be perfect, it should still provide a reasonable depiction.

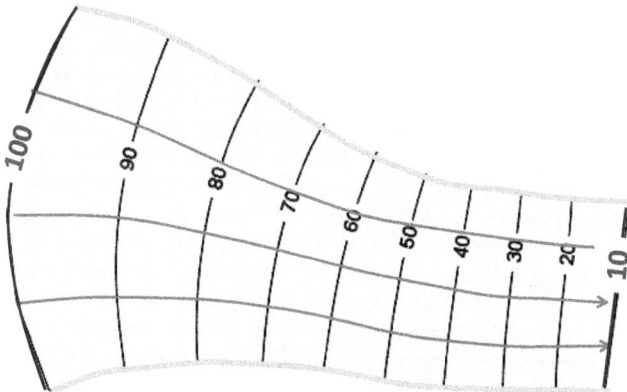

FIGURE 7.5   Flow net for a relatively complex-shaped aquifer.

In this specific case, the respective values of N_t, h_1, h_2, and N are as follows: $N_t=4$, $h_1=100$ m, $h_2=10$ m, and $N=9$, resulting in a Δh value of 10 m. Given a K value of 10, the value of Q will be

$$Q = 4 \times 10 \times 10 = 400 \frac{m^3}{day} \text{ per meter of aquifer thickness.}$$

Example 7.3: Aquifer Case with Wells

Example 7.4 has been modified to include two scenarios: one with a pumping well and another with two pumping wells, which naturally impacts the flow field. In the case of one well (as shown in Figure 7.6), eight tubes are depicted, with five being captured by the well and three escaping towards the right-side boundary. The dotted line provides a rough representation of how the flow through the tubes is divided between the well and the downstream boundary. Water within the area enclosed by the dotted lines will be collected by the well, while the remaining water will be directed towards the stream. For a Δh value of 10 m and a hydraulic conductivity K of 10 m/day, the flow rate of a single tube is 100 m³/day/m. Consequently, the flow rates to the well and the boundary are 500 and 300 m³/day/m, respectively.

This flow net design serves two preliminary analysis purposes. First, it can be employed to optimize the allocation of flow between the well for public water usage and the stream to support ecological benefits. Second, it can be utilized to safeguard the stream against contamination introduced upstream from the well, such as potential leakage from storage tanks or landfills. The well will effectively capture the contaminant if it is located anywhere within the area defined by the dotted line. However, it's important to acknowledge the limitations of using flow nets, as real-world applications can be considerably more complex and may necessitate extensive evaluations. One limitation, which will be discussed in Chapter 8 pertains to contaminant transport by mixing, which can result in dispersion outside the flow tubes. In such cases, the contaminant may not be effectively captured by the well.

Figure 7.7 extends this analysis to the scenario involving two wells. In this case, there are a total of seven tubes, and the flow rate for a single tube remains 100 m³/day/m, with a hydraulic conductivity K of 10 m/day. The flow is divided among the two wells and the stream, with each well receiving two tubes (200 m³/day/m)

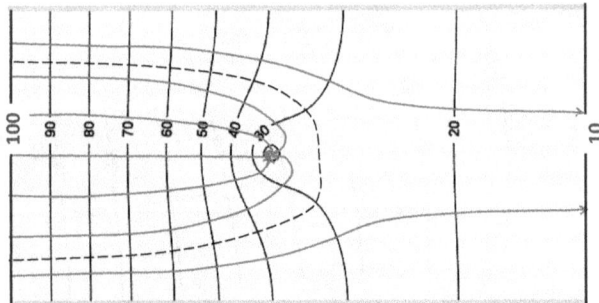

FIGURE 7.6 Flow net for a rectangular aquifer with one well.

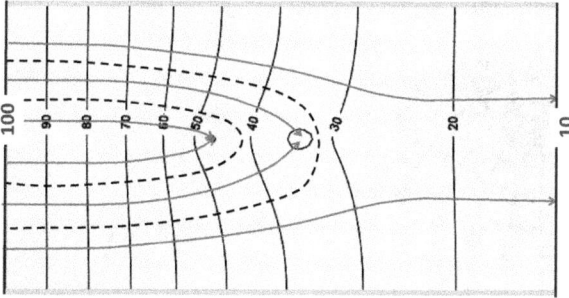

FIGURE 7.7   Flow net for a rectangular aquifer with two wells.

and three tubes directed towards the stream (300 m³/day/m). Once again, the dotted lines illustrate the partitioning of tubes among the various recipients of water. As with the case of one well, this analysis can be applied to manage groundwater resources effectively by optimizing the distribution of flow between wells and the stream or to protect the stream from potential contamination.

Example 7.4: Aquifer Cross Section

The flow net depicted in Figure 7.8 illustrates the flow between the upper aquifer boundary, where the equipotential value is 90, and the lower left side where the value reaches zero. Several crucial observations can be made concerning the aquifer in this figure.

Firstly, it's evident that the elements of the net are smaller near the exit boundary, indicating higher water velocity. This occurs because the available area for flow, represented by the width of the tubes, decreases in size. Secondly, it is essential to ensure that all flow lines exit through the specified-head boundary where all the equipotential lines converge (as seen in Figure 7.9a). The line above the specified-head boundary denotes a no-flow condition, and no flow lines can exist.

Lastly, the no-flow boundary condition along the right and bottom sides results in a sharp, 90-degree turn for the flow line, as depicted in Figure 7.8. This configuration aligns with theoretical assumptions of frictionless flow. However, in reality,

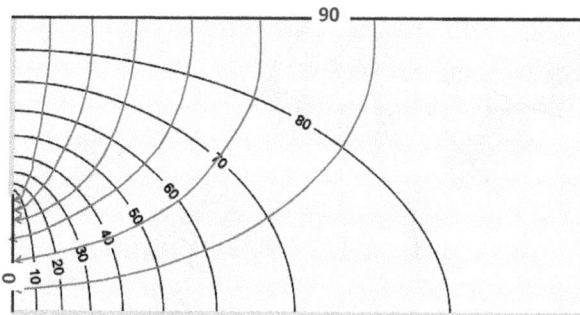

FIGURE 7.8   Flow net for an aquifer cross section.

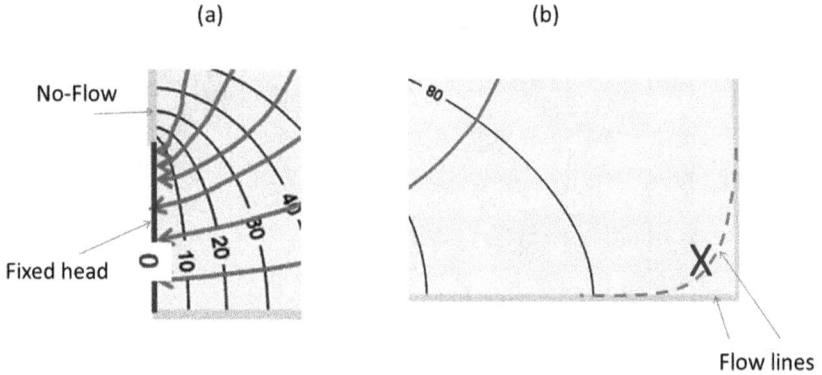

FIGURE 7.9   Notes for Example 7.15: (a) Flow lines exist through the correct boundary; (b) No-flow lines should intersect.

water in the corner area may become stagnant. Thus, caution is warranted to prevent drawing a flow line along the vertical no-flow line that subsequently curves to meet the no-flow section at the bottom (as shown by the dotted line in Figure 7.9b). Such a depiction is not acceptable, as it results in the intersection of two flow lines.

Example 7.5: Aquifer Cross Section with Sheet Piles 1

The example shown in Figure 7.10 illustrates a practical engineering problem related to either draining a construction site or controlling river flow. The figure depicts a cross-section of an aquifer where water accumulates up to 10 m behind a structure designed to obstruct the flow, assuming a completely dry condition behind the structure. Naturally, water will move within the aquifer, as indicated by the flow net. This water would need to be promptly removed, for example, through a pumping and disposal process. The flow net assumes that the structure stands without penetrating the ground, possibly through above-ground support. This may be impractical, if not

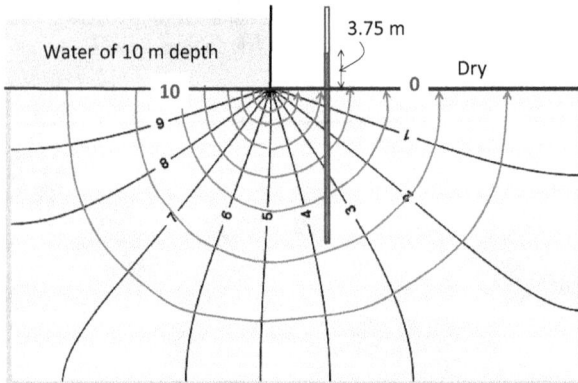

FIGURE 7.10   Flow net for flow in an aquifer's cross section where water is accumulating behind a structure to keep the downstream side dry.

entirely impossible. However, the objective here is to estimate the rate at which water must be removed under these conditions. Based on the calculation scheme described above, $\Delta h = 1$ m, and $N_t = 9$. Assuming K is 10 m per day, the total flow rate would be 90 cubic meters per day per meter, which should be the rate of water removal.

If an observation well is added downstream from the structure, as shown in the figure, the water would rise in the well to a level approximately 3.75 m above the ground surface. This inference is based on the value of the hydraulic head at the bottom of the well, which falls between the head contours of 3 and 4 m.

It is also important to note that the velocity of water exiting the structure is relatively high due to the narrower width of the flow tubes in this area. As mentioned earlier, the flow rate is the same for all tubes; therefore, the velocity should be greater in the smaller-sized tubes. Such high velocity can potentially lead to scouring problems through a process known as "piping," where water erodes soil particles in the aquifer. This can negatively impact the stability of the structure.

Example 7.6: Aquifer Cross Section with Sheet Piles 2

This example illustrates a more practical scenario in which a structure, composed of sheet piles is inserted into the ground to provide support. These sheets are typically constructed from steel and feature interlocking edges, making them effective at retaining soil or water. However, their most crucial function lies in significantly reducing both the flow rate and exit-water velocity behind the sheet.

It's important to note that the sheet itself is impervious and, therefore, is treated as a no-flow boundary. As depicted in the flow net in Figure 7.11, water within the aquifer will move downward and exit by circumventing the sheet. The initial flow line coincides with the left face of the sheet, descending and then ascending along the right face. Notably, there are key differences compared to the previous example: a reduction in the number of flow tubes and an increase in flow line spacing near the sheet.

Under this design, only four tubes are present, compared to nine in the previous example, marking a remarkable 44% reduction in the total flow rate. When comparing the sizes of the flow tubes around the sheet in both examples, it becomes evident that there is a significant decrease in velocities in the current design. This decrease in velocity also leads to a reduction in the head in the monitoring well.

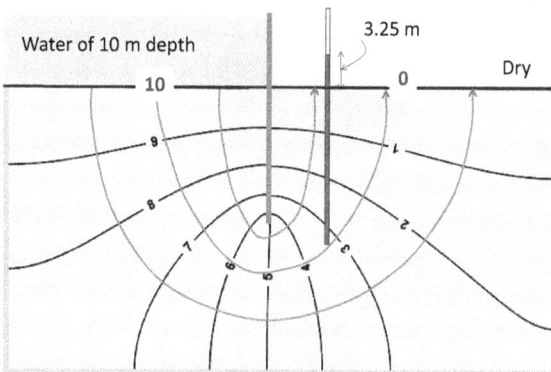

FIGURE 7.11 Flow net for the case of a deep metal sheet used to dry out the area downstream from the sheet.

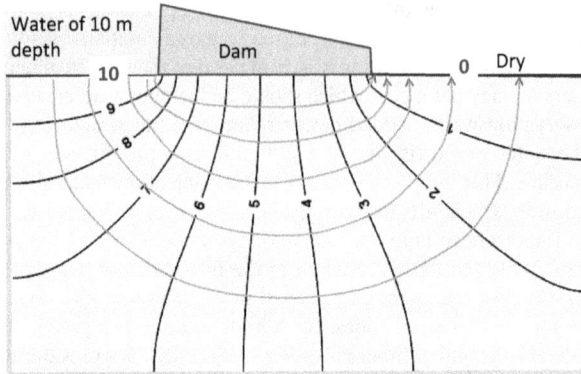

FIGURE 7.12 Flow net for an aquifer's cross section where water accumulates behind a large structure (a dam) to keep the downstream side dry.

Example 7.7: Aquifer Cross Section with a Dam

Sheet piles are a suitable choice for construction purposes, but dams are more effective for controlling river flows. In addition to their water flow restriction capabilities, their larger size, as illustrated in Figure 7.12, also helps decrease both the flow rate and the velocities below the dam. While Figure 7.12 depicts a completely dry downstream condition, it's important to note that this is not always necessary, as will be demonstrated in the following example.

For six tubes, Δh of 1 m, and K of 10 m/day, the flow rate is 60 m³/day/m.

Example 7.8: Aquifer Cross Section with a Dam and Sheet Piles

It is possible to integrate a dam with sheet piles to further reduce flow rates and velocities, as shown in Figure 7.13. Compared to the previous example, the number of tubes is reduced from six to four, resulting in a 33% decrease in flow rate. The sizes of the tubes have been increased around the dam's body, which leads to reduced water velocities in the aquifer. As mentioned earlier, this is crucial for preventing piping that can damage the dam's foundation.

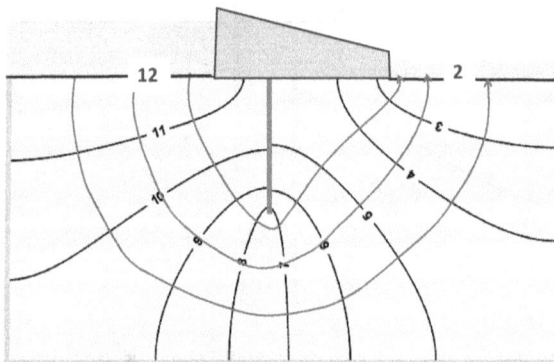

FIGURE 7.13 Flow net for a case involving a dam and a sheet pile.

7.2 CALCULATION OF WATER VELOCITY AND TRAVEL TIMES

Figure 7.14 depicts four cells within a flow net, each meticulously represented as "squares" by enclosing them with circles. As previously discussed, the flow net operates within a two-dimensional framework, with flow estimates made on a per-unit-width basis. The primary goal here is to estimate the water velocity within each of these cells and predict the total time it takes for water to traverse from the center of cell 1 to the center of cell 4. The small circles within the cells serve as markers for their respective centers.

To achieve this, we will first calculate the velocities within each cell and then estimate the incremental travel times between the centers of successive cells. These calculations will be executed using equations (7.5) through (7.7), as described below.

1. Estimate the area of the tube perpendicular to the flow direction:

$$A = d \text{ per unit width} \tag{7.5}$$

where d is the diameter of the circle, which equals both the tube width and the equipotential-increment width for a perfect square.

2. Calculate the seepage (or average linear) velocity via

$$v = \frac{\Delta Q}{An} = \frac{\Delta Q}{dn} \tag{7.6}$$

where ΔQ is the tube flow rate and n is porosity. This average velocity will be estimated for each cell and is assumed to be that at the cell's center.

3. Estimate the travel time between successive cells using the expression:

$$t = \frac{L}{v} \tag{7.7}$$

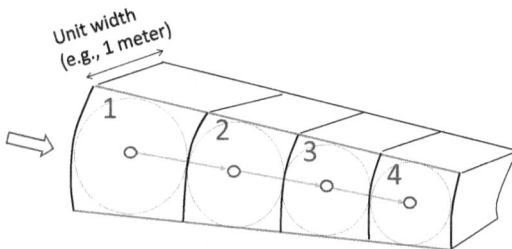

FIGURE 7.14 A three-dimensional schematic display of four cells from a flow net. The small solid circles represent the centers of the cells. The large dotted circles are used to guide the drawing of perfect "square" cells. The lines connecting the circles show the travel distances between successive cells.

where L is the distance between the centers of successive cells and should equal the sum of the radii of the two relevant circles (the lines connecting the cells in Figure 7.14).

4. The calculations are repeated for successive cells, and the times are added to estimate the total travel time.

Example 7.9: Velocity and Travel Time

Given the following data, estimate the velocity within each of the four cells in Figure 7.14 and the total travel time.

- The tube flow rate $\Delta Q = 23 \, m^3/day/m$
- The cell dimensions are 7, 5.5, 4, and 2.5 m for the four cells, respectively.
- Porosity is 0.3.

For cell 1,

$$v = \frac{\Delta Q}{dn} = \frac{23}{7 \times 0.3} = 10.95 \text{ m/day}$$

$$L = \frac{7 + 5.5}{2} = 6.25 \text{ m}$$

$$t = \frac{L}{v} = \frac{6.25}{10.95} = 0.57 \text{ day}$$

For cell 2,

$$v = \frac{\Delta Q}{dn} = \frac{23}{5.5 \times 0.3} = 13.93 \text{ m/day}$$

$$L = \frac{5.5 + 4}{2} = 4.75 \text{ m}$$

$$t = \frac{L}{v} = \frac{4.75}{13.93} = 0.34 \text{ day}$$

For cell 3,

$$v = \frac{\Delta Q}{dn} = \frac{23}{4 \times 0.3} = 19.17 \text{ m/day}$$

$$L = \frac{4 + 2.5}{2} = 3.25 \text{ m}$$

$$t = \frac{L}{v} = \frac{3.25}{19.17} = 0.17 \text{ day}$$

The total travel time will be the sum of 0.57, 0.34, and 0.17, or 1.08 days. Note that, as expected, the velocity increases and the incremental time decreases as the tube area ($A = d$ per unit width) gets smaller.

7.3 FLOW NETS FOR NON-IDEAL AQUIFER SYSTEMS

The principles discussed in the previous examples can be generalized to construct flow nets for non-ideal conditions, such as flow in anisotropic and layered aquifers. As explained in The Groundwater Project (2024a), the process begins by transforming the geometry of the system into an isotropic system, where a conventional flow net is constructed. The flow net is then transformed back to the original anisotropic system.

For layered systems, as described by The Groundwater Project (2024b), groundwater flowlines refract when they cross a geological boundary between two formations with different hydraulic conductivities. This refraction is similar to how light reflects when passing from one medium to another. However, unlike light, which follows a sine law, groundwater refraction obeys a tangent law.

7.4 NUMERICAL CONSTRUCTION OF FLOW NETS FOR NON-IDEAL AQUIFER SYSTEMS

Chapter 11 of this text discusses the numerical modeling of groundwater systems, which involves using specialized software to simulate more realistic scenarios that are closer to field conditions. In such cases, the graphical approaches discussed earlier would not be suitable due to violations of the underlying assumptions. These situations may include complex boundary conditions, variability in hydraulic conductivity across flow domains, the dominance of transient conditions, and the presence of partially saturated zones.

As an illustration, Figure 7.15 shows the plan view of a confined aquifer with dimensions of 1,000×1,000 ft and two zones that may have different hydraulic conductivity values. The aquifer thickness is set at 100 m. Hydraulic head values are defined at the corners, as shown, and are assumed to gradually and linearly change between these values along the left and right sides. No-flow conditions are assumed for the top and bottom boundaries. The water flow simulations were conducted using the MODFLOW simulation model (Harbaugh et al., 2000), while the pathline simulations utilized the MODPATH model developed by Pollock (2012). Calculations and result presentations were completed using GMS (AQUAVEO, 2023). The following three cases were simulated.

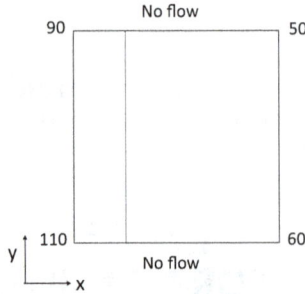

FIGURE 7.15   A schematic plot of an aquifer used in the numerical simulation of a flow net.

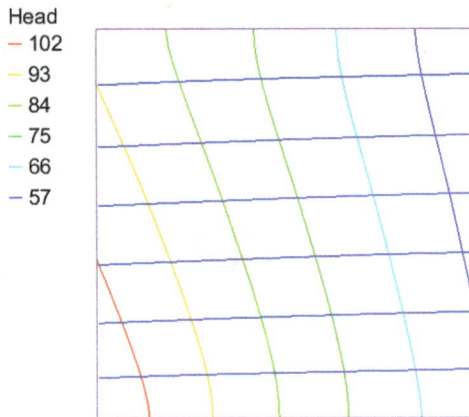

FIGURE 7.16   Results of a numerical simulation of a flow net for a homogeneous/anisotropic aquifer system.

7.4.1 A Homogeneous/Anisotropic Aquifer System

A single aquifer is represented here with a homogeneous hydraulic conductivity value (K_x) of 100 ft/day and an anisotropy ratio of 0.1 (the K_y to K_x ratio, where x and y represent the axes shown in the figure). The resulting flow net is displayed in Figure 7.16. The main difference between this case and a homogeneous and isotropic case is the lack of orthogonality between the flow lines and equipotential lines.

7.4.2 Two Contrasting Aquifer Systems

The aquifer consists of two zones, each of which is homogeneous and isotropic. The hydraulic conductivity values are 300 ft/day for the left side and 100 ft/day for the right side. The simulation results are shown in Figure 7.17. In this case, orthogonality is preserved, but a clear refraction of the flow lines at the interface between the two zones is evident.

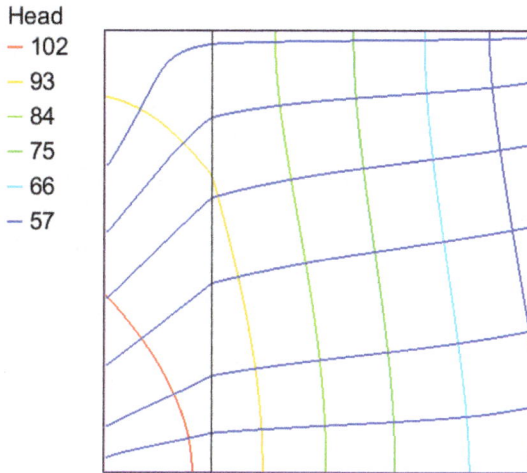

FIGURE 7.17 Results of a numerical simulation of a flow net for two contrasting aquifer zones.

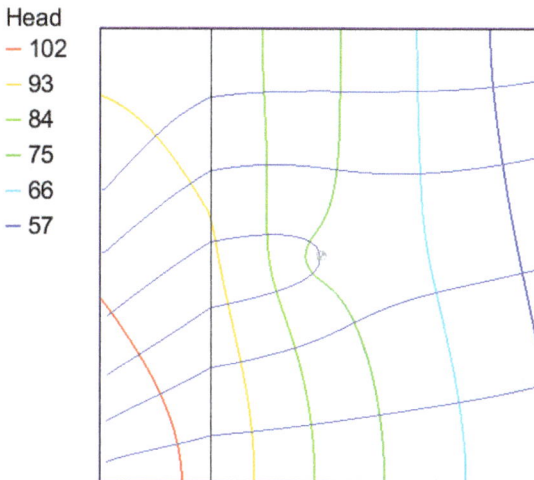

FIGURE 7.18 Results of a numerical simulation of a flow net for two contrasting aquifer systems with a pumping well.

7.4.3 Two Contrasting Aquifer Systems with a Pumping Well

The final case is similar to the one in Section 7.4.2, but includes a pumping well at the center of the aquifer, operating at a rate of 100,000 ft³/day. The results in Figure 7.18 show the refraction of the flow lines, along with the capture of some of these lines due to the pumping action.

7.5 ASSIGNMENTS

1. Draw a reasonably accurate flow net for the aquifer in Figure 7.19. Gray in the middle represents a no-flow area. Estimate the discharge rate based on head values in meters. The hydraulic conductivity is 10 m per day. In this and all subsequent cases, the gray lines indicate a no-flow condition, while the labeled lines represent a specified (fixed) head condition.
2. For the two cases in Figures 7.20 and 7.21, draw a reasonably accurate flow net. Estimate the discharge rate. Head values are in meters and hydraulic conductivity is 10 m per day. Estimate the water level in the observation well.

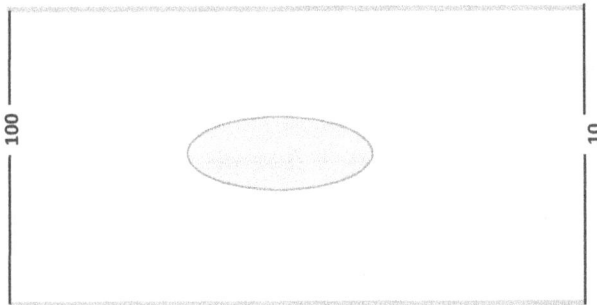

FIGURE 7.19   Problem 1: flow net for an area with a no-flow zone within a domain.

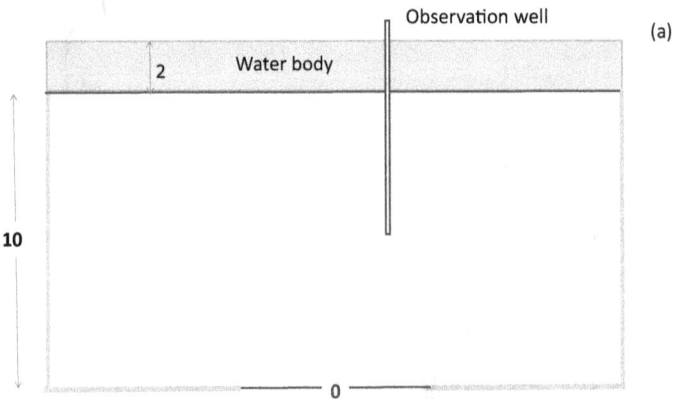

FIGURE 7.20   Problem 2: flow net for an area with an accumulating water body.

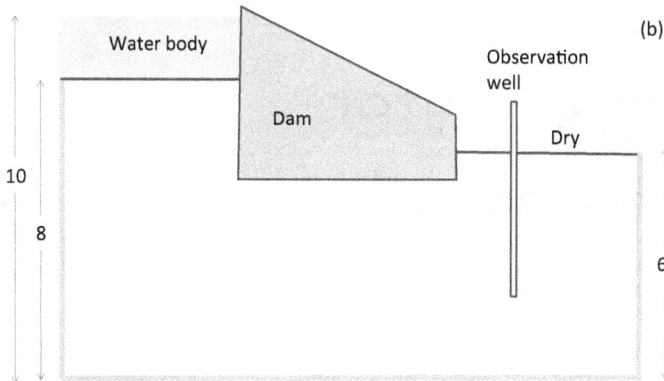

FIGURE 7.21   Problem 3: flow net for an area with an accumulating water body and an earth dam.

REFERENCES

AQUAVEO. (2023). Groundwater modeling system. Retrieved from https://www.aquaveo.com/software/gms-groundwater-modeling-system-introduction

Bear, J. (1972). *Dynamics of Fluids in Porous Media.* New York, NY: American Elsevier Publishing Company.

Freeze, R. A., & Cherry, J. A. (1979). *Groundwater.* Englewood Cliffs, NJ: Prentice-Hall.

Fetter, C. W. (2001). *Applied Hydrogeology* (4th ed.). Upper Saddle River, NJ: Prentice-Hall.

Harbaugh, A. W., Banta, E. R., Hill, M. C., & McDonald, M. G. (2000). MODFLOW-2000, the U.S. Geological Survey modular ground-water model; user guide to modularization concepts and the ground-water flow process. *U.S. Geological Survey Open-File Report 00–92,* 121 pp. https://doi.org/10.3133/ofr200092

Poeter, E., & Hsieh, P. (2020). Graphical construction of groundwater flow nets. *Groundwater Project.* Retrieved from https://gw-project.org/books/graphical-construction-of-groundwater-flow-nets/

Pollock, D. W. (2012). *User guide for MODPATH version 6—A Particle-tracking Model for MODFLOW (U.S. Geological Survey Techniques and Methods 6-A41).* Reston, VA: U.S. Geological Survey. https://pubs.usgs.gov/tm/6a41/

The Groundwater Project. (2024a). Box 5 – Drawing flow nets for anisotropic systems. Retrieved from https://books.gw-project.org/graphical-construction-of-groundwater-flow-nets/chapter/box-5-drawing-flow-nets-for-anisotropic-systems/

The Groundwater Project. (2024b). Flow nets: 5.1 Flow nets by graphical construction. Retrieved from https://fc79.gw-project.org/english/chapter-5/

Groundwater Contamination

8

8.1 INTRODUCTION

Water contamination refers to the introduction of chemical substances or microorganisms into surface or subsurface waters. Water pollution occurs when the quality of water is degraded by contaminants, rendering it toxic to humans or the environment. While water cannot be entirely free of contaminants, efforts are necessary to protect against pollution by keeping contaminant levels below unacceptable thresholds that can lead to human diseases or even death. Water is highly susceptible to contamination and can dissolve more substances than any other liquid. Contamination can originate from various sources, including farms, towns, and numerous facilities, generating harmful substances that readily dissolve and mix with water. Some contaminants fully dissolve in water, like salt, while others, such as petroleum products (referred to as immiscible liquids), can either float or sink in bulk, forming contamination plumes. However, the parts of these plumes in contact with water are subject to dissolution, releasing dissolved product components.

Serious and deadly cases of water pollution affecting millions of people worldwide have been well-documented by organizations such as the EPA (2021), The Lancet (2022), Centers for Disease Control and Prevention (2025), and the United Nations (2023), among others. Contamination can result from accidental or illegal releases from various sources. Pathogens such as bacteria and viruses spread from sewage treatment facilities, farms, and urban areas, causing diseases such as cholera, giardia, typhoid, and Legionnaires' disease. Chemical pollutants from numerous waste disposal sites also impact water supplies, potentially leading to gastrointestinal illnesses, negative effects on the nervous system or reproductive health, and chronic diseases such as cancer.

In addition to human health issues, environmental damage is widespread in ecological systems that encompass waterways and lakes. Agricultural and other human activities can generate nutrients, specifically nitrogen and phosphorus, which may end up in surface water bodies through surface runoff and subsurface flows. The accumulation of nutrients can lead to eutrophication, a natural process resulting in the growth of unsightly scum on the water surface, diminishing recreational value and contributing to other environmental problems. This process can also deplete dissolved oxygen from the water due to bacterial activity, sometimes causing fish kills. Furthermore, chemicals and heavy metals from industrial and municipal wastewater can reduce the lifespan and reproductive abilities of certain organisms. Large fish, such as tuna, are prone to accumulating high quantities of toxins, like mercury, through the food chain.

Among water bodies, chemical waste and climate change are causing oceans to become more acidic, making it more challenging for shellfish and coral to survive. Pollution impacts the ability of shellfish and other species to build shells and may affect the nervous systems of marine life, including sharks. Marine debris, such as plastic products and soda cans, can strangle, suffocate, and starve marine animals.

The transport and fate of contaminants in the surface and subsurface domains are primarily controlled by physical, chemical, and biological processes. Physical processes, mainly influenced by flow velocity, carry and spread contaminants in flowing water. Chemical and biological degradation processes can reduce chemical concentrations or change their form. However, the nature and timescales of these processes can vary between surface and subsurface environments. Unlike surface water contamination, the movement of contaminants through the subsurface is complex and difficult to assess, predict, and remediate. Different types of contaminants react differently with soils, sediments, and other geologic materials, often traveling along intricate flow paths at varying velocities. Groundwater moves very slowly, and incidents of contamination can persist for many years before detection, resulting in environmental and health damage.

8.2 GROUNDWATER CONTAMINANT TRANSPORT PROCESSES

The following are the main terms and definitions discussed in the following sections.

- **Adsorption**: adherence of molecules in solution to the surface of solids.
- **Convection (Advection)**: the process by which solutes are transported by moving groundwater at a rate equal to the average linear velocity.
- **Contaminant Transport (Convective) Velocity**: the rate at which contamination moves through an aquifer.
- **Diffusion**: a process describing molecules moving randomly from areas of high solute concentrations to areas of low concentration in the direction of the solute concentration gradient, also referred to as molecular diffusion.
- **Dispersivity**: a scale-dependent property of an aquifer that determines the degree to which a dissolved chemical will spread in flowing groundwater. Dispersivity comprises three directional components: longitudinal, transverse, and vertical.
- **Dispersion**: causes solute molecules to travel faster or slower than the average linear velocity; spreading of the solute in the direction of groundwater

flow (longitudinal dispersion) or perpendicular to groundwater flow (transverse or vertical dispersion).

- **Dispersion Coefficient (Mechanical)**: a measure of the spreading of a flowing substance due to the nature of the porous medium, with its interconnected channels distributed randomly in all directions.
- **Dispersion Coefficient (Hydrodynamic)**: the sum of the coefficients of mechanical dispersion and molecular diffusion.
- **Flow Path**: the subsurface course a water molecule or solute would follow in a given groundwater velocity field.
- **Porosity (Total)**: the ratio of the volume of void spaces in a rock or sediment to the total volume of the rock or sediment.
- **Porosity (Effective)**: the ratio of the total volume of voids available for field transmission to the total volume of the porous medium.
- **Retardation**: a process that causes chemicals to move slower than the average convective velocity, resulting in less spreading. It is caused by chemical adsorption.
- **Source of Contaminants**: the physical location and spatial extent of the source contaminating the aquifer.

Contaminants can infiltrate groundwater aquifers either through percolation through soils or via direct deep well injection. Examples of these contaminants include agricultural chemicals and substances leaking from surface and subsurface storage tanks. Unlike deep injection sources, such as waste disposal wells, various processes that influence the destiny and movement of contaminants in near-surface soils are crucial as they impact the rates and concentrations of contaminants that eventually reach the groundwater system. Some contaminants, like salt or nitrate, are highly soluble and can swiftly migrate from surface soils to the saturated materials located below the water table, often occurring during and after rainfall events. Another example involves the chemical transformation of nitrogen from fertilizers, ultimately generating nitrate, which serves as a significant source of contamination.

Contaminants are categorized based on their solubility and other characteristics that dictate whether they behave conservatively or non-conservatively. Numerous physical and chemical processes take place in the subsurface, as depicted in Figure 8.1. Besides the physical processes, only a few chemical processes will be addressed in subsequent sections.

The following sections will cover both conservative and non-conservative soluble chemicals, as well as density-dependent chemicals. The latter category encompasses cases of saltwater intrusion and the transport of immiscible liquids. These immiscible liquids, also referred to as non-aqueous phase liquids, include substances such as gasoline, diesel, and crude oils.

Detailed information on the physical and chemical processes contributing to the fate and transport of contaminants can be found in groundwater hydrology books, including works by Bear (1979), Schwartz and Zhang (2003), Fetter and Kreamer (2022), and Bair and Lahm (2006).

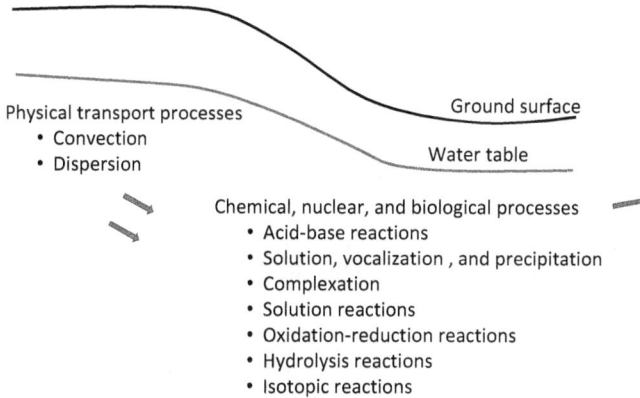

Physical transport processes
- Convection
- Dispersion

Ground surface

Water table

Chemical, nuclear, and biological processes
- Acid-base reactions
- Solution, vocalization , and precipitation
- Complexation
- Solution reactions
- Oxidation-reduction reactions
- Hydrolysis reactions
- Isotopic reactions

After Domenico and Schwartz (1998)

FIGURE 8.1 Physical and chemical transport processes. (After Domenico and Schwartz, 1998.)

8.2.1 Conservative Contaminant Transport

In the absence of chemical or biological activities, physical processes govern the transport and fate of conservative chemicals. These processes include convection, which involves the movement of chemicals through flowing water, and dispersion, which arises from variations in velocity within porous media. A demonstration of the physical transport of conservative chemicals can be found in a video by Fox (2014). A summary of these processes is provided in Science in the Courtroom (2021).

8.2.1.1 Convective (or advective) groundwater transport

Convective transport, also known as advective transport, describes the process by which soluble contaminants, or solutes, are carried by flowing groundwater. Convective flow velocities are linked to Darcy's flux, which relies on the hydraulic conductivity of the aquifer material and the hydraulic gradient that drives the flow. As previously discussed in Section 6.3, Darcy's law serves as the foundation for quantifying the rate of fluid flow through subsurface materials. The estimation of Darcy's (or specific) flux q is accomplished by

$$q = \frac{Q}{A} = KS \tag{8.1}$$

In this equation, Q, K, and S represent the total discharge through an area A, hydraulic conductivity, and water head gradient, respectively. It is important to note that specific flux cannot be used to assess contamination fate and transport in porous media. The reason for this limitation is that q is estimated as the average flow rate across the entire area of the tube in Darcy's setup. Consequently, Darcy's law is considered to be based on a macroscopic scale, quantifying flow across the entire media of the tube, encompassing both grains and pores. In reality, however, contaminants move exclusively between the soil grains or through the pores and voids that constitute material porosity. This movement occurs on a much smaller, microscopic scale and is exceedingly challenging, if not impossible, to quantify accurately. Nevertheless, the analysis of contaminant transport must be conducted on macroscopic or field-scale levels. In practical terms, the macroscopic scale corresponds to dimensions on the order of field-sampling sizes, such as those for soil cores, typically around a foot or so (as illustrated in Figure 8.2). Furthermore, the assessment of contaminant transport requires consideration of sites with dimensions of hundreds or even thousands of feet. Therefore, the process of upscaling from the micro-scale to the macro-scale, as depicted in the figure, is a complex and often insurmountable task.

To address this challenge, the micro-scale velocity is averaged over the effective area available for water flow, which is the total area multiplied by porosity, expressed as follows:

$$A_e = An \tag{8.2}$$

And v is thus estimated by

$$v = \frac{Q}{A_e} = \frac{Q}{An} = \frac{KS}{n} \tag{8.3}$$

In the literature, the velocity v is often referred to as the average linear velocity. This term is used because the actual distribution of velocities is too complex to fully characterize, so it is simplified as a linear function of all controlling variables. In simpler terms, v serves as a macroscopic variable that represents the microscopic velocities, which are more challenging to describe in detail. As an illustration, Example 8.1 provides calculations of travel times.

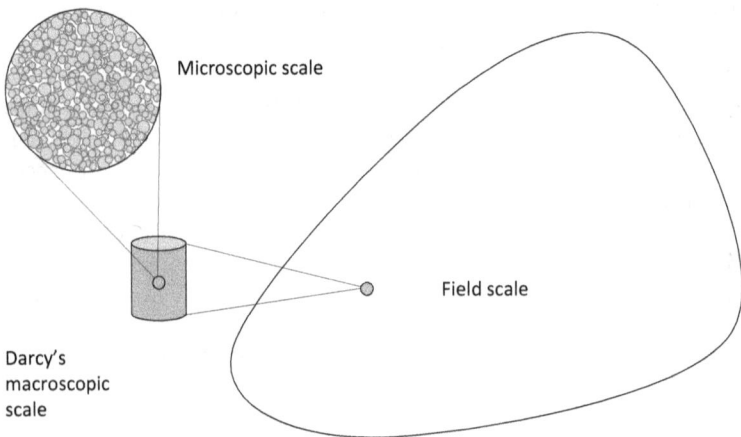

Microscopic scale

Darcy's macroscopic scale

Field scale

FIGURE 8.2 Different scales related to contamination transport ranging from the microscopic to field scales.

Example 8.1: Contaminant Travel Time

A contaminated site is represented by the dark gray area in the Figure 8.3, where three wells, A, B, and C, are located. The numbers next to each well indicate the water level in each well above the same datum. Sampling from Well A has shown that the contamination has reached the top of the clay layer. To estimate the travel time for the contamination to reach Well D, located 1,000 m away from the con-tamination site, we'll need to consider the following information:

- Well A measures the water level immediately above the clay layer.
- Well B measures the water level immediately below the clay layer.
- The thickness of the clay layer is 8 m.
- Wells C and D measure the water levels in the sand layer.
- Conductivity K and porosity n are given for in Figure 8.3 for the two layers.

The water velocity is estimated from

$$v = \frac{KS}{n}$$

The slope S is estimated by using the difference between water levels in two wells divided by the separating distance between them. Wells A and B are roughly located at the same location. The water level in these wells indicates a downward vertical flow across the top layer. The slope is calculated as follows:

$S = \dfrac{24\text{m} - 20\text{m}}{8\text{m}} = 0.5$. The respective vertical velocity through the top layer will be

$$v = \frac{KS}{n} = \frac{0.01 \text{ m/day} \times 0.5}{0.01} = 0.5 \frac{\text{m}}{\text{day}}$$

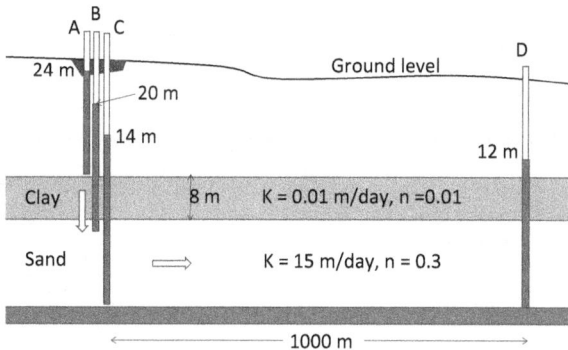

FIGURE 8.3   Sketch for an aquifer used in estimating contaminant travel time.

And the vertical time of travel

$$t = \frac{\text{distance}}{\text{velocity}} = \frac{8}{0.5} = 16 \text{ days}$$

Similarly, for the bottom layer, the slope is calculated as follows:

$$S = \frac{14 \text{ m} - 12 \text{ m}}{1,000 \text{ m}} = 0.002.$$

The respective horizontal velocity through the bottom layer will be:

$$v = \frac{KS}{n} = \frac{15 \text{ m/day} \times 0.002}{0.3} = 0.1 \text{ m/day}$$

And the horizontal time of travel:

$$t = \frac{\text{distance}}{\text{velocity}} = \frac{1,000}{0.1} = 10,000 \text{ days (about 27 years)}.$$

The total travel time is dominated by the travel in the lower layer.

8.2.1.2 Contamination by convection

Consider the sand column in the lower plot of Figure 8.4, where a contaminant is continuously injected from the left-hand side, while the water flows toward the right. The figure illustrates a gray region representing the extent of contamination at a specific time. At that moment, the lower portion of the column, depicted in white, remains uncontaminated, and a distinct boundary separates the contaminated (gray) and uncontaminated (white) sections. Due to convective transport, there is no spreading of the contaminant, and all molecules move at the same velocity, resulting in the sharp interface seen in the figure. As time advances, the contaminated area will progressively

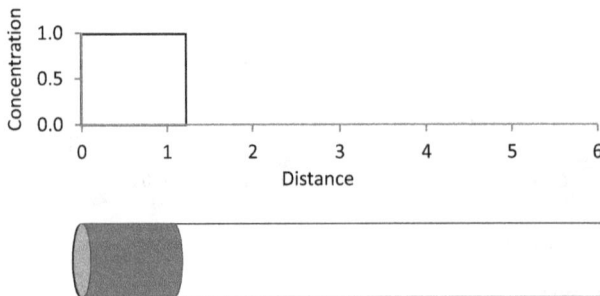

FIGURE 8.4 Top: Concentration profile versus distance at a specific time for a column flow with continuous contaminant injection. Bottom: Contaminant distribution in the soil column at the same specific time.

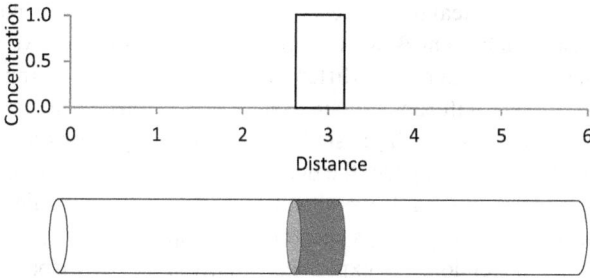

FIGURE 8.5 Top: Concentration profile versus distance at a specific time for a column flow with limited time contaminant injection. Bottom: Contaminant distribution in the soil column at the same specific time.

extend to the right, while the uncontaminated section diminishes until the entire column becomes contaminated.

Contamination is typically quantified in terms of mass per unit volume, such as milligrams per liter (mg/L). Assuming that the initial concentration of the injected contaminant was 1.0 mg/L, the entire gray area should also have a concentration of 1.0 mg/L, as shown in the figure. The uncontaminated portion of the column will maintain a concentration of zero, and once again, there will be a well-defined sharp interface between the contaminated and uncontaminated regions.

Figure 8.5 depicts an alternative scenario involving the brief introduction of a contaminant into a flowing water system, followed by a subsequent flush with clean water. The gray-shaded region at the bottom of the illustration represents the contaminant's position at a specific point in time, which will gradually move from left to right as time elapses. As the contamination exits the column, the entire column will return to a clean state.

In the upper part of the diagram, the concentration within the column is depicted as either 1.0 mg/L for the contaminated section or zero for the clean section. Once again, there is a distinct and abrupt transition in concentration between the contaminated and clean sections on both sides of the column.

8.2.1.3 The effect of dispersion (spreading)

In the discussion of convective transport described above, an assumption was made that all contaminant molecules were moving at the same velocity, resulting in a sharp change in concentrations between the contaminated and clean sections of the column. However, in most realistic scenarios, spreading will occur because molecules will move at different microscopic velocities. These differences in microscopic velocities, relative to the average value v, will cause some molecules to end up ahead or behind the average location at any given time.

The disparity between transport with convection alone and transport with both convection and spreading can be likened to a marathon competition. Imagine participants gathered at a starting point A, all running toward point B, where the race finishes. If all runners are moving at the same pace, they will only occupy a specific section of the road at any given time, leaving the rest of the road clear. In this scenario, all runners will simultaneously reach point B, which represents transport by convection only.

However, in reality, runners have different paces, leading to varying densities along the course. Most runners will run at a rate close to the average pace, arriving at a given time, where the crowd density is highest. As you move away from the average location of those running at the average pace, the densities decrease. In this more realistic case, winners can be clearly identified. If we substitute people with molecules and concentration with densities, this analogy helps illustrate the importance of pore–space mixing, formally known as dispersion. As explained later, neglecting dispersion can significantly impact contaminant first arrival and concentration distributions.

Figure 8.6 illustrates the case of a continuous injection in a sand column, accounting for dispersion, and compares it with the case of convection only. Instead of a sharp concentration change, as seen in the case of convection alone, dispersion results in a gradual change from the maximum concentration at the injection point to zero as you move farther away. Over time, more parts of the column become contaminated, but again, with a gradual change in concentration. With dispersion, the contaminant covers a larger volume compared to the case with convection only. As shown in the figure, the first arrival reaches about 2.0 and 1.0 m for the two cases, respectively. However, concentrations are lower with dispersion in the first meter of the tube.

It can be stressed that accounting for dispersion is crucial when considering the extent of contamination. Conversely, dispersion is less important when the concern is the level of contamination, such as in cases related to health issues.

Figure 8.7 illustrates similar displays, this time focusing on a limited-time injection scenario. Once again, dispersion is responsible for a gradual transition from maximum concentration at the center of the contaminated volume to zero concentration farther away on both sides. As time progresses, the contaminated volume shifts to the right, exhibiting a gradual concentration change until it is entirely cleared by the introduction of clean water. Notably, due to dispersion, the contaminant occupies a larger volume compared to the case with convection alone. However, it is worth noting that concentrations within the contaminated area will be lower. Just as with continuous injection, it is

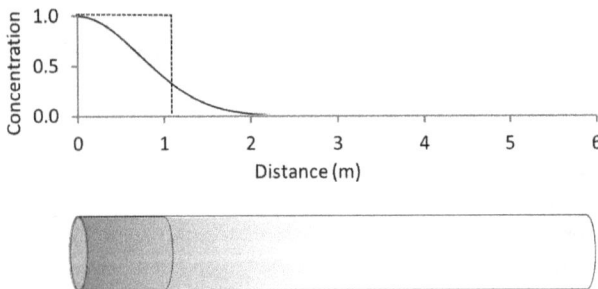

FIGURE 8.6   Top: The black line shows the concentration profile in a column at a specific time (continuous contaminant injection) with convection and dispersion. Bottom: Column view of contaminant shows concentration values represented by grey color intensity. The dotted lines shows the same information with convection only.

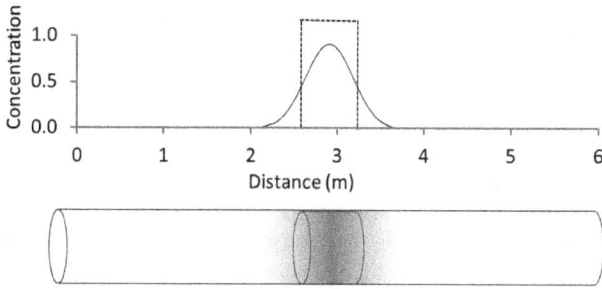

FIGURE 8.7 The top black line shows the concentration profile in a column at a specific time (limited time injection) with convection and dispersion. The bottom column view of the contaminant shows concentration represented by grey color intensity. The dotted lines shows the same information with convection only.

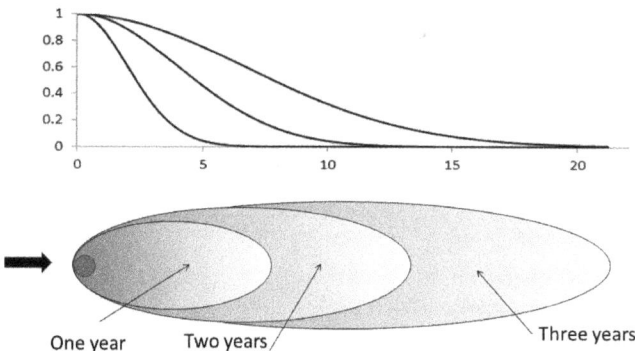

FIGURE 8.8 Continuous contaminant injection with dispersion.

again crucial to account for dispersion when there are concerns regarding the extent of contamination, whereas it is less significant when the primary concern is the contamination level.

The same discussion regarding the sand column, where water flow is one-dimensional, can be extended to encompass cases involving two- and three-dimensional spreading. Figures 8.8 and 8.9 illustrate the two-dimensional scenarios for continuous and limited-time injections, respectively. These figures depict the time progression of contaminated areas and concentration levels. In Figure 8.8, continuous injection results in an increase in mass within the aquifer, with rising concentrations bounded by the concentration of the injected contaminant. Over time, the plume continues to expand, covering additional portions of the aquifer.

Conversely, Figure 8.9 demonstrates that the volume of the injected contaminant remains constant when no chemical or biological activities are occurring. As the size of the plume grows, concentrations decrease, ultimately leading to complete dilution of the contamination.

The plots in Figures 8.8 and 8.9 can be regarded as projections for three-dimensional scenarios. In this case, the plumes take on a three-dimensional shape, roughly resembling an American football.

In Figures 8.8 and 8.9, the plumes should spread in the direction of flow and extend in the direction perpendicular to it. Longitudinal dispersion, which occurs more prominently due to the flow, pertains to dispersion in the direction of flow. As we will explain later, despite the one-dimensional nature of the flow, the contaminant can also spread in other directions known as transverse dispersion. This dispersion arises from variations in aquifer microscopic velocities, causing solute molecules to disperse in all directions. This principle remains applicable in three-dimensional spreading, where transverse dispersion occurs to a lesser extent in directions perpendicular to the flow, causing uneven distribution of the contaminant.

The spatial distribution of the contaminant in Figure 8.10 differs from the ideal case depicted in Figure 8.8. Several potential factors contribute to this disparity, including aquifer characteristics, such as the spatial distribution of hydraulic conductivity, and the nature of the flow field, which governs water velocities.

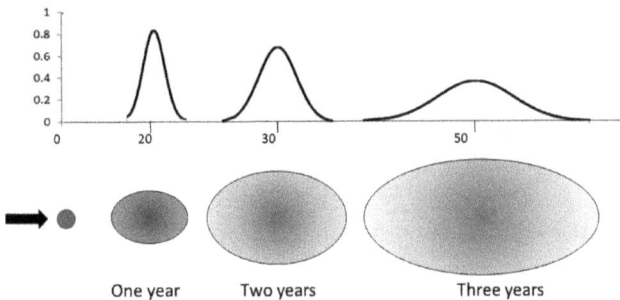

FIGURE 8.9 Limited-time contaminant injection with dispersion.

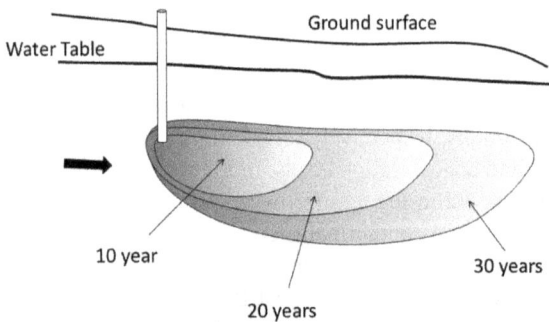

FIGURE 8.10 Contamination due to deep-well injection of a dissolved contaminant, such as a wastewater chemical.

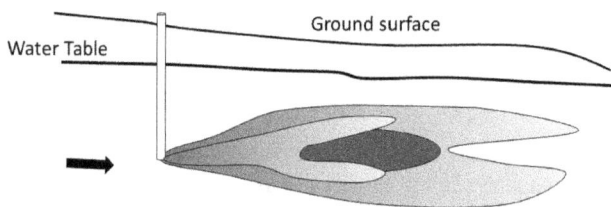

FIGURE 8.11 Plume configuration depends on aquifer characteristics. Here the distribution is affected by the existence of the dark-gray zone of low conductivity that locally slows down the spread of the contamination.

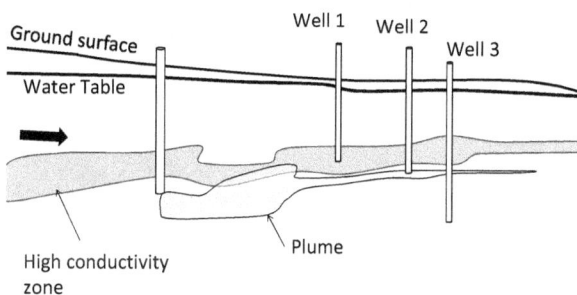

FIGURE 8.12 Effect of a high conductivity zone on plume configuration. Well 2 will be contaminated sooner than Well 1, although Well 1 is closer to the contamination source.

Aquifer heterogeneities can significantly influence the configuration of the contaminant plume. For instance, as illustrated in Figure 8.11, the concentration profile is influenced by the presence of a low conductivity zone that locally retards the spread of contamination. Consequently, concentrations within such a zone would be higher due to the entrapment of the contaminant in that area. Similarly, the influence of a high conductivity zone on the plume pattern can have significant consequences, as demonstrated in Figure 8.12. In this example, it is evident that Well 2 will become contaminated sooner than Well 1, despite Well 1's closer proximity to the contamination source, owing to preferential flow through the high conductivity zone.

8.2.1.4 Causes of dispersion

Contaminant movement is governed by molecular diffusion and mechanical dispersion, both of which contribute to the phenomenon known as hydrodynamic dispersion. Molecular diffusion involves the movement of contaminants within the water-filled pore space and is influenced by the concentration gradient. It can occur even in stagnant water. This process can be likened to the behavior of a drop of colored dye in a glass of clear water, where the gradual change in color represents the diffusion of the dye into areas with cleaner water.

The diffusive solute flux is quantified using Fick's law, Fick (1855):

$$F_m = D_m S_c \tag{8.4}$$

In this context, F_m represents the flux, D_m stands for the molecular diffusion coefficient, and S_c denotes the concentration gradient, which signifies the rate of change of concentration with distance at a given time. For instance, when considering concentrations at two distinct points, the movement of a chemical species occurs from the point of higher concentration to the point of lower concentration, with a flux rate that is directly proportional to the difference in concentrations and inversely proportional to the separation distance between the two points.

Mechanical dispersion is a microscopic process that takes place on both the soil particle and pore scales, as illustrated in Figures 8.13 and 8.14. In Figure 8.13, the local velocities within the pores vary based on their sizes, and this variability can even occur within a single pore. The velocity distribution within the pore follows a bell-shaped

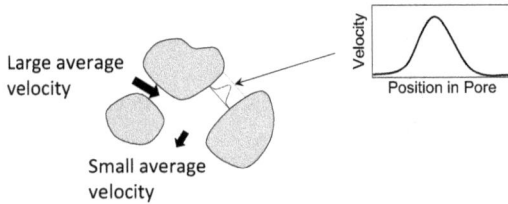

FIGURE 8.13   A microscopic view of the grain-pore system. The gray color shapes represent soil grains. Change in velocity inside the pore spaces is due to (1) variations in average velocities within pores of different sizes, and (2) variations in velocities within the same pore.

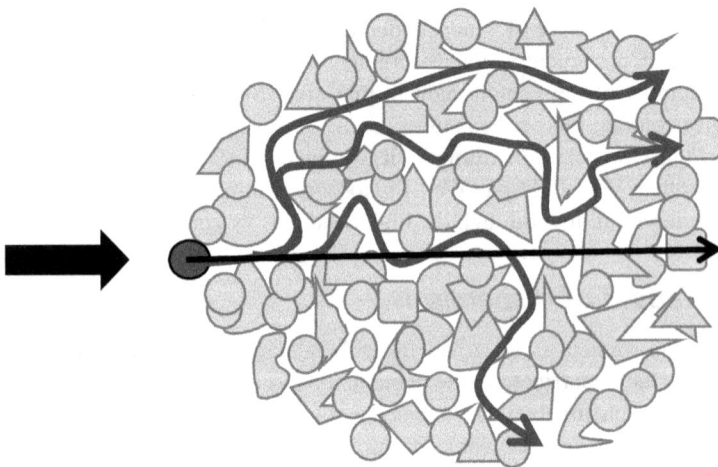

FIGURE 8.14   Illustration of the dispersion process where solute particles follow different pathways to end up at different locations despite originating from the same spot.

curve, as depicted in the figure, with lower velocity near the grain surface and peak velocity at the center of the pore, where friction forces are minimal. The average velocity within different pores will depend on their relative sizes, meaning that the velocity of a particular molecule is influenced by its specific location.

Furthermore, molecular travel can be influenced by factors such as the shape of particles, their size distribution, and the potential tortuosity of the pore network, as demonstrated in Figure 8.14. As defined by Bear (1979), tortuosity is the ratio of the actual flow path length to the straight-line distance between the ends of the flow path. In Figure 8.14, contaminant molecules injected at the small gray circle will each move at different velocities and in different directions through the pores. Over time and with many injected molecules, the contaminant will spread, covering a larger area as it advances. The straight black line in Figure 8.14 represents the direction of convective flow that three molecules would follow, assuming no dispersion occurs.

Mechanical dispersion results in the movement of contaminant molecules at varying speeds relative to the average convective velocity, both along and across the hydraulic gradient that drives water flow. This leads to the dispersion or spreading of the contaminant plume in different directions relative to the convective front.

The model FEFLOW (Diersch, 2013) was used in several simulations to illustrate contaminant transport in an aquifer, both with and without dispersion. Figure 8.15 depicts the scenario of transport in the aquifer driven solely by convection, without accounting for dispersion. Chemical molecules are introduced at three different locations. In the absence of dispersion, all injected particles follow the flow or streamlines, as explained in the flow-net discussion in Chapter 7. To address dispersion, it becomes necessary to assess the microscopic-scale transport of chemical particles, which is practically unfeasible. One approach to tackle this challenge is to simulate random pathways for the injected particles, as depicted in Figure 8.16. Molecules move in a stochastic manner, forming a plume or cloud. The extent of spreading depends on the characteristics of the media and the flow velocity.

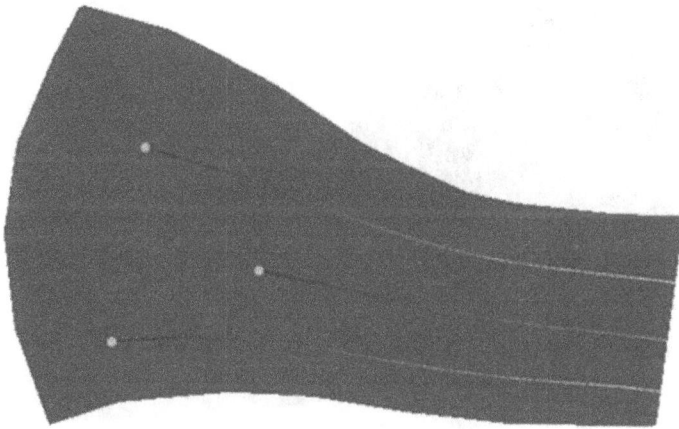

FIGURE 8.15 Contaminant molecules moving with convection only in the direction of flow lines.

However, generating these pathways does not inherently replicate the exact microscopic details and can produce varying pathways due to their random nature. Therefore, it is crucial to verify predictions against measured contamination data to ensure accuracy and build confidence in the results.

When comparing the extent of contamination in Figures 8.15 and 8.16, it becomes evident that neglecting dispersion can significantly underestimate the spread of contaminants. The broader dispersion pattern shown in Figure 8.16 can have substantial implications for health and other related concerns. Remediation efforts are also substantially impacted by the nature of the contamination distribution.

As previously discussed, the configuration of the contaminant plume is influenced by the spatial distribution of aquifer characteristics. Comparing Figures 8.16 and 8.17

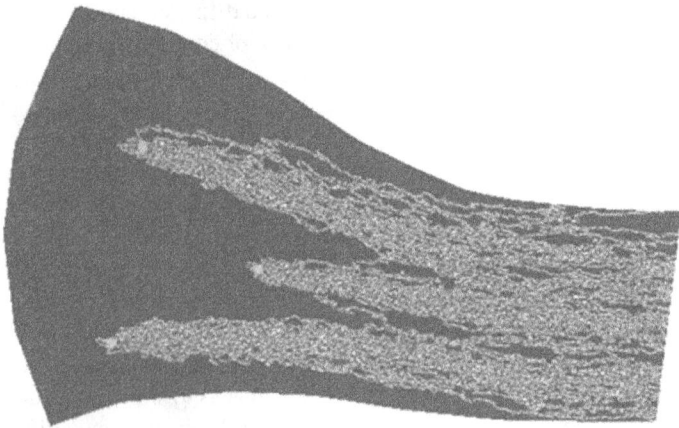

FIGURE 8.16   Contaminant molecules moving with convection and dispersion.

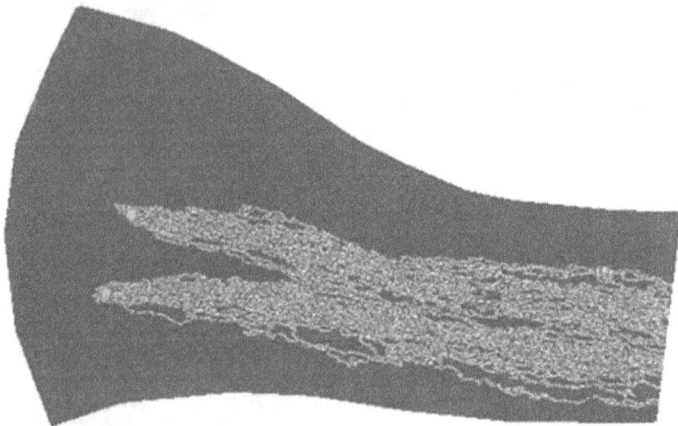

FIGURE 8.17   Spreading in a homogeneous aquifer with the same conductivity at all locations.

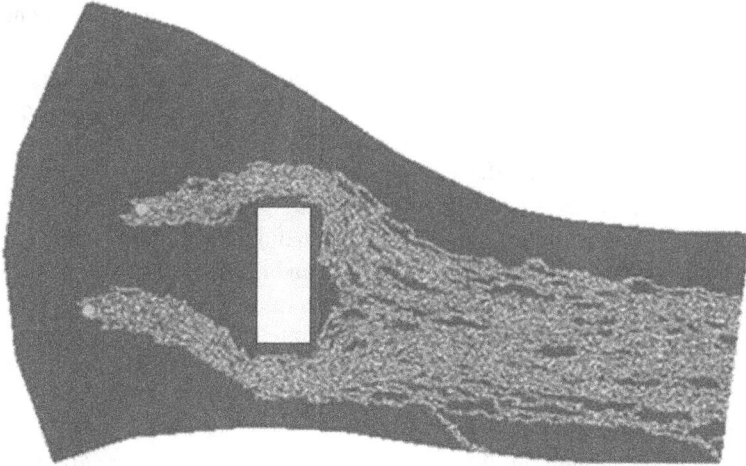

FIGURE 8.18 Spreading in an aquifer with a low conductivity zone downstream from the sources of injection.

illustrate the impact of a low conductivity zone located immediately downstream from the injection points, as depicted in the latter figure. In this scenario, the contaminant tends to circumvent the area with lower conductivity, resulting in increased lateral spreading (Figure 8.18).

As discussed earlier, assessing velocity on the microscopic scale is a complex task. This complexity also applies to mechanical dispersion, which is a microscopic process reliant on velocity. Consequently, the solution was to upscale to a macroscopic level for characterizing dispersive transport and its associated parameters. The initial step involved adapting equation (8.4), originally formulated for molecular diffusion, to encompass mechanical dispersion. In other words, it is assumed that the flux is directly proportional to the concentration gradient, introducing a constant of proportionality known as the mechanical dispersion coefficient, denoted as D_d:

$$F_d = D_d S_c \tag{8.5}$$

where F_d is the mechanical dispersion flux. The mechanical dispersion coefficient is evaluated by relating it to the average velocity and a media characteristic parameter called dispersivity (α). For example, the relationship is expressed as follows for one-dimensional transport:

$$D_d = \alpha v \tag{8.6}$$

Finally, the total flux F due to molecular diffusion and mechanical dispersion would be:

$$F = (D_m + D_d)S_c = DS_c \tag{8.7}$$

In equation (8.7), D is referred to as the hydrodynamic dispersion coefficient, which combines the effects of both diffusive and dispersive processes. D_m has a significantly smaller value compared to D_d, indicating that molecular diffusion becomes significant only in stagnant or near-stagnant water conditions. Otherwise, mechanical dispersion dominates.

Equation (8.5) is based on the assumption of a constant D_d. However, field studies have demonstrated the scale-dependent nature of D_d due to the challenges in upscaling from microscopic to macroscopic levels. As the contaminant plume expands, it interacts with more variability in porous media, leading to an increase in D_d values. Gelhar et al. (1992) compiled comprehensive data showing D_d ranging from a few centimeters in laboratory experiments to thousands of meters in large field studies. The main challenge lies in the inability to directly measure D_d, making it necessary to estimate values based on field-measured contamination data.

Besides the aquifer's characteristics, the extent of spreading depends on the convective velocity of groundwater flow. Therefore, it is important to recognize that neither convection nor mechanical dispersion occurs in stagnant water because both processes depend on the magnitude of flow velocity. In general, any realistic assessment of dispersion requires the accurate evaluation of the aquifer's flow field and resulting velocities.

8.2.2 Non-Conservative (Reactive) Contaminant Transport

The transport and fate of contaminants can be significantly influenced by chemical and biological activities that result in the transfer of contaminant mass between the liquid and aquifer solid materials or the conversion of dissolved chemicals from one form to another. These processes have a direct impact on the mass balance of the chemicals and the spatial distribution of their concentrations. Chemicals can be broadly categorized into two groups: organic and inorganic types (ATSDR, 2023; SCIENCE NOTES, 2023).

Organic chemicals encompass a group of human-made compounds, including pesticides, gasoline, dry-cleaning solvents, and degreasing agents. This group includes volatile organic chemicals, which are substances containing carbon that evaporate at room temperature, as well as synthetic organic chemicals. Organic contaminants can either be adsorbed onto surfaces, or degraded through microbiological processes, occurring under both aerobic and anaerobic conditions, which correspond respectively to the presence and absence of oxygen.

Inorganic substances, on the other hand, consist of chemicals that lack carbon atoms. Examples include ammonia, hydrogen sulfide, metals like sodium and aluminum, and most elements, such as carbon and nitrogen. Inorganic contaminant levels can be attenuated through various processes, including precipitation, oxidation–reduction, and adsorption. Precipitation reactions result in the formation of insoluble solids known as precipitates (LibreTexts, 2023a). These reactions are useful in determining the presence of specific elements in a solution. For instance, the presence of lead in water sources can be indirectly tested by introducing a certain chemical and monitoring the formation of a precipitate.

Oxidation and reduction describe the addition or removal of oxygen or hydrogen from a compound. Oxidation occurs when a chemical species gains oxygen or loses hydrogen, while reduction is defined by the opposite process. For example, hydrogen can be added to reduce nitrogen, producing ammonia.

Adsorption, a phenomenon applicable to organic chemicals as well, refers to the adherence of molecules in a solution to the surface of solids. For instance, organic chemicals can adsorb onto particles of organic carbon that may be present in the aquifer material. Adsorption leads to a retardation process, reducing the velocity of a contaminant relative to the convective groundwater velocity. The retardation factor, representing the ratio of these two velocities, can reach values as high as 10, effectively immobilizing the contaminant.

Some chemical reactions occurring in this context can be beneficial, such as natural chemical degradation induced by bacterial activities, which can reduce chemical concentrations and slow down contamination progression. Conversely, other reactions can be detrimental, generating chemicals that undergo minimal or no transformations, persisting in the environment, as seen in processes generating nitrate through nitrogen transformations.

Assessing and remediating contamination incidents is inherently challenging due to the complex and dynamic nature of groundwater systems. Chemical processes are continually influenced by the redistribution of dissolved species, affected by dispersion processes, and subject to the complexities of aquifer properties.

8.2.3 Approximate Equations for Assessing Contaminant Fate and Transport

The governing equation for the transport and fate of contaminants can be intricate, often requiring complex solutions to address realistic aquifer conditions. Nevertheless, simplified solutions do exist, and they can be valuable for initial analyses. These simplified solutions entail restricting conditions, such as assuming uniform-flow fields and homogeneous states. For instance, solutions are available to address scenarios involving the limited-time release of a contaminant volume into an aquifer, such as those resulting from accidental spills. In this scenario, a chemical mass denoted as M is presumed to be released at a known location within an initially uncontaminated aquifer. The flow is considered to be one-dimensional, solely along the x-axis. As described by LMNO Engineering, Research, and Software (2023), if the mass M is uniformly injected across both the width and height of the groundwater aquifer, dispersion will occur exclusively in the x-direction, making it one-dimensional. However, if M is uniformly injected only across the width of the groundwater aquifer, dispersion will manifest in both the x and y directions, resulting in two-dimensional dispersion. Lastly, when M is injected at a specific point within the groundwater, dispersion will occur in all three dimensions: x, y, and z. (Note that x in this context represents a symbol for distance, distinct from the multiplication operator.)

Three distinct solutions are available to estimate the solute concentration C for these three cases, as outlined in Bear (1972).

For a one-dimensional case, the solution at any location x and time t:

$$C(x,t) = \frac{M}{2nHW\sqrt{\pi D_x t}} \exp\left[-\frac{(x-vt)^2}{4D_x t}\right] \tag{8.8}$$

For a two-dimensional case, the solution at any location x, y and time t:

$$C(x,y,t) = \frac{M}{4nW\pi t\sqrt{D_x D_y}} \exp\left[-\frac{(x-vt)^2}{4D_x t} - \frac{y^2}{4D_y t}\right] \tag{8.9}$$

For a three-dimensional case, the solution at any location x, y, z and time t:

$$C(x,y,z,t) = \frac{M}{4n(\pi t)^{1.5}\sqrt{D_x D_y D_z}} \exp\left[-\frac{(x-vt)^2}{4D_x t} - \frac{y^2}{4D_y t} - \frac{z^2}{4D_z t}\right] \tag{8.10}$$

where

$$D_x = \alpha_x v \tag{8.11}$$

$$D_y = \alpha_y v \tag{8.12}$$

$$D_z = \alpha_z v \tag{8.13}$$

The variables in the above equations are as follows.

α_x, α_y, α_z=dispersivities in x, y, and z directions, respectively.
C=chemical concentration.
D_x, D_y, D_z=dispersion coefficients in x, y, and z directions, respectively
M=chemical mass injected.
n=porosity [percentage].
t=time.
v=pore or linear water velocity in the x-direction.
H=aquifer thickness (for one-dimensional case).
W=aquifer width (for one or two-dimensional cases).
x, y, z=location distances; x is the direction of groundwater flow, y is the vertical distance from the centerline of the plume, and z is the lateral distance.

Calculation software is available from LMNO Engineering, Research, and Software (2023). Other solutions for different aquifer and transport features are also documented by that online site.

It is important to note that the location of the peak in the x direction is at $x=vt$ and the peak value is located at $y=0$ for the two-dimensional case and $y=z=0$ for the three-dimensional case. The respective expressions for the values of the peak for the three cases will be as follows:

$$C(x,t)=\frac{M}{2nHW\sqrt{\pi D_x t}}$$ (8.14)

$$C(x,y,t)=\frac{M}{4nW\pi t\sqrt{D_x D_y}}$$ (8.15)

$$C(x,y,z,t)=\frac{M}{4n(\pi t)^{1.5}\sqrt{D_x D_y D_z}}$$ (8.16)

Example 8.2: Transport and Fate Simulation

Estimate the concentrations along the x-axis for $x=0$ to 60 m, with an increment of 4 m, at locations $y=z=0$, using the following information: the injected mass $M=18,000$ kg, time $t=1,000$ days, velocity $v=0.024$ m/day, $n=35\%$, x-direction dispersivity $\alpha_x=1$ m, and $\alpha_y=\alpha_z=0.1$ m. (Data are slightly modified from LMNO Engineering, Research, and Software, 2023.)

Equation (8.10) is applied. Consistent units should be used. Here are the meter-day-kilogram:

$D_x=1\times0.024=0.024$ m; $D_y=D_z=0.1\times0.024=0.0024$ m

$$C(x,y,z,t)=\frac{M}{4n(\pi t)^{1.5}\sqrt{D_x D_y D_z}}\exp\left[-\frac{(x-vt)^2}{4D_x t}-\frac{y^2}{4D_y t}-\frac{z^2}{4D_z t}\right]$$

$$C(x,y,z,t)=\frac{1,800}{4\times35\times(3.14\times1,000)^{1.5}\sqrt{0.024\times0.0024\times0.0024}}$$
$$\exp\left[-\frac{(x-0.024\times1,000)^2}{4\times0.024\times1,000}\right]$$

$$C(x,y,z,t)\sim1.8\times\exp\left[-\frac{(x-24)^2}{96}\right]$$

The results are shown in Table 8.1. They can be plotted as illustrated in the graph in Figure 8.19, which shows the C distribution in the x direction with a peak at v multiplied by t, which equals 24 m. As discussed in Section 8.2.1.3, as time progresses, the peak will decrease, and the contamination plume will expand, moving away from the source at $x=0$.

TABLE 8.1   Results of Example 8.2

DISTANCE X	CONCENTRATION C
0	0.004
4	0.028
8	0.125
12	0.402
16	0.924
20	1.524
24	1.800
28	1.524
32	0.924
36	0.402
40	0.125
44	0.028
48	0.004
52	0.001
56	0.000
60	0.000

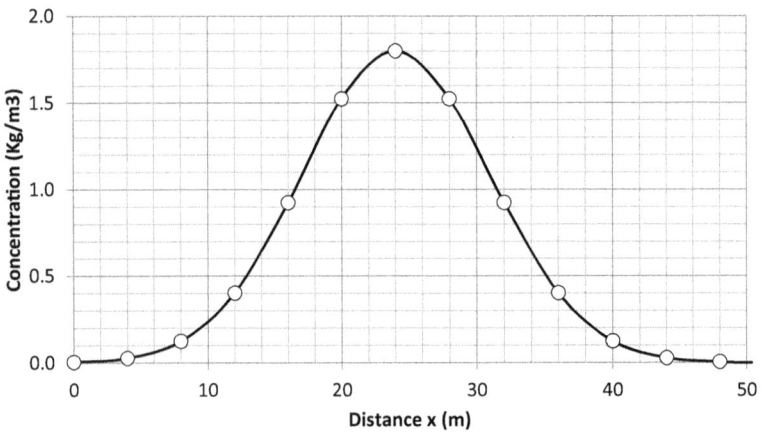

FIGURE 8.19   Concentration-distance profile for transport in an aquifer.

8.2.4 Non-Aqueous Phase Liquid Contamination

Non-aqueous phase liquids (NAPLs) are a class primarily composed of hydrocarbons that exist as immiscible phases when in contact with water and air. Nevertheless, some constituents of these liquids can still dissolve in water at the interface between the water

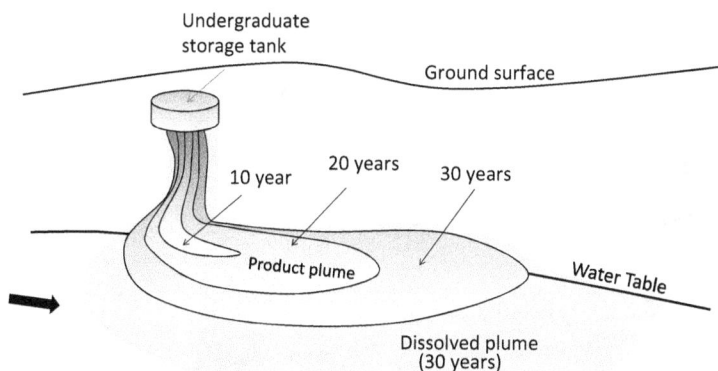

Undergraduate
storage tank

Ground surface

10 year   20 years   30 years

Product plume

Water Table

Dissolved plume
(30 years)

FIGURE 8.20   Contamination due to leakage from an underground storage tank of a light non-aqueous phase liquid (such as diesel or gasoline).

phase and the NAPL phase. Further details on NAPLs are available in EPA (1995) and Sethi and Di Molfetta (2019).

Differences in the physical and chemical properties of water and NAPLs create a physical interface between the two liquids, preventing them from mixing. Depending on their density relative to water, NAPLs are typically categorized as either light NAPLs (LNAPLs) or dense NAPLs (DNAPLs). When LNAPLs spill at the surface or in the subsurface, they penetrate the unsaturated zone, accumulate, float on the water table, spread in the direction of moving water, and gradually release their soluble fraction into the groundwater (see Figure 8.20). Consequently, two primary plumes emerge: a free product plume and a dissolved plume. Additionally, the contaminant may also exist in gaseous and adsorbed phases.

Conversely, DNAPLs vertically infiltrate the aquifer. Depending on the volume of the release, DNAPL contamination may impact the entire saturated thickness and migrate along the bottom of the aquifer (refer to Figure 8.21). Similar to LNAPLs, DNAPLs act as a continuous source of contamination by releasing their soluble fraction into the groundwater. As with LNAPLs, both free product and dissolved plumes exist, along with the possibility of the contaminant being present in gaseous and adsorbed phases.

The unsaturated zone contains both water and air. Therefore, along the periphery of the infiltrated NAPL product, a portion of the hydrocarbon will be retained by capillary forces due to the unsaturated condition, remaining as a residual product within the soil pores. When infiltrating rainwater comes into contact with residual NAPL, it dissolves soluble components. Additionally, volatilization can spread contamination into the air.

In general, NAPLs can exist in four phases within the subsurface: as a free product (NAPL), dissolved in water, in a gaseous state, and adsorbed onto soil or aquifer materials (see Figure 8.22). Different NAPL phases may partition or transition from one phase to another, depending on environmental conditions. For example, soluble components may dissolve from the NAPL into the passing groundwater. The same molecule may

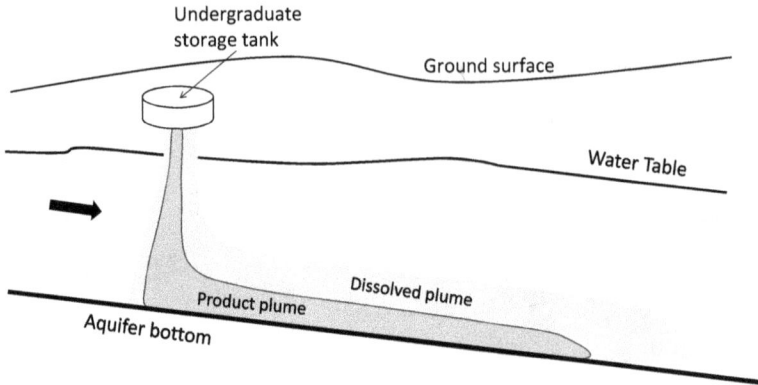

FIGURE 8.21 Contamination due to leakage from an underground storage tank of a dense non-aqueous phase liquid (such as crude oil).

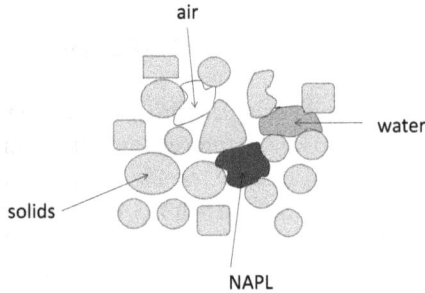

FIGURE 8.22 Contamination may exist in four phases: gas, adsorbed to soil materials, dissolved in water, or as NAPL.

adsorb onto a solid surface and subsequently desorb into the passing groundwater. As a result, remediation and cleanup efforts become complex because they must address all NAPL phases. Treatments would involve pumping out the majority of the product; however, these efforts would not effectively remove the residual or adsorbed NAPL, which may necessitate specialized removal methods, such as soil vapor extraction systems. Contaminated water containing dissolved-phase NAPL in the saturated zone would be pumped out for treatment.

8.2.5 Ocean Saltwater Intrusion

Aquifer saltwater contamination can result from various factors, including seawater intrusion in coastal areas, historical saltwater infiltration into aquifers, leaks from geological formations, and the presence of tidal lagoons. Additional sources of contamination include irrigation return flows and the disposal of saline waste by humans. In this section, we will focus primarily on seawater intrusion in coastal aquifers, as it poses the most significant threat. This issue can be highly challenging for communities that rely primarily on fresh groundwater for various purposes, including drinking and agriculture. Factors such as droughts, excessive groundwater pumping, and rising sea levels are the primary contributors to the encroachment of seawater into freshwater groundwater sources. The combination of rising sea levels and declining freshwater levels creates steeper gradients that allow saltwater to intrude farther inland. This intrusion reduces the availability of freshwater in aquifers, rendering the resource unsuitable for use and potentially leading to the abandonment of wells.

The movement of freshwater toward the sea acts as a natural barrier to prevent saltwater from encroaching into freshwater aquifers. However, under certain conditions, such as extended dry spells and excessive groundwater extraction, this encroachment can intensify. Therefore, the interface between freshwater and saltwater, as depicted in Figure 8.23, can shift either seaward or landward, depending on the extent of encroachment.

The figure illustrates the presence of a freshwater lens with an assumed sharp interface, indicating no mixing between the two zones. Under hydrostatic conditions, meaning no water flow, a simplified formula known as the Ghyben–Herzberg equation can estimate the position of this sharp interface. In this equation, it is important to note that the formation of the freshwater lens is primarily attributed to density considerations, as freshwater is less dense than saltwater. The equation establishes a relationship between the depth of saltwater (h_s) and the freshwater head (h_f), as shown in Figure 8.23:

$$h_s = \frac{\rho_s - \rho_f}{\rho_s} h_f \tag{8.17}$$

where ρ_s and ρ_f are the densities of saltwater and freshwater, respectively. This equation may also be applicable for the case of horizontal flow toward the ocean in areas away

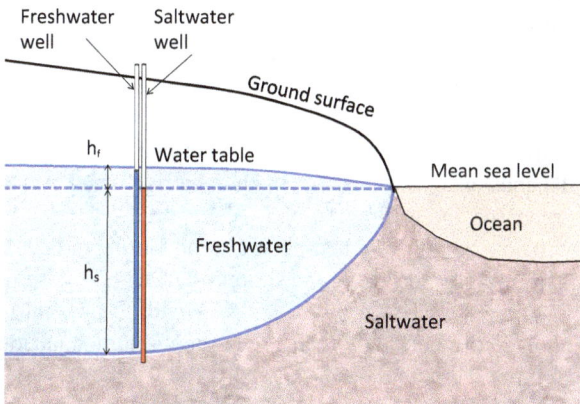

FIGURE 8.23 Saltwater intrusion and the Ghyben–Herzberg relation.

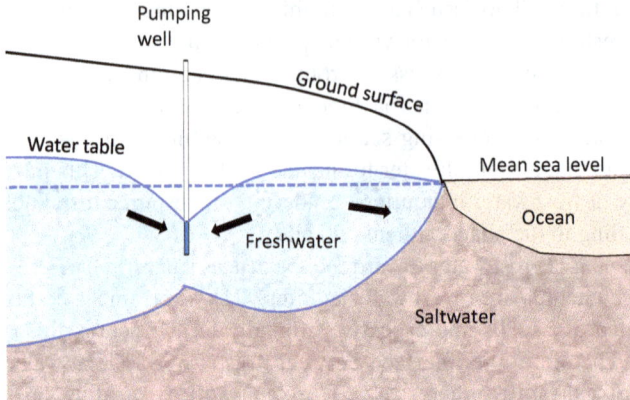

FIGURE 8.24 Pumping decreases the fresh hydraulic head (water level) and causes up-coning, i.e., the rise of the saltwater towards the well.

from the shoreline. The values of ρ_s and ρ_f are about 1.025 and 1.0 g/cm³ at a temperature of 20°C, leading to the following expression:

$$h_s = 40h_f \tag{8.18}$$

This equation simply states that, under the assumptions considered, the depth of the interface below the mean sea level is 40 times the freshwater head above the mean sea level.

Groundwater pumping can reduce the size of the freshwater lens and cause saltwater up-coning toward the pumping well (Figure 8.24), increasing the chance of contaminating the water supply. To minimize up-coning, rather than using a single well, it is recommended to distribute the pumping rate over multiple wells. However, the overall decrease in freshwater storage cannot be avoided.

Unfortunately, in real-world systems, the sharp interface assumption is not valid. Saltwater and freshwater mix, creating a dispersion or transition zone (refer to Figure 8.25). Salinity varies between that of saltwater and freshwater within a mixing zone whose size depends on aquifer dispersion characteristics and hydraulic conditions, including aquifer recharge and well pumping.

It's important to note that the Ghyben–Herzberg equation predicts a zero-depth (h_s) for the saltwater–freshwater interface at the shoreline when h_f is zero. However, this is unrealistic because some space is required for freshwater to flow into the ocean. The schematic in Figure 8.25 provides a more accurate representation, accommodating this requirement and aligning with field observations.

Additionally, the plot illustrates a broader mixing zone due to the influence of tides and wave actions, particularly when combined with the relatively thin freshwater lens.

Note that the distribution of the mixing zone is site-specific and can vary significantly from what is depicted in Figure 8.25, depending on the hydrological and

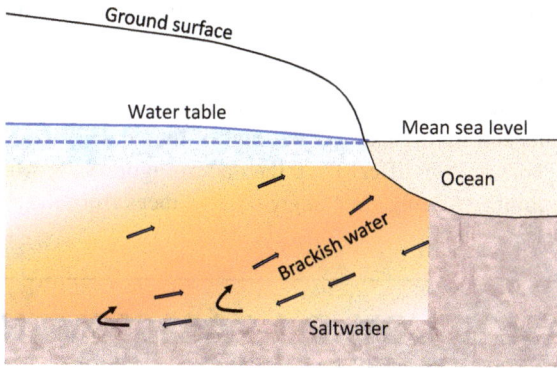

FIGURE 8.25 Effects of mixing between saltwater and freshwater forming a brackish zone. The arrows show the flow and circulation of water within various zones.

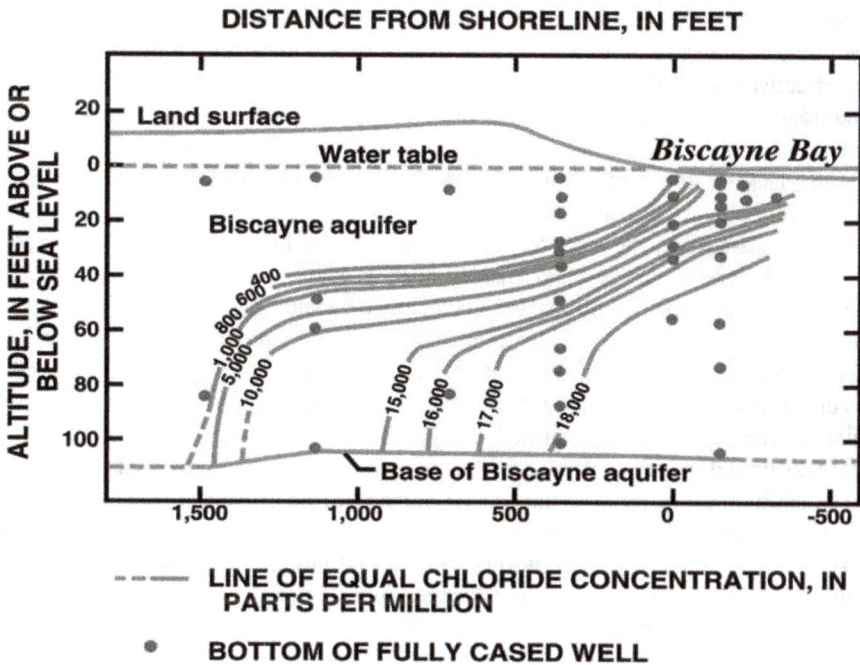

LINE OF EQUAL CHLORIDE CONCENTRATION, IN PARTS PER MILLION

BOTTOM OF FULLY CASED WELL

FIGURE 8.26 Saltwater intrusion for the Biscayne aquifer near Miami, Florida (USGS, 2017).

geological conditions of the site. For example, Figure 8.26 illustrates saltwater intrusion in the Biscayne aquifer near Miami, Florida (USGS, 2017). This plot illustrates the freshwater lens and the transition or mixing zone above the saltwater zone. The values for chloride concentrations (measures of salinity), measured using wells, are expressed in parts per million, equivalent to milligrams per liter (mg/L). In the figure, the mixing zone exhibits variability in chloride concentration, ranging from 400 to 18,000 mg/L. Generally, 250 mg/L is considered an acceptable limit for concentration in freshwater, while the value for ocean water is 19,000 mg/L. In the plot, the size of the mixing zone is significantly influenced by the base (or bottom) of the aquifer, which contributes to additional mixing. This profile differs from the one in Figure 8.25, which lacks such a base, resulting in a thinner mixing zone at greater distances from the shoreline.

8.3 CASE STUDY: RISK ASSESSMENT OF ON-SITE SEWAGE DISPOSAL SYSTEMS

Cesspools and septic tanks are typically necessary in situations where a connection to a centralized municipal sewer service is not practical or available, particularly in rural areas with low-density housing where establishing a treatment facility is not feasible. In septic systems, solid and liquid waste can be separated for appropriate disposal. The liquid wastewater flows into a leach field, where it undergoes a filtration process. The effectiveness of these systems in protecting the environment depends on certain conditions being met, such as maintaining a sufficient distance from the subsurface water table and the ocean. In contrast, a cesspool is essentially a pit lined with cement or stone, lacking the ability to filter waste. As a result, cesspools can lead to soil and groundwater contamination.

Studies assessing the risks to human health posed by wastewater contamination include the work of Hrudey and Hrudey (2007), who examined waterborne disease outbreaks in developed countries and documented 75 such cases. In 40 of these cases, wastewater contamination was identified as the primary cause. One example is an outbreak at the Washington County Fair in New York State in 1999, which resulted in two deaths (Centers for Disease Control and Prevention, 1999, Novello, 2000). In other countries, Said et al. (2003) identified sewage effluent as a source of waterborne disease outbreaks linked to private drinking water supplies in England and Wales.

In addition to the adverse impacts on human health, effluent from on-site sewage disposal systems (OSDS) can harm the environment by increasing biological productivity in streams and nearshore waters. Nitrate and phosphate are the most common limiting nutrients in these waters, and excessive concentrations of these ions can lead to overproduction of plant matter, displacing native plants and causing toxic algal blooms (Rabalais, 2002).

A study by Whittier and El-Kadi (2009) assessed the human health and environmental risks associated with OSDS on the island of Oahu, Hawaii. The relative risk is a crucial consideration when addressing potential issues with limited financial resources. Oahu serves as the state's major urban center, with most of the population concentrated along the southern coastal plain. Figure 8.27 displays the distribution of OSDS across the island, with higher concentrations primarily found near the coast. As shown in the figure, smaller urban areas and rural communities rely on on-site wastewater disposal.

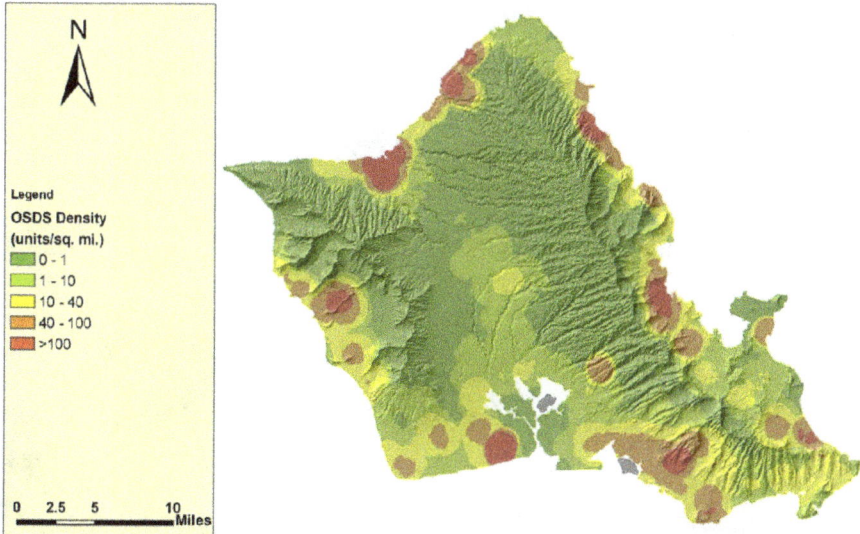

FIGURE 8.27   OSDS spatial island-wide distribution represented by the number of units in a square mile (Whittier and El-Kadi, 2009).

The study conducted by Whittier and El-Kadi identified several key factors contributing to contamination risks. These factors included the quantity, distribution, and types of OSDS, the effluent loads, critical receptors such as drinking water sources, streams, and near-shore waters, as well as the required buffers around potential risk areas.

The study's findings revealed the presence of 14,606 OSDS units, collectively releasing nearly 10 million gallons per day (mgd) of sewage into the environment. A significant portion of this discharge, approximately 77%, was attributed to cesspools, releasing an estimated 7.2 mgd of untreated sewage effluent. Remarkably, cesspools were responsible for nearly 96% of the potential nitrogen release from OSDS, accounting for 1,660 kg/day out of a total of 1,732 kg/day.

To assess the spatial distribution of OSDS risk, a weighted approach was employed that considered various risk factors, including the OSDS's proximity to receptors, their type, and the volume of waste they generated. The results are depicted in Figure 8.28. It is important to note that the values on the plot represent relative risk rather than absolute values. The study's results indicated that the majority of high-risk areas were located along the shoreline, posing a significant threat to the ocean. This problem could further exacerbate the potential for near-surface areas to experience flooding due to sea-level rise. Inland areas were also a cause for concern, particularly in terms of groundwater quality. Nitrate contamination originating from OSDS could elevate groundwater concentrations to levels exceeding the maximum contamination level of 10 mg/L, rendering the water unsuitable for consumption. Figure 8.29 displays the locations of OSDS units with the highest risk scores, highlighting areas requiring special attention and mitigation efforts.

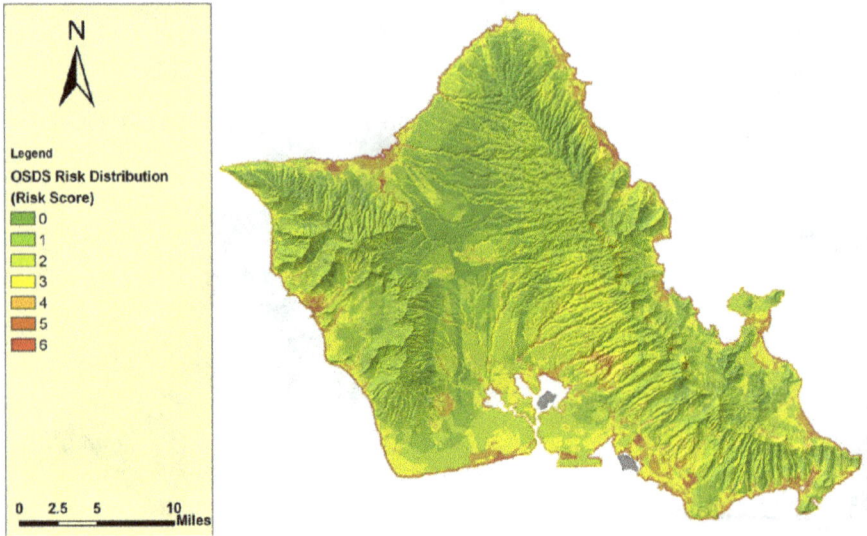

FIGURE 8.28   Spatial island-wide OSDS risk distribution (Whittier and El-Kadi, 2009).

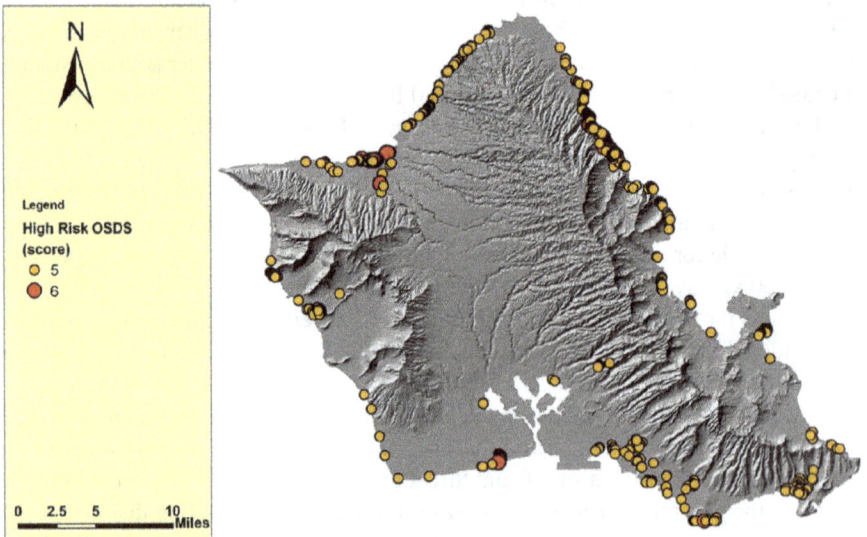

FIGURE 8.29   Location of high-risk OSDS (Whittier and El-Kadi, 2009).

8.4 CASE STUDY: A WATERSHED PROTECTION PLAN

Nawiliwili Bay on the island of Kauai, Hawaii, is nourished by three primary streams originating from the Nawiliwili Watershed: Huleia Stream, Puali Stream, and Nawiliwili Stream (refer to Figure 8.30). As per Section 303(d) of the Federal Clean Water Act (EPA, 2022), the Hawaii Department of Health designates Nawiliwili Bay, Nawiliwili Stream, and Huleia Stream as water bodies experiencing impaired water quality due to excessive turbidity. Furthermore, Nawiliwili Bay is also listed as impaired due to excessive nutrients and microorganisms.

El-Kadi et al. (2004) conducted a study aimed at fulfilling the mandates set by the federal government, as represented by the EPA, to formulate a comprehensive restoration and protection plan for the Nawiliwili Watershed. This proposed plan encompassed specific elements stipulated by the EPA, which include identifying contamination sources necessitating load reductions, establishing precise load reduction targets, and defining milestones for their implementation. Additionally, the plan should outline remediation measures, provide estimates for technical and financial support requirements, and detail associated costs and resource allocations. Lastly, the plan mandates the creation of a public information and education initiative.

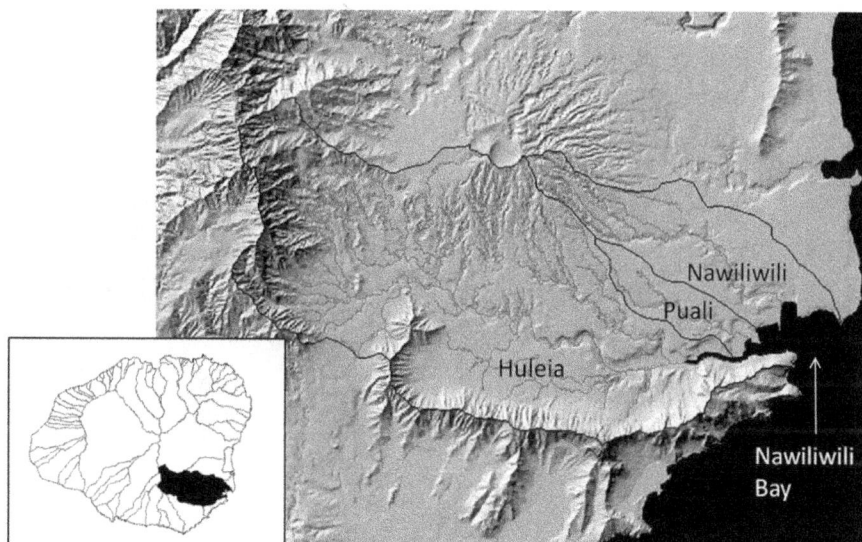

FIGURE 8.30 The three main streams in the Nawiliwili Watershed, which combines the three stream basins. Topography defines the boundaries of the watersheds or basins. The inset map shows the location of the Nawiliwili Watershed on the island of Kauai, Hawaii (El-Kadi et al., 2004).

Based on a field assessment, the study proposed specific restoration and protection activities. These include preventing soil erosion from agricultural lands and controlling non-native and invasive plant and animal species. Equally important are the recommendations to develop an accurate water budget for the watershed, eliminate cesspools, and implement low-impact development practices.

Figure 8.31 illustrates the recommended management measures for restoring the Nawiliwili Watershed. Among the recommended actions, best management practices (BMPs) aim to reduce the entry of chemical pollutants and sediment into water resources. They consist of both structural and non-structural practices, with the latter providing preventive benefits and being relatively cost-effective. Examples include preserving open spaces and naturally vegetated areas, which can mitigate erosion and support natural infiltration and evapotranspiration processes. Creating or maintaining riparian buffers, like wetlands, can reduce pollutant loads by slowing water flow, trapping sediment, and encouraging natural infiltration processes.

On the other hand, structural BMPs are designed to treat stormwater runoff and can be categorized into four types: source control BMPs, source filtration BMPs, regional detention and treatment systems, and pollution prevention practices (El-Kadi et al., 2004).

Agricultural BMPs are measures aimed at reducing the entry of fertilizers, pesticides, animal waste, and other pollutants into water resources while maintaining agricultural production. These include chemical application and irrigation

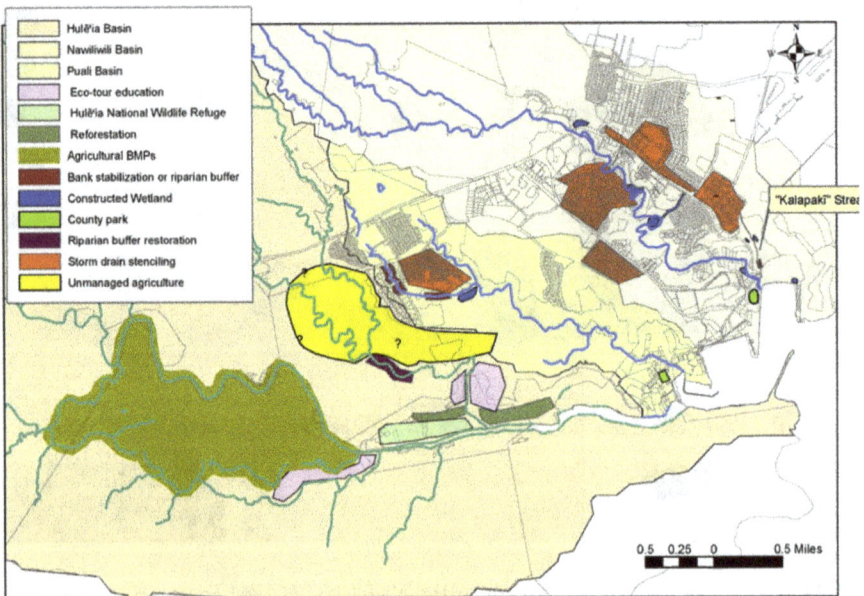

FIGURE 8.31 Approximate locations of management measures proposed for the restoration of the Nawiliwili Watershed. Question marks indicate approximate boundaries (El-Kadi et al., 2014).

management to minimize water and chemical losses to the environment. Moreover, water bodies can be protected by employing buffers, setbacks, and swales to minimize or prevent the transport of sediments and chemicals from agricultural production areas.

Lastly, the study delves into the economic implications of the restoration plan, considering potential expenditures in agriculture, recreation and tourism, and households. These changes are expected to have a ripple effect throughout the economy of the island of Kauai, where the watershed is located. The study also addresses the cost of restoration versus the expected benefits and potential sources of funding for the restoration and protection plan. The estimated total cost of restoration is approximately $6 million (2004 dollars) for a five-year plan.

8.5 SUPERFUND SITES

The Comprehensive Environmental Response, Compensation, and Liability Act (CERCLA) was enacted by Congress on December 11, 1980. It granted the U.S. Environmental Protection Agency (EPA) the authority to regulate hazardous substances at contaminated waste sites across the nation (EPA, 2023a). Additionally, the act empowered the EPA to hold individuals or companies responsible for contamination, compelling them to cover the costs of site cleanup. CERCLA also established the Superfund program, a federal initiative for managing contaminated sites categorized by priority and the extent of waste contamination. This program is commonly referred to as the National Priorities List (NPL).

Superfund sites are those identified as posing significant risks, primarily to humans, necessitating intervention by the U.S. Federal Government, represented by the EPA. These sites encompass a range of facilities, including oil refineries, mines, waste disposal sites, chemical storage and processing facilities, and industrial areas. To determine priority rankings, the EPA employs a hazard ranking system, which involves site inspections and the collection of all available information. According to federal law, the EPA must update the NPL annually, and site rankings determine the order of remediation. A site must be listed on the NPL before federal funds can be allocated for its cleanup. Once the cleanup process commences, the EPA oversees the remediation efforts to restore the sites to practical use. Funding for these cleanup activities is derived from the Superfund Trust Fund, which also supports legal actions against potentially responsible site owners and operators. For sites under federal government jurisdiction, agencies managing them are obliged to fund cleanup actions from their budgets.

As of May 2023, there are 1,894 Superfund sites among the 50 states (EPA, 2023a). New Jersey has the highest number of Superfund sites, with 152, followed by Pennsylvania with 126 sites and California with 114 sites. Conversely, states with the fewest Superfund sites include North Dakota (two sites), Nevada (three sites), and South Dakota (four sites).

8.5.1 The General Steps Included in Site Remediation

Although they can be site-specific, the steps and procedures involved in addressing a Superfund site are well established. The following is a list of most of the acronyms used in Superfund procedures and reporting.

- AOC: Administrative Order of Consent: An agreement between the company and EPA; the company agrees to pay for damages and to cease activities that caused the damages
- ARARs: Applicable or Relevant and Appropriate Requirements
- BRA: Baseline Risk Assessment
- CERCLA: Comprehensive Environmental Response, Compensation, and Liability Act of 1980
- COCs: Constituents of Concern
- FS: Feasibility Study
- NPL: National Priorities List (Superfund List)
- O&M: Operation and Maintenance
- RAOs: Remedial Action Objectives
- RI: Remedial Investigation
- Record of Decision (ROD): A public document that explains the remediation plan for the cleanup of a Superfund site
- TI: Technical Impracticability: Allows a company to obtain a waiver

Assessing and remediating Superfund sites depend on site conditions and the level and type of contaminants. The process starts with identifying the problem and concludes with closing the site after remediation is complete. Despite the variations among sites, the general steps in handling the assessment and cleanup processes are somewhat similar, which can be summarized as follows.

1. Accidental spill or discovery of serious concerns
2. Water use discontinued from the affected sources
3. EPA's Preliminary Assessment/Site Investigation.
4. Initiation of EPA's Hazard Ranking Scoring process.
5. If warranted, the site is added to the National Priority List (NPL)
6. Order of Consent (AOC) signed by EPA and the potential responsible party
7. Cleanup investigation begins
8. Cleanup investigation is completed
9. EPA proposes a final remedy for the site
10. ROD is signed
11. EPA selects the final remedy; construction begins
12. Public comments requested for the proposed cleanup settlement
13. Construction is complete

14. EPA begins the first five-year review of cleanup, with reports to follow every five years
15. The site is closed and removed from the NPL.

The following steps for site remediation for a Superfund site (steps 7–9 above) or any contaminated site are usually adopted.

1. Efforts are initiated to pinpoint the source of contamination, which can sometimes be ambiguous due to potential overlaps between different sources. In certain cases, extensive and costly fieldwork is required.
2. Chemical data are collected from water samples obtained from observation wells, which are often dug at significant expense, especially in deep and large aquifers.
3. Field data are used to develop a conceptual model for the site (refer to Chapter 10). This model encompasses factors such as the geological nature of the aquifer, its extent, conditions at its boundaries, contaminant sources and characteristics, climate data, and the history of land use and land cover changes.
4. Subsequently, a simulation model is created. This model serves as a predictive tool that integrates all available site information with equations describing water flow and contaminant transport within the aquifer (as detailed in Chapter 10). It can be used to forecast future contaminant levels and evaluate potential cleanup strategies, including financial considerations.
5. Cleanup operations are executed, and observation wells are employed to monitor progress in the cleanup efforts.

In the following section, we will illustrate Superfund procedures using an example from a site in Hawaii.

8.5.2 Case Study: Del Monte Superfund Site, Oahu, Hawaii

Del Monte Fresh Produce (Hawaii) Inc. cultivated and processed pineapples on a plantation located in Oahu, Hawaii, from approximately 1946 until November 2006 (EPA, 2023b). During this time, several pesticides, specifically soil fumigants, were applied to control nematodes that attack pineapple roots. Unfortunately, these chemicals were stored, mixed, and, regrettably, some spills occurred in close proximity to the Kunia drinking water well, which is concerning.

In April 1977, there was an accidental spill of 500 gallons of pesticides at the site. As a result, the fumigants contaminated the shallow subsurface soil, as well as perched and deep basal groundwater. Figure 8.32 provides a map of the area including contaminated site.

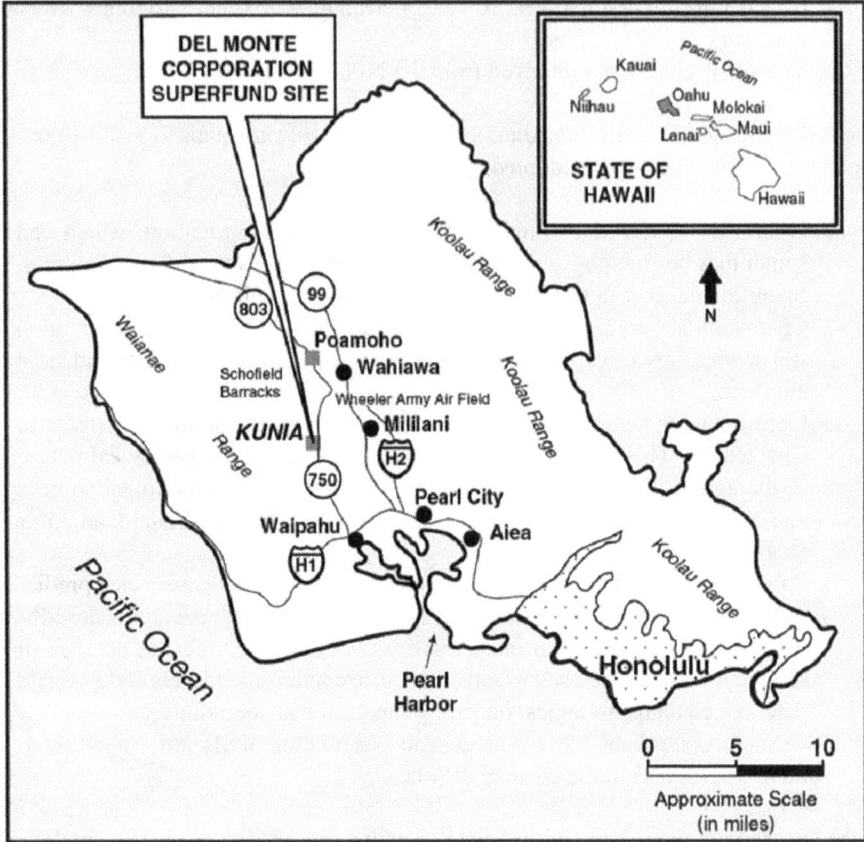

FIGURE 8.32   Location map of the Superfund site, Oahu, Hawaii (EPA, 2023b).

The constituents of concern (COCs) include ethylene dibromide (EDB), 1,2-dibromo-3-chloropropane (DBCP), 1,2-dichloropropane (1,2-DCP), and 1,2,3-trichloropropane (1,2,3-TCP). The soil and shallow groundwater were found to be contaminated with EDB, DBCP, DCP, solvents, TCP, benzene, and the pesticide lindane. Meanwhile, the deep groundwater was contaminated by EDB, DBCP, and TCP. Concerns arise from the fact that individuals who come into contact with or ingest this contaminated groundwater or soil are at risk.

8.5.2.1 Initial actions

- **1980**: The state ordered the Kunia well removed from service.
- **1981 and 1983**: Del Monte excavated 2,000 and 16,000 tons of soil and spread it on nearby fields.
- **1980–1994**: Del Monte pumped deep and shallow groundwater from three extraction wells; water was used to control dust on roads and sprayed on unplanted pineapple fields.

- **1994**: EPA requested that Del Monte discontinue this practice since it constituted unlawful disposal of a hazardous substance under the Resource Conservation and Recovery Act (RCRA) and the Superfund statute. Del Monte complied.
- **1990**: EPA started a Preliminary Assessment/Site Investigation.
- EPA determined that COCs had been released into soil and perched groundwater at the site and that a substantial threat of release to basal groundwater existed.
- **1992**: EPA completed the Hazard Ranking Scoring process.
- **1994**: The site was added to the National Priorities List (NPL).
- **1995**: The Del Monte Company, EPA, and the state of Hawaii signed an administrative order of consent (AOC) for a Remedial Investigation/ Feasibility Study and Engineering Evaluation and cost analysis.
- **2003**: The ROD was signed. The selected remedy addressed soil and groundwater contamination.
- **March 2003**: EPA issued a Proposed Plan for remedy

The remediation plan addressed various contaminated areas with specific procedures. For the perched aquifer and deep soil in the source area, remediation involves pumping contaminated water for treatment. Phytoremediation is utilized to treat the contaminated groundwater, including water extraction and diversion into a closed-loop treatment cell planted with Koa haole plants. These plants break down toxic compounds into non-toxic ones. A vegetated soil cover (cap) is placed over the contaminated soil in the spill area to reduce rainwater infiltration, which carries contaminants down to the basal aquifer. Land use would be restricted to prevent damage to the cap. Finally, soil contaminants will be removed using a soil vapor extraction system.

For the deep basal groundwater, the selected remedy's objective was to eliminate potential future exposure to contaminants and restore the basal aquifer's drinking water quality. A phased pump-and-treat approach was designed to clean contaminated groundwater, starting at the Kunia Well, the source area of contaminants. Monitoring wells would serve to (1) characterize the extent of contaminated groundwater, (2) assess the effectiveness of groundwater pumping, and (3) evaluate whether natural processes (natural attenuation) can reduce contaminant concentrations to meet drinking water standards in the remainder of the aquifer. If natural breakdown was not evident, additional pumping wells would be added to ensure comprehensive capture and treatment of the plume. Contaminated groundwater would be treated to meet drinking water standards using air stripping and carbon filtration, and the treated water would be used for irrigation. Land use restrictions should be enforced to prevent activities that may interfere with groundwater extraction and monitoring.

In 2008, construction of the selected remedies was completed, and the Preliminary Close Out Report was signed. In 2010, the first five-year report found that the remedy was constructed in accordance with the ROD requirements and is functioning as intended. However, it was noted that groundwater in the basal aquifer has background levels of COCs above Hawaii's maximum contamination levels due to historical pesticide application in the area, aside from the accidental spill of COCs. Additionally, restoring the groundwater in the basal aquifer to drinking water levels proved to be more challenging than expected. Therefore, it was recommended that the remedial action objectives and cleanup levels for the basal aquifer be re-evaluated at a later date. Nevertheless, the report concluded that the remedy is protective of human health and the environment because there is no exposure to untreated basal aquifer groundwater. The last five-year report (EPA, 2020) supported these conclusions but indicated that there is an impact on the basal aquifer from nonpoint agricultural sources up-gradient. It was recommended that background concentrations be re-evaluated annually to detect any trends.

8.5.2.2 Summary (typical of superfund sites)

- **1977**: Accidental spill
- **1980**: Kunia well disconnected
- **1990**: EPA's Preliminary Assessment/Site Investigation.
- **1992**: EPA's Hazard Ranking Scoring process.
- **1995**: Site added to NPL
- **1995**: Order of Consent (AOC) signed
- **1997**: Cleanup Investigation To Begin
- **1999**: Cleanup Investigation Completed
- **2003**: U.S. EPA Proposed Final Remedy for Site
- **2003**: ROD signed
- **2003**: Poamoho Section Proposed for Removal from Superfund List
- **2005**: EPA selects final remedy—Construction underway
- **2005**: Public Comments Requested for Del Monte Cleanup Proposed Settlement
- **2008**: Construction complete
- **2010**: EPA Begins First Five-Year Review of Cleanup with reports to follow every 5 years
- **2010**: EPA Finds Del Monte Site Remedy Protective of the environment
- **2015**: Second Five-Year Report
- **2020**: Third Five-Year Report. The 2020 Site Status (EPA, 2020) is listed as follows:

Contamination at the site is currently being treated through a dual-phase soil vapor and groundwater treatment system. A total of 18,000 tons of contaminated soil was removed in two short-term removal actions in 1981 and 1983. EPA delisted the Poamoho section of the site from the Superfund program's National Priorities List (NPL) in January 2004. Investigation results indicated that the area poses no significant threat to human health or the environment.

8.6 ASSIGNMENTS

1. Describe the differences between dispersivity, dispersion coefficient, and diffusion coefficient.
2. Example 8.2 dealt with assessing the three-dimensional concentrations following a sudden release of a contaminant. The calculations were

made at locations along the x-axis for $x=0$–60 m, with an increment of 4 m. Locations along the x-axis were chosen at $y=z=0$. The injected mass M was 18,000 kg, the time for calculations $t=1,000$ days, velocity $v=0.024$ m/day, $n=35\%$, x-direction dispersivity $\alpha_x=1$ m, and $\alpha_y=\alpha_z=0.1$ m. (Data are slightly modified from LMNO Engineering, Research, and Software, 2023).

A. Repeat the calculations for dispersivities $\alpha_x=2$ m and $\alpha_y=\alpha_z=0.2$.
B. Repeat the calculations for time $t=2,000$ days.
C. The term velocity v multiplied by time t ($x=vt$) provides the location of the peak. For the same data, calculate the peak values and locations for times 500, 1,000, 1,500, 2,000, and 2500.

3. Select a Superfund site and complete a timeline of handling the site as described in Section 8.5.1, following the example in Section 8.5.2. The site can be selected from the list on the website https://www.epa.gov/superfund.

REFERENCES

ATSDR. (2023). Inorganic substances. Retrieved from https://wwwn.cdc.gov/tsp/substances/ToxChemicalListing.aspx?toxid=37

Bair, E. S., & Lahm, T. D. (2006). *Practical Problems in Groundwater Hydrology*. Upper Saddle River, NJ: Pearson Prentice Hall.

Bear, J. (1972). *Dynamics of Fluids in Porous Media*. New York: American Elsevier.

Bear, J. (1979). *Hydraulics of Groundwater*. New York: McGraw-Hill.

Centers for Disease Control and Prevention. (1999). Outbreak of Escherichia coli O157 and Campylobacter among attendees of the Washington County Fair—New York, 1999. *MMWR. Morbidity and Mortality Weekly Report*, 48(36), 803–805. https://pubmed.ncbi.nlm.nih.gov/10499785/

Centers for Disease Control and Prevention. (2025, May 29). *Waterborne disease in the United States*. Retrieved July 2, 2025, from https://www.cdc.gov/healthy-water-data/waterborne-disease-in-us/index.html

Diersch, H. -J. (2013). *FEFLOW—Finite Element Modeling of Flow, Mass and Heat Transport in Porous and Fractured Media*. Springer. https://doi.org/10.1007/978-3-642-38739-5

Domenico, P. A., & Schwartz, F. W. (1998). *Physical and Chemical Hydrogeology* (2nd ed.). New York: John Wiley & Sons Inc.

El-Kadi, A. I., Mira, M., Dhal, S., & Moncur, J. E. T. (2004). Assessment and protection for the Nawiliwili watershed: Phase 3—Restoration and protection plan for the watershed. (Rep. No. WRRC-2004–05). Water Resources Research Center, University of Hawai'i. Retrieved from https://health.hawaii.gov/cwb/files/2013/05/PRC_WatershedNawiliwiliBay.pdf

EPA. (1995). Light nonaqueous phase liquids. Retrieved from https://www.epa.gov/sites/default/files/2015-06/documents/lnapl.pdf

EPA. (2020). Third five-year review report for Del Monte Corporation (Oahu Plantation) Superfund site Honolulu, Hawaii. Retrieved from https://semspub.epa.gov/work/09/100021407.pdf

EPA. (2021). Drinking water. Retrieved from https://www.epa.gov/report-environment/drinking-water

EPA. (2022). Summary of the clean water act. Retrieved from https://www.epa.gov/laws-regulations/summary-clean-water-act

EPA. (2023a). Superfund. Retrieved from https://www.epa.gov/superfund

EPA. (2023b). Del Monte Corp. (Oahu Plantation) Kunia, Hi. Retrieved from https://cumulis.epa.gov/supercpad/cursites/csitinfo.cfm?id=0902876

Fetter, C. W., & Kreamer, D. (2022). *Applied Hydrogeology* (5th ed.). Long Grove, IL: Waveland Press.

Fick, A. (1855). On liquid diffusion. *Philosophical Magazine, 10*(63), 30–39. https://doi.org/10.1080/14786445508641925

Fox, G. (2014). Basics of groundwater hydrology, Lab 5 groundwater model 1. Retrieved from https://www.youtube.com/watch?v=AtJyKiA1vcY

Gelhar, L. W., Welty, C., & Rehfeldt, K. R. (1992). A critical review of data on field-scale dispersion in aquifers. *Water Resources Research, 28*(7), 1955–1974. https://doi.org/10.1029/92WR00607

Hrudey, S. E., & Hrudey, E. J. (2007). Published case studies of waterborne disease outbreaks – Evidence of a recurrent threat. *Water Environment Research, 79*(3), 233–245. https://doi.org/10.2175/106143006X95483

LibreTexts. (2023a). Precipitation reactions. Retrieved from https://chem.libretexts.org/Bookshelves/Inorganic_Chemistry/Supplemental_Modules_and_Websites_(Inorganic_Chemistry)/Descriptive_Chemistry/Main_Group_Reactions/Reactions_in_Aqueous_Solutions/Precipitation_Reactions

LMNO Engineering, Research, and Software. (2023). Groundwater contaminant transport: 3-D pulse (slug) injection. Retrieved from https://www.lmnoeng.com/Groundwater/transport-Pulse.php#Variables

Novello, A. (2000). *The Washington County Fair Outbreak Report*. Albany, NY: New York State Department of Health. 199 pages.

Rabalais, N. N. (2002). Nitrogen in aquatic ecosystems. *Ambio, 32*(2), 102–112. https://doi.org/10.1579/0044-7447-31.2.102

Said, B., Wright, F., Nichols, G. L., Reacher, M., & Rutter, M. (2003). Outbreaks of infectious disease associated with private drinking water supplies in England and Wales 1970–2000. *Epidemiology and Infection, 130*, 469–479. https://www.ncbi.nlm.nih.gov/pmc/articles/pmc2869983/

Schwartz, F. W., & Zhang, H. (2003). *Fundamentals of Groundwater*. New York, NY: John Wiley & Sons.

Science in the Courtroom. (2021). Module 8: Movement of TCE to wells G and H. Retrieved from https://serc.carleton.edu/woburn/student-modules/contaminants/index.html

SCIENCE NOTES. (2023). List of metals. Retrieved from https://sciencenotes.org/list-metals/

Sethi, R., & Di Molfetta, A. (2019). Transport of immiscible fluids. In: *Groundwater Engineering*. Springer Tracts in Civil Engineering. https://doi.org/10.1007/978-3-030-20516-4_14

The Lancet. (2022). Pollution and health: A progress update. Retrieved from https://www.thelancet.com/journals/lanplh/article/PIIS2542-5196(22)00090-0/fulltext

United Nations. (2023). *Water*. Retrieved from https://www.un.org/en/global-issues/water

U.S. Geological Survey. (2017). Freshwater-saltwater interactions along the Atlantic Coast. Retrieved from https://water.usgs.gov/ogw/gwrp/saltwater/fig3.html

Whittier, R., & El-Kadi, A. I. (2009). Human and environmental risk ranking of onsite sewage disposal systems. Final Draft, Submitted to State of Hawai'i Department of Health, Safe Drinking Water Branch, Honolulu, Hawaii. Retrieved from https://health.hawaii.gov/wastewater/files/2015/09/OSDS_OAHU.pdf

Surface Water Contamination

9

9.1 OPEN-WATER BODY MECHANISMS

Within the water body, several physical fate and transport processes occur: advection, diffusion, and dispersion. These processes can be defined similarly to how they were in Chapter 8, which discussed parallel groundwater processes. Advection occurs when a contaminant is transported by fluid motion without significant spreading. On the other hand, diffusion is defined as the spreading of contaminants due to a concentration gradient, moving from areas of high concentration to areas of lower concentration. This process results from random mixing and can be quantified using Fick's law (1855), as discussed in Section 8.2 [equation (8.4)]. In Fick's law, the mass flux in a specific direction is directly proportional to the concentration gradient in that direction, with the proportionality constant known as the diffusion coefficient.

Dispersion is a process similar to diffusion but is primarily driven by variations in water velocity. It is a much more efficient spreading process compared to diffusion and contributes to the mixing of contaminants within a water body.

In the water body, the primary chemical and biological processes include precipitation and dissolution, biodegradation, and photodegradation. Precipitation involves the removal of a chemical species from water through the formation of solid minerals. Dissolution, on the other hand, is the reverse process where ionic species diffuse away from solids into the water. Biodegradation results from microorganism activities that transform or break down organic materials. This process can occur in the presence of oxygen, known as aerobic biodegradation, or without oxygen, known as anaerobic biodegradation. Additional details about these chemical and biological processes are provided in Section 8.2 of Chapter 8.

Photodegradation, or photolysis, refers to the chemical degradation caused by the radiant energy of light. It occurs when a compound absorbs sunlight. The extent of photodegradation depends on the amount of sunlight available and the degree to which the chemical can absorb sunlight. Contaminants near the water surface are particularly vulnerable to solar radiation and subsequent photodegradation.

9.2 WATER–AIR AND WATER–SOIL MECHANISMS

Across the water–air interface, wind and hydraulic turbulence can induce the transfer of airborne contaminants into the water or release them into the atmosphere through mechanical processes. When the concentration of a gas in the atmosphere surpasses that in the water, the gas tends to be absorbed into the water, and vice versa. Nevertheless, the effectiveness of this transport process may be hindered by the formation of gas and water films at the interface between the water body and the atmosphere.

Section 9.1 above delves into chemical and biological mechanisms at the water–soil interface, and some of these mechanisms are also outlined in Section 8.2 (Chapter 8). Another notable process is adsorption, which involves the attachment of molecules in a solution to the surface of soils. It's worth noting that photodegradation is improbable in deeper systems within the soil zone.

9.3 WATER QUALITY PROBLEMS OF MAJOR WORLD RIVERS

Water quality in rivers is negatively affected by several factors, including agricultural practices that involve the use of chemicals such as fertilizers, pesticides, and herbicides that run off into the river. Factories and industries in urban areas may directly discharge untreated or partially treated wastewater into rivers, containing harmful chemicals, heavy metals, and other toxic substances. Other sources of pollution include the discharge of untreated sewage and solid waste, which introduces pathogens, organic waste, and debris into the water. Additionally, rivers used as transportation routes often see vessels discharging waste, oil, and other pollutants directly into the water.

These pollutants from various sources harm aquatic life; for example, excess nutrients can lead to eutrophication, harmful algal blooms, and decreased oxygen levels. Pollutants can also create health hazards, necessitating costly treatment before the water is suitable for drinking and other domestic uses.

In some cases, water salinity increases due to drainage from irrigation fields and the introduction of high-salinity pollutants. Saltwater intrusion at the river's outlet, where ocean or sea water mixes with river water, also contributes to elevated salinity levels. Increases in climate-related sea level rise would certainly worsen the problem.

Sedimentation from soil erosion increases turbidity, reducing light penetration and negatively impacting aquatic plants and organisms. Accumulating sediment can also disrupt water flow and decrease a river's reservoir capacity.

The introduction of non-native plant and fish species further compounds water quality issues. For example, water hyacinth can spread extensively, clogging waterways, reducing oxygen levels, and increasing transpiration loss, which disrupts ecosystems and complicates water treatment.

As discussed in Section 1.8, climate change is altering precipitation patterns, potentially reducing river flows while increasing the frequency and severity of storms. Lower flow rates concentrate pollutants, worsening water quality. Additionally, reduced flow can heighten conflicts over water resources among countries sharing a river basin. More intense storms exacerbate flood risks and sedimentation problems. Floodwaters can introduce additional pollutants into the river, such as those from waste disposal sites.

This section addresses the Nile, the Amazon, and the Mississippi—significant water supply sources with substantial economic and environmental importance. These rivers are globally recognized as some of the longest, yet they face considerable environmental degradation due to issues like overuse, waste disposal, and climate change. Furthermore, they contend with conflicting demands that transcend national or state borders. A summary of key information is provided in Tables 9.1 and 9.2.

TABLE 9.1 Information summary of covered rivers (set 1)

RIVER	LOCATION	LENGTH. RANK	PATHWAY
The Nile	Eastern and northeastern Africa	6,650 km (4,130 miles). Longest.	Flows generally north through the Democratic Republic of the Congo, Tanzania, Burundi, Rwanda, Uganda, Kenya, Ethiopia, Eritrea, South Sudan, Republic of the Sudan, and Egypt
Amazon	Northern South America	6,400 km (4,000 miles). Second longest	Originates in the Peruvian Andes Mountains and flows northern Brazil into the Atlantic Ocean
Mississippi	Covers 31 U.S. states and 2 Canadian provinces	5,970 km (3,710 miles). Fourth longest.	It collects water from 41% of the contiguous United States. Emptying into the Gulf of Mexico

TABLE 9.2 Information summary of covered rivers (set 2)

RIVER	TRIBUTARIES	USE	CONCERNS
The Nile	Ghazal River, the Blue Nile, and the Atbara River	Irrigation and almost all water supply. The Aswan High Dam provides flood protection, hydroelectric power.	Concern over erosion, contamination, conflict between Egypt, the Sudan, and Ethiopia
Amazon	More than 1,000. Major are Ucayali, Tahuyo, Tamshiyaçu, Itaya, Nanay	Trade. Contains the world's most extensive rainforest and is home to an extraordinary diversity of birds, mammals, and other wildlife.	Since the 1960s, the effects of economic exploitation on the region's ecology and the destruction of the rainforest have generated worldwide concern.
Mississippi	Major: Arkansas, Illinois, Missouri, Ohio, and Red rivers	Transportation, industry, and recreation.	Pollution concern from industry in addition to insecticides and fertilizers

9.3.1 The River Nile

The River Nile is not only the longest river in the world but has also played a significant role in the lives of its inhabitants. Sources of hydrological information about the Nile include Britannica (2023c), Shahin (2002), and Sutcliffe and Parks (1999). Useful information and additional publications can be found in Wikipedia Nile (2023) and No Water No Life (2023). The following hydrological summary is based on various resources listed above.

Figure 9.1 depicts the River Nile Basin, encompassing the entire length of the river, spanning approximately 6,650 km (4,130 miles). The basin covers an expansive area of about 3,254,555 km^2 (1,256,591 mile^2), roughly constituting 10% of Africa's total landmass. This extensive basin spans across 11 countries: the Democratic Republic of the Congo, Tanzania, Burundi, Rwanda, Uganda, Kenya, Ethiopia, Eritrea, South Sudan, the Republic of Sudan, and Egypt.

As shown in Figure 9.1, the water in the Nile is sustained by the Blue Nile, the Atbara, and the White Nile. The discharges and water levels of the main Nile stream and its tributaries vary with the seasons, typically rising during the summer due to heavy tropical rains in Ethiopia. In South Sudan, the flood begins as early as April, while its effects reach Aswan, Egypt, in July and peak in October. The river's levels are usually at their lowest from March to May. While the flood is a relatively regular phenomenon, it can occasionally vary in both volume and timing. Before the implementation of water management structures, consecutive years of high or low floods resulted in crop failure, famine, and disease.

Moving northward through Sudan, the river enters Lake Nasser, which spans parts of Egypt and Sudan. The peak of the flood typically enters Lake Nasser in July or August, with an average daily inflow of approximately 25.1 billion cubic feet. Contributions from the Blue Nile, Atbara, and the White Nile account for about 70%, 20%, and 10%, respectively. During low water levels in early May, the discharge is approximately 1.6 billion cubic feet per day, primarily sourced from the White Nile with some input from the Blue Nile. On average, roughly 85% of the water in Lake Nasser originates from the Ethiopian Plateau. This man-made lake, one of the largest in the world, can cover an area of 2,600 mile^2 (6,700 km^2) and has a substantial storage capacity of more than 40 mile^3 (168 km^3). However, due to the region's hot and arid conditions, the lake can lose up to 10% of its volume to evaporation when full, decreasing to around 3% at its minimum capacity.

The last stretch of the river flows north to Cairo and then into the delta, which is formed by silt deposited primarily by the river during flood seasons. Unfortunately, the absence of annual flood waters after the construction of the High Aswan Dam has adversely affected the delta's fertility, leading to increased salinity. The delta extends about 100 miles from north to south and approximately 155 miles at its widest point between Alexandria and Port Said. The river eventually flows into the Mediterranean Sea through two branches: the Rosetta Branch to the west and the Damietta Branch to the east.

FIGURE 9.1   The Nile Basin. Adapted from the Nile basin map by Hel-hama, licensed under the Creative Commons Attribution 3.0 Unported License. (Available at: https://commons. wikimedia.org/wiki/File.png. License details: https://creativecommons.org/licenses/by/3.0/.)

The Nile is the primary water source for Egypt, Sudan, and South Sudan, supporting agriculture and fishing practices. It was historically a vital waterway for transportation, particularly before the widespread use of motor transport. However, with the development of air, rail, and highway infrastructure in the 20th century, dependency on the river for transport has significantly diminished. In Egypt, the Nile also serves as a tourist attraction, with sailing ships used for this purpose.

The construction of the Ethiopian Renaissance Dam has become a source of conflict between Egypt and Sudan on one side and Ethiopia on the other (Climate Diplomacy, 2023; Roussi, 2019). Egypt is especially concerned that the dam will reduce the amount of water it receives from the Nile, particularly given increased demands from population growth and urbanization (Climate Diplomacy, 2023). Egypt claims historical rights to Nile water, whereas Ethiopia asserts its rights based on the fact that 85% of the water originates from highland sources within its territory. Whittington et al. (2014) proposed a strategy to resolve the issue by agreeing on rules for filling the dam's reservoir, operating rules during drought periods, and Ethiopia's rights to develop its water resources infrastructure for the benefit of its people. However, the dam's construction and reservoir filling have progressed to their final stages while negotiations have not been successful. Egypt appears to be in a weaker position, relying on the goodwill of the Ethiopian government without securing substantial international support.

Despite the critical importance of the Nile for its basin countries, water quality deterioration is a major concern. Kipsang et al. (2024) conducted an extensive review revealing that the river has been contaminated by numerous pollutants, including toxic heavy metals and organic contaminants. The water quality is believed to be at or above the World Health Organization's acceptable guidelines for drinking water, agricultural irrigation, and aquatic life support.

Water quality studies primarily focus on analyzing available data to assess contamination levels, identify potential sources of pollutants, and recommend mitigation actions. Commonly assessed parameters include pH, dissolved oxygen, nitrate and phosphate concentrations, salinity (or electrical conductivity), heavy metal concentrations (e.g., lead and mercury), biological oxygen demand, and total suspended solids.

The River Nile is particularly vital for Egypt, providing essential water resources for agriculture, industry, and domestic use. However, this crucial resource faces significant pollution challenges due to industrial discharges, agricultural runoff, and untreated sewage. For example, Abdel-Satar et al. (2017) evaluated several water quality parameters in the river using water quality indices. Samples were collected and analyzed from various locations along a reach, examining physicochemical parameters and heavy metals (including cadmium, lead, and copper). The results indicated significant pollution levels near industrial and agricultural zones.

A similar study by Elnazer et al. (2018) concluded that the samples were unsuitable for drinking due to unacceptable concentrations of calcium, chlorine, cadmium, copper, and lead, as well as chromium. The water was considered suitable for irrigation, except for some samples that contained elevated concentrations of arsenic and chromium. Additionally, Taher et al. (2021) studied spatial variation in the Nile's Damietta branch and found that all water quality parameters were below the permissible limits according to World Health Organization standards, except for turbidity, biochemical oxygen demand[1], and chemical oxygen demand[2] based on Egyptian regulations. The authors used water quality indicators for a set of pollutants to suggest that the branch represents a good to excellent source of drinking water before entering secondary treatment.

The study by Hassan et al. (2023) addressed another contaminant, specifically microplastics, which are tiny plastic particles that either result from the breakdown of larger plastic debris or are manufactured for specific uses. Their research revealed that microplastics were abundant in the water, sediments, fish, and crayfish throughout the study sites. They stressed the need for decision-makers to take appropriate measures to mitigate these risks.

Osman et al. (2015) examined a different form of contamination by screening for multiple hormonal activities in water and sediment using in vitro bioassays and gonadal histology. The results indicated significant negative activities along the studied Nile's course, with downstream sites experiencing more serious pollution than upstream sites due to industrial and anthropogenic activities.

Virtually all studies, including those by Elnazer et al. (2018) and Abdel-Satar et al. (2017), emphasize the necessity of strictly enforcing protection laws by national and international authorities to safeguard the Nile's water quality. These studies highlight the importance of continuous monitoring and pollution control measures aimed at preserving the river's ecosystem and ensuring water safety for various uses. Coordinated efforts are needed to maintain the river's water quality for current and future needs, alongside integrated water management strategies to sustain the river's ecological health. Without immediate action, the degradation of water quality will pose serious risks to biodiversity, human health, and sustainable development, especially in Egypt.

9.3.2 The Amazon

Several publications are available about the Amazon Basin, including Goulding et al. (2003), Britannica (2023a), Chimu (2023), and Wikipedia Amazon (2023). Figure 9.2 illustrates the Amazon Basin, encompassing the river and its main tributaries. The Amazon River is the largest river in South America and boasts the world's largest drainage system in terms of both flow discharge and basin area. Its total length is at least 4,000 miles (6,400 km), making it slightly shorter than the Nile River. Nevertheless, recent advancements in global positioning systems and satellite imagery have led to debates over its exact length, with some suggesting it may be longer than the Nile. The final determination of its length remains a subject of ongoing discussion.

The westernmost source of the Amazon River is located high in the Andes Mountains, within 100 miles (160 km) of the Pacific Ocean, while its mouth empties into the Atlantic Ocean. The Amazon basin, also known as Amazonia, covers an area of approximately 2.7 million square miles (7 million square km). At its widest point, the basin stretches about 1,725 miles (2,780 km) from north to south, encompassing regions in Brazil, Peru, Colombia, Ecuador, Bolivia, and Venezuela (see Figure 9.2). However, the majority of the Amazon's main stream and a significant portion of its basin are located within Brazil.

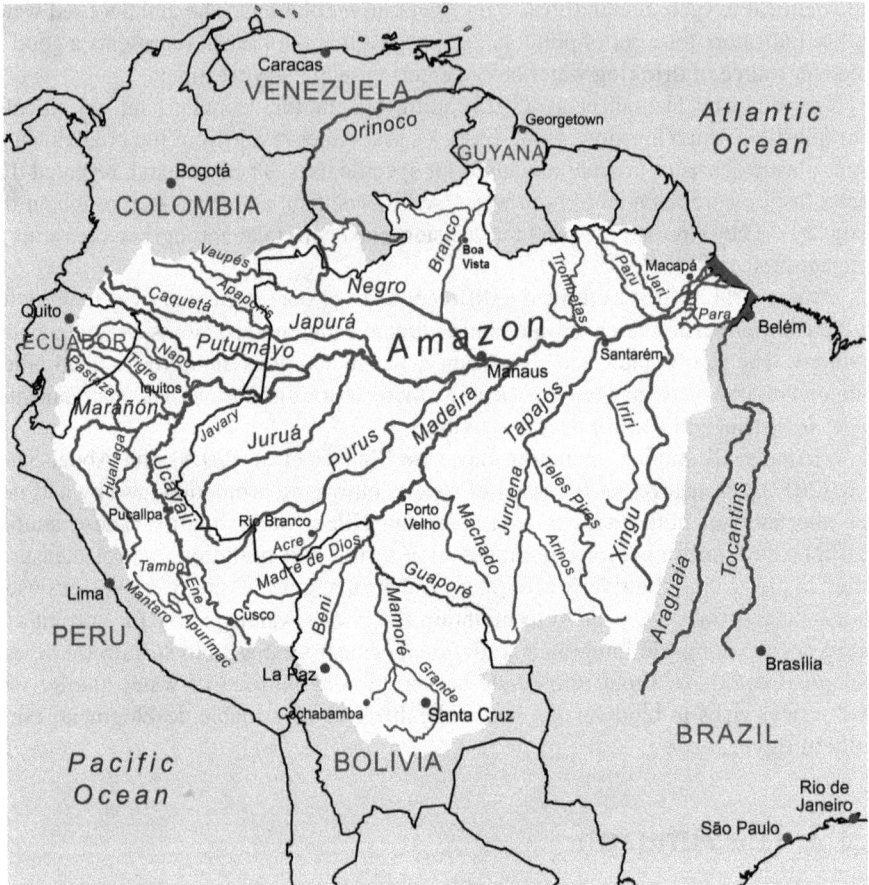

FIGURE 9.2 The Amazon Basin map. Adapted from "Amazonrivermap_fr" by Shannon1, licensed under the Creative Commons Attribution-ShareAlike 4.0 International License. Available at: https://commons.wikimedia.org/wiki/File:Amazonrivermap_fr.svg. License details: https://creativecommons.org/licenses/by-sa/4.0/

The river experiences an average flood-stage discharge of about 49,000 mile^3 (200,000 m^3) per second at its mouth, which is roughly four times that of the Congo River and ten times the volume carried by the Mississippi River. This copious freshwater discharge extends the dilution of the ocean's saltiness for more than 100 miles (160 km) from the shoreline. Sediment from the Amazon River has played a role in the formation of the 200-mile-long (320-km) island of Tupinambarana. Ships, depending on their sizes, can travel upstream as far as 1,300 miles (2,090 km) at any time of the year.

The ecological significance of the river is remarkable, with over two-thirds of the basin covered by an expansive rainforest, representing approximately half of the Earth's remaining rainforests and constituting the single largest reserve of biological resources.

The Amazon River is fed by more than 1,000 tributaries originating from the Guiana Highlands, the Brazilian Highlands, and the Andes. Notably, six of these tributaries, including the Japurá, Juruá, Madeira, Negro, Purus, and Xingu rivers, each exceed 1,000 miles (1,600 km) in length. The Negro River, the largest of all the Amazon tributaries, contributes around one-fifth of the Amazon's total discharge. Its tributaries include the Branco, Vaupés, and Guainía rivers. In its lower reaches, the Negro River widens and becomes filled with islands, reaching widths of up to 20 miles (32 km) in certain areas. The Madeira River, with a discharge of approximately two-thirds that of the Negro River, joins the Amazon below Manaus.

It is estimated that the Amazon deposits 1.3 million tons of sediment per day in the ocean. However, the river is not building a delta because most of that sediment is transported northward by coastal currents and deposited along the coasts of northern Brazil and French Guiana.

The lowland areas of the main river and its tributaries are subject to annual flooding. Fortunately, these floods are not characterized by catastrophic events. This is mainly due to the large size of the basin, the gentle water gradient, and the significant storage capacity of both the floodplain and the estuaries of the river's tributaries. In the upper course of the Amazon, there are two annual floods caused by the rainy seasons from the Peruvian Andes (from October to January) and from the Ecuadorian Andes (from March to July). As we move downstream, these two seasons gradually merge into one, and the river's rise progresses slowly downstream in an enormous wave from November to June. Subsequently, the waters recede by the end of October. Flood levels can reach from 40 to 50 ft (12 to 15 m) above their low levels.

The Amazon basin has long been relatively uninhabited, with areas populated by indigenous groups who have predominantly relied on hunting and small-scale agriculture with low yields. The native population has dwindled over the years due to development and commercial exploitation of their lands, resulting in the breakdown of native life. Many indigenous groups were enslaved by early European explorers, especially during organized raids from the 16th to the 18th century. Additionally, many succumbed to European diseases such as influenza, measles, and smallpox. Survivors fled into increasingly remote and inaccessible sections of the Amazon basin.

Since World War II, the economic development of the Amazon basin has included road construction, oil discoveries, and timber harvesting. Commercial exploitation has targeted tropical hardwoods, river fish, and clandestinely produced cocaine, along with livestock operations. These activities have led to the widespread displacement of indigenous groups, who have either been forced onto new reserves or left to barely survive. In the mid-1950s, Brazil decided to construct a new inland capital, Brasília, leading to plans for massive road and highway construction and efforts to create new settlements. However, these goals were not fully realized.

Developments also include upland small plantations for crops such as rice, manioc, corn, cacao, coffee, nuts, and black pepper. Corporate farming and agro-forestry operations have had limited success, primarily due to poor soil fertility in some locations. Cattle pastures dominate land use on cleared parts of the Amazon basin. Successful operations involve high-quality timbers like mahogany, Amazonian cedar, and many other species. Other trees yield perfumes, flavorings, and pharmaceutical ingredients. Rubber, a valuable commodity, is obtained from both wild trees and those grown in small plantations.

Mining, as seen in the rich mineral complex of the Serra dos Carajás area west of the town of Marabá, Brazil, is highly profitable but has had harmful effects on the environment. This site is considered one of the world's largest and richest iron ore deposits, also producing gold, copper, nickel, manganese, and tin. Unfortunately, the site requires the clearing of thousands of acres of forest annually to provide charcoal for the mining operation. Additionally, large amounts of mercury used to extract gold in Brazilian sites are released into the river, leading to health-hazard fish contamination. Consequently, mercury contamination has grown among Amazonian peoples, especially those in more isolated groups that consume large amounts of fish.

Hydroelectric power stations were constructed to meet the energy requirements of the Brazilian Carajás development and the cities of Belém and Manaus. Growing concerns about the negative environmental effects of constructing large dams have placed other projects on hold.

The principal oil developments within Amazonia have occurred in the Oriente regions of the Andean countries. Necessary oil pipelines run through Colombia, Ecuador, northeastern Peru, and end at export terminals on the Pacific coast. Developments have had minimal consequences within the Brazilian and Bolivian portions of the basin.

Unfortunately, human activities have increasingly threatened the equilibrium of the forest's complex ecology. Deforestation has accelerated, and mineral discoveries have opened the basin to new settlers and corporations. Notable deforestation has occurred in areas of Brazil, Colombia, and Ecuador. The cultivation of coca for the illicit production of cocaine continues to stimulate such activities in western Amazonia. However, it is challenging to quantify the basin-wide extent and rate of deforestation in some instances due to the difficulty of distinguishing between regenerating secondary vegetation and undisturbed areas. More recently, the use of radar has improved the precision of investigations, which can aid future conservation plans.

The Amazon Basin plays a crucial role in the global carbon cycle and climate system (Rosen et al., 2024). Fisher et al. (2008) examined the impacts of persistent deforestation and climate change on the region's carbon and water cycles. Changes in the Amazon Basin's carbon balance and hydrological processes have significant implications for global climate regulation (Esteves, 1998; Barros et al., 2014). In this regard, Rosen et al. (2024) concluded that the Brazilian Amazon has been a net carbon source during recent climate extremes and that the south-eastern Amazon was a net land carbon source over their study period. Overall, the results point to increasing human-induced disturbances (deforestation and forest degradation by wildfires) and reduction in the old-growth forest sink during drought. Moura et al. (2016) further emphasized major alterations in carbon sequestration and release patterns, with critical consequences for regional and global climate dynamics.

In general, most available studies address the hydrological aspects of the Amazon River and its water quality. Issues such as land use, climate change, and the assessment of specific contaminants are prominent in this research. Many studies highlight the need for sustainable land management practices by providing critical information to policymakers and environmental managers, aiming to balance development with environmental protection. For example, the study by McClain et al. (1997) investigates the sources, composition, and transport of dissolved organic matter in the Amazon River. This research integrates hydrology, biogeochemistry, and ecology to explore how organic carbon and nutrients are transferred from land to streams and rivers, considering both natural processes and anthropogenic influences. Similarly, Moreira-Turcq et al. (2003) examine the characteristics and behavior of organic matter in the mixing zone of the Rio Negro and Rio Solimões in the Amazon Basin. Unlike McClain et al., this study also considers the composition of particulate matter. The study findings emphasize the important role of water mixing processes on biogeochemical dynamics and the overall carbon cycle in the Amazon River system.

Recognizing that the Amazon River is one of the largest rivers in the world in terms of discharge and sediment transport, the study by Filizola and Guyot (2011) examines sediment load and its impact on water quality, focusing on suspended sediment. By analyzing records from monitoring stations, the study emphasizes the relationship between sediment, the hydrological cycle, climate events, and human activities. The results show significant seasonal variations, with peak sediment loads occurring during the rainy season. Regional differences in sediment transport are observed, influenced by tributary input and land use changes.

International concern about the ecological consequences of continuing deforestation and other development-related activities has been growing. Consequently, efficiently absorbing carbon dioxide would decrease contributions to global warming through the greenhouse effects. Such activities may also reduce the region's evapotranspiration, disrupting the hydrologic cycle by increasing surface runoff and reducing groundwater recharge. Further, deforestation can negatively affect the unique gene pool of the Amazon Rainforest, which houses perhaps two-thirds of the world's known organisms. Particular concern has been placed on the threat to biodiversity and the potential loss of the basin's as-yet-unknown and unexploited resources, such as pharmaceuticals. Finally, the survival and welfare of the region's indigenous peoples, who are an integral part of the rainforest's ecosystem, are also at stake.

International concern about deforestation led to the United Nations Conference on Environment and Development, held in Rio de Janeiro in 1992 (UN, 1992). The concern was based on the view that the Amazon basin is a global resource, serving as a control mechanism for the world's climate and as a genetic repository for the future. Understandably, the countries of the region viewed such calls as a challenge to their national sovereignty. It is hoped that future efforts will secure more support and cooperation from all concerned parties.

9.3.3 The Mississippi

Publications related to the Mississippi River include Morris (2012), Kunkel et al. (1994), Galat and Frazier (1996), Wiener et al. (1984), Barry (1993), and Wiener et al. (1984). Online resources are also available, such as Kammerer (May 1990), Dempsey (2018), EPA (2023), Nature Conservancy (2023), National Park Service (2023a, 2023b), and Britannica (2023b). The following summary is based on these resources.

FIGURE 9.3 The Mississippi Basin. Adapted from Mississippiriver-new-01by Shannon1, licensed under the Creative Commons Attribution-Share Alike 4.0 International License. Available at https://commons.wikimedia.org/wiki/File:Mississippiriver-new-01.png. License details https://creativecommons.org/licenses/by-sa/4.0/

Figure 9.3 illustrates the Mississippi River Basin, which includes the river and its primary tributaries. The river is recognized as the fourth-longest river in the world. Although there are some discrepancies in the exact length, it is approximately 3,710 miles (5,970 km) long, stretching from Lake Itasca in northern Minnesota to its delta in southern Louisiana, where it empties into the Gulf of Mexico. The average discharge near Baton Rouge, Louisiana, is approximately 593,000 cubic feet per second (16,800 cubic meters per second). The basin encompasses more than 1,245,000 mile^2 (490,000 km^2) and includes all or parts of 31 states and two Canadian provinces.

The lower alluvial valley of the river is a relatively flat plain covering about 35,000 mile^2 (13,700 km^2) along the river. This valley starts just below Cape Girardeau, Missouri, is approximately 600 miles (375 km) in length, with varying widths ranging from 25 to 125 miles (15.6–78 km). It encompasses parts of seven states, including Missouri, Illinois, Tennessee, Kentucky, Arkansas, Mississippi, and Louisiana.

Major tributaries of the Mississippi River include the following rivers: St. Croix, Wisconsin, Rock, Illinois, Kaskaskia, Ohio, Yazoo, Big Black, Minnesota, Des Moines, Missouri, White, Arkansas, Ouachita, Red, and Atchafalaya.

Historically significant events include the floods of 1849 and 1850, which caused extensive damage in the Mississippi River Valley. Consequently, in 1879, the US Congress established the Mississippi River Commission and tasked it with safeguarding the riverbanks, enhancing safety and navigation, facilitating commerce and trade, and protecting against catastrophic floods. However, more major floods occurred in 1912, 1913, and 1927, with the latter being the most devastating in Lower Mississippi Valley history. It resulted in levee breaches, inundating cities, towns, and farms, destroying crops, paralyzing industries and transportation, and incurring substantial monetary losses. This calamity claimed over 200 lives and displaced over 600,000 people. In response, the Flood Control Act of 1928 was enacted, committing the federal government to a concrete flood control program. This legislation authorized the Mississippi River and Tributaries Project, the nation's first comprehensive flood control and navigation initiative.

From an economic perspective, river navigation saw gradual improvements, culminating in a channel depth of nine feet (3 m) by 1930. This was followed by the construction of 26 locks and dams spanning from Minneapolis, Minnesota, to St. Louis, Missouri. The river now annually supports approximately $400 billion in commercial activity and sustains 1.3 million jobs. About 500 million tons of goods are transported through the river each year, a substantial portion being corn and soybeans from Iowa and other Midwestern farms. Roughly 60% of all grain exported from the U.S. is shipped via the Mississippi River. The river also serves as a source of drinking water for millions of inhabitants in 50 cities. Moreover, the river basin plays a pivotal role in providing essential habitat for numerous fish and wildlife species.

Over the years, the Mississippi River's characteristics have undergone significant changes due to extensive engineering practices. The river's meanders and floodplains have also been altered because of extensive agricultural and urban development activities. Many adjacent surface water bodies, including wetlands, riparian zones, and tributaries, have been disconnected from the river due to these engineering modifications. This has had a detrimental impact on the ecology of the basin by reducing habitat for native plants and animals, consequently diminishing the biological productivity of the entire river basin. Furthermore, the channelization has diminished the capacity of the coastal marshes in Louisiana to mitigate erosion and the impact of Gulf storms, as they typically absorb some of the flow energy of the water.

Various activities have also negatively affected the basin's ecology, leading to increased pollution, bank erosion, turbidity, sediment resuspension, and the disruption of native species' habitats. The Mississippi River is regarded as one of the most polluted rivers in the United States, largely due to contaminants from agricultural runoff, sewage treatment plants, and industrial facilities. Agricultural runoff and partially treated or untreated human waste introduce pesticides, nutrients, other chemicals, and pathogens into the river. Industrial facilities release a variety of pollutants, including heavy metals and toxic chemicals. Water quality standards for mercury, bacteria, sediment, polychlorinated biphenyl, and nutrients are exceeded in some stretches of the Mississippi River. Additionally, newly introduced contaminants such as microplastics and pharmaceuticals have been discovered in the river, rendering the water unsuitable for fishing, swimming, and drinking. These pollutants also contribute to habitat loss.

A host of studies have addressed the water quality of the Mississippi River, particularly concerning chemicals of concern that significantly affect this quality. These studies examine sources of contaminants, levels of contamination, and potential negative impacts on health and the environment.

Meade (1995) identifies various contaminants, including pesticides, herbicides, heavy metals, and industrial chemicals. The study also identifies the sources of these contaminants, their concentrations in different parts of the river, and trends over time. Battaglin and Goolsby (1994) take a different approach by focusing on agricultural chemical use, land use, and cropping practices in the Mississippi River Basin. Their data are used to assess patterns of chemical pollution in the river that correlate with agricultural activities. These findings can assist in managing chemical inputs and mitigating their impacts on water quality.

Pereira et al. (1996) also focus on agricultural practices, specifically on the occurrence and transport of herbicides such as atrazine, cyanazine, and alachlor, and their degradation. The study compiles data on chemical concentrations, distribution, and persistence patterns in water and sediment samples in the lower Mississippi River and its tributaries. It also highlights the seasonal variations of various chemicals and the factors influencing their transport, underscoring the potential risks to water quality and aquatic life, particularly in areas with intensive agricultural activity.

Turner and Rabalais (1991) expand the scope by including nutrient contamination, particularly nitrogen and phosphorus, in addition to pesticides and herbicides, all related to intensified agricultural practices. Goolsby and Battaglin (2000) and Alexander et al. (2000) assess nitrogen sources and the amount of nitrogen transported to the Gulf of Mexico from the Mississippi River, which delivers nearly all of the fresh water flowing into the Gulf. These sources include fertilizers, atmospheric deposition, and animal waste. Turner and Rabalais (1991) also address the ecological consequences of increased nutrient concentrations, such as the development of hypoxic zones in the Gulf of Mexico. Hypoxia refers to a condition where there is a deficiency of oxygen in the water, leading to "dead zones" where oxygen levels are so low that most marine life cannot survive. This phenomenon has a detrimental impact on marine life, including the lucrative commercial shrimp fishery.

Rabalais et al. (2002) extensively discuss hypoxia in the Gulf of Mexico, focusing on its ecological impacts and the challenges of translating scientific knowledge into policy. Like Turner and Rabalais (1991), the authors advocate for comprehensive management strategies to reduce contaminant inputs, emphasizing the need for collaboration between scientists, policymakers, and stakeholders.

Climate change further exacerbates the environmental challenges facing the river. Aging infrastructure is ill-equipped to handle the shifts in precipitation patterns driven by climate change. This leads to more frequent and severe flooding along the Mississippi and its floodplain, carrying excess sediment and chemicals into the river.

States like Iowa have adopted nutrient reduction strategies to reduce fertilizer runoff into the Mississippi River. For example, in 2018, Iowa lawmakers allocated \$282 million over 12 years to fund water quality initiatives, including measures such as growing cover crops, building bioreactors, and restoring wetlands to help farmers reduce nutrient loss to the river (Des Moines Register, 2022).

9.4 ASSIGNMENT

Rivers can be classified based on various characteristics, including length, size of the drainage area, average discharge, the number of countries in the drainage basin, and their economic, ecological, cultural, and religious significance. Global influence, as well as the severity of conflicts within or across borders, are also important factors. Major rivers and their main tributaries include the Yangtze (China), Congo (spanning the Republic of Congo and several other African nations), Mekong (China and other Asian countries), Rhine (Germany and several European countries), and Brahmaputra (India, China, and other Asian countries).

Select one of these rivers, or another of your choice, and prepare a report following the same structure as the sections above. Your discussion should cover the geography and hydrology of the river, its significance and ecological importance, specific water quality issues, and other challenges.

NOTES

1 A measure of the amount of oxygen that microorganisms in water will consume as they break down organic matter. High BOD levels suggest a high concentration of biodegradable organic pollutants, which can lead to oxygen depletion in aquatic environments, harming fish and other organisms.
2 A measure of the total amount of oxygen required to chemically oxidize both organic and inorganic compounds in water, providing a broader indication of water quality.

REFERENCES

Abdel-Satar, A. M., Ali, M. H., & Goher, M. E. (2017). Indices of water quality and metal pollution of Nile River, Egypt. *Egyptian Journal of Aquatic Research*, *43*(1), 21–29. https://doi.org/10.1016/j.ejar.2016.12.006.

Alexander, R. B., Smith, R. A., & Schwarz, G. E. (2000). Effect of stream channel size on the delivery of nitrogen to the Gulf of Mexico. *Nature, 403*(6771), 758–761. https://www.nature.com/articles/35001562

Barros, N., Cole, J. J., Tranvik, L. J., Prairie, Y. T., Bastviken, D., Huszar, V. L. M., de Araújo Lima, M., & Roland, F. (2014). Water quality and seasonal variations in the Amazon River and its tributaries. *Journal of Hydrology, 519*, 2430–2445. https://doi.org/10.1016/j.jhydrol.2014.08.056

Barry, J. M. (1993). *Rising Tide: The Great Mississippi Flood of 1927 and How It Changed America*. New York, NY: Simon & Schuster.

Battaglin, W. A., & Goolsby, D. A. (1994). Spatial data in geographic information system format on agricultural chemical use, land use, and cropping practices in the United States. U.S. Geological Survey Water-Resources Investigations Report 94-4176. https://doi.org/10.3133/wri944176.

Britannica. (2023a). Amazon River. Retrieved from https://www.britannica.com/place/Amazon-River

Britannica. (2023b). Mississippi River. Retrieved from https://www.britannica.com/place/Mississippi-River

Britannica. (2023c). Nile River. Retrieved from https://www.britannica.com/place/Nile-River

Chimu. (2023). 15 Facts About the Amazon River That'll Blow Your Mind. Retrieved from https://www.chimuadventures.com/blog/2022/09/amazon-river/

Climatic Diplomacy. (2023). Dispute over Water in the Nile Basin. Retrieved from https://climate-diplomacy.org/case-studies/dispute-over-water-nile-basin

Dempsey, C. (2018). Geography facts about the Mississippi Watershed. Retrieved from https://www.geographyrealm.com/geography-facts-mississippi-watershed/

Des Moines Register. (2022). Pollution and habitat loss make Mississippi River among nation's most endangered. *Des Moines Register*. Retrieved from https://www.des-moinesregister.com/story/money/agriculture/2022/04/19/mississippi-river-map-endangered-american-rivers-list/7332940001/

Elnazer, A. A., Mostafa, A., Salman, S. A., Arafa, H. M., & Alharbi, A. D. (2018). Temporal and spatial evaluation of the River Nile water quality between Qena and Sohag Cities, Egypt. *Bulletin of the National Research Centre, 42*, 3, https://doi.org/10.1186/s42269-018-0005-6.

EPA. (2023). The Mississippi/Atchafalaya River Basin (MARB). Retrieved from https://www.epa.gov/ms-htf/mississippiatchafalaya-river-basin-marb

Esteves, F. A. (1998). *Fundamentos da Limnologia*. Rio de Janeiro: Editora de Universidade.

Fick, A. (1855). On liquid diffusion. *Philosophical Magazine, 10*(63), 30–39. https://www.tandfonline.com/doi/abs/10.1080/14786445508641925

Filizola, N., & Guyot, J. L. (2011). Suspended sediment yields in the Amazon Basin: A first estimation. *IAHS-AISH Publication, 367*, 60–68.

Fisher, J. B., Tu, K. P., & Baldocchi, D. D. (2008). The importance of soil moisture in the Amazon rainforest. *Geophysical Research Letters*, *35*(18), L18405. https://doi.org/10.1029/2008GL034706

Galat, D. L., & Frazier, A. G. (1996). Overview of river-floodplain ecology in the Upper Mississippi River Basin. In: Kelmelis, J. A. (ed.), *Science for Floodplain Management into the 21st Century* (Vol. 3). U.S. Government Printing Office. https://observatoriopantanal.org/wp-content/uploads/crm_perks_uploads/5cb0f734750a11456042675850236/2019/08/1996_Overview_of_River_Floodplain_Ecology_in_the_Upper_Mississippi_River_Basin.pdf

Goolsby, D. A., & Battaglin, W. A. (2000). Nitrogen in the Mississippi Basin—Estimating sources and predicting flux to the Gulf of Mexico. U.S. Geological Survey Fact Sheet 135–00. Retrieved from https://pubs.usgs.gov/fs/2000/0135/report.pdf

Goulding, M., Barthem, R., & Ferreira, E. J. G. (2003). *Smithsonian Atlas of the Amazon*. Smithsonian Books. https://www.scirp.org/reference/referencespapers?referenceid=1892083

Hassan, Y. A. M., Badrey, A. E. A., Osman, A. G. M., & Mahdy, A. (2023). Occurrence and distribution of meso- and macroplastics in the water, sediment, and fauna of the Nile River, Egypt. *Environmental Monitoring and Assessment, 195*(9), 1130. https://doi.org/10.1007/s10661-023-11367-7

Hel-hama. (2018). Nile basin map [Map]. Wikimedia Commons. https://commons.wikimedia. org/wiki/File:Nile_basin_map.png. Licensed under the Creative Commons Attribution 3.0 Unported License. https://creativecommons.org/licenses/by/3.0/

Kammerer, J. C. (1990). Largest Rivers in the United States. *U.S. Geological Survey.* https://pubs. usgs.gov/of/1987/ofr87-242/

Kipsang, N. K., Kibet, J. K., & Adongo, J. O. (2024). A review of the current status of the water quality in the Nile water basin. *Bulletin of the National Research Centre, 48*, 30. https://doi. org/10.1186/s42269-024-01186-2

Kunkel, K. E., Changnon, S. A., & Angel, J. R. (1994). Climatic aspects of the 1993 Upper Mississippi River Basin flood. *Bulletin of the American Meteorological Society, 75*(5), 811–822. https://doi.org/10.1175/1520-0477(1994)075<0811:CAOTUM>2.0.CO;2

McClain, M. E., Richey, J. E., Brandes, J. A., & Pimentel, T. P. (1997). Dissolved organic matter and terrestrial–lotic linkages in the central Amazon basin of Brazil. *Global Biogeochemical Cycles, 11*(3), 295–311. https://agupubs.onlinelibrary.wiley.com/doi/pdf/10.1029/97GB01056

Meade, R. H. (1995). Contaminants in the Mississippi River, 1987–92. *U.S. Geological Survey Circular* 1133. https://pubs.usgs.gov/circ/circ1133/

Moreira-Turcq, P., Seyler, P., Guyot, J. L., & Etcheber, H. (2003). Exportation of organic carbon from the Amazon River and its main tributaries. *Hydrological Processes, 17*(7), 1329–1344. https://doi.org/10.1002/hyp.1287

Morris, C. (2012). *The Big Muddy: An Environmental History of the Mississippi and its Peoples from Hernando de Soto to Hurricane Katrina.* Oxford University Press. https://drew.locate. ebsco.com/instances/3d136644-a2a2-409f-a67c-8b4b02042e0d?option=subject&query= Mississippi%20River%20Valley--History

Moura, M. A., dos Santos, V. M., & Almeida, R. M. (2016). Long-term water quality trends in the Amazon River Basin. *Environmental Monitoring and Assessment, 188*(8), 1–14. https://doi. org/10.1007/s10661-016-5470-5

National Park Service. (2023a). Mississippi River facts. Retrieved from https://www.nps.gov/ miss/riverfacts.htm

National Park Service. (2023b). Water quality in the Mississippi River. Retrieved from https:// www.nps.gov/miss/learn/nature/waterquality.htm

Nature Conservancy. (2023). Mississippi River basin. Retrieved from https://www.nature.org/ en-us/about-us/where-we-work/priority-landscapes/mississippi-river-basin/

No Water No Life. (2023). Nile River Basin. Retrieved from https://nowater-nolife.org/ nile-river-basin/

Pereira, W. E., Domagalski, J. L., & Hostettler, F. D. (1996). Occurrence and transport of herbicides and their degradation products in the lower Mississippi River and its tributaries. *Environmental Science & Technology, 30*(2), 275–283. https://pubs.usgs.gov/publication/70015779

Rabalais, N. N., Turner, R. E., & Scavia, D. (2002). Beyond science into policy: Gulf of Mexico hypoxia and the Mississippi River. *BioScience, 52*(2), 129–142. https://doi.org/10.1641/00 06-3568(2002)052[0129:BSIPGO]2.0.CO;2

Rosan, T. M., Sitch, S., O'Sullivan, M., ... et al. (2024). Synthesis of the land carbon fluxes of the Amazon region between 2010 and 2020. *Communications Earth & Environment, 5*, 46. https://doi.org/10.1038/s43247-024-01205-0

Roussi, A. (2019). Gigantic Nile dam prompts clash between Egypt and Ethiopia. *Nature, 574*, 159–160. https://doi.org/10.1038/d41586-019-02987-6

Shahin, M. (2002). *Hydrology and Water Resources of Africa.* Springer. https://www.scirp.org/ reference/referencespapers?referenceid=2959836

Shannon1. (2008). Amazonriverbasin basemap [Map]. Wikimedia Commons. https://commons. wikimedia.org/wiki/File:Amazonriverbasin_basemap.png. Licensed under the Creative Commons Attribution-Share Alike 4.0 International License. https://creativecommons.org/ licenses/by-sa/4.0/

Shannon1. (2016). Mississippiriver-new-01 [Map]. Wikimedia Commons. https://commons.wikimedia.org/wiki/File:Mississippiriver-new-01.png. Licensed under the Creative Commons Attribution-Share Alike 4.0 International License. https://creativecommons.org/licenses/by-sa/4.0/

Sutcliffe, J. V., & Parks, Y. P. (1999). *The Hydrology of the Nile* (IAHS Special Publication No. 5, p. 161). IAHS Press. https://www.scirp.org/reference/referencespapers?referenceid=2223261

Taher, M. E. S., Ghoneium, A. M., Hopcroft, R. R., & ElTohamy, W. S. (2021). Temporal and spatial variations of surface water quality in the Nile River of Damietta Region, Egypt. *Environmental Monitoring and Assessment, 193*(3), 128. https://doi.org/10.1007/s10661-021-08919-0

Turner, R. E., & Rabalais, N. N. (1991). Changes in Mississippi River water quality this century. *BioScience, 41*(3), 140–147. https://doi.org/10.2307/1311453

Osman, A. G. M., AbouelFadl, K. Y., Krüger, A., & Kloas, W. (2015). Screening of multiple hormonal activities in water and sediment from the river Nile, Egypt, using in vitro bioassay and gonadal histology. *Environmental Monitoring and Assessment, 187*(6), 1–16. https://doi.org/10.1007/s10661-015-4553-z

United Nations. (1992). United Nations Conference on Environment and Development, Rio de Janeiro, 1992. Retrieved from https://www.un.org/en/conferences/environment/rio1992

Whittington, D., Waterbury, J., & Jeuland, M. (2014). The Grand Renaissance Dam and prospects for cooperation on the Eastern Nile. *Water Policy, 16*(4), 595–608. https://doi.org/10.2166/wp.2014.011b

Wiener, J. G., Anderson, R. V., & McConnville, D. R. (eds.). (1984). *Contaminants in the Upper Mississippi River: Proceedings of the 15th Annual Meeting of the Mississippi River Research Consortium*. Butterworth Publishers. https://pubs.usgs.gov/publication/85491

Virtual Laboratory Experiments

10

10.1 INTRODUCTION

This chapter presents virtual laboratory experiments corresponding to various topics covered in this textbook. Each experiment includes setup instructions, procedures, and example data sets. This information is also useful for establishing a physical laboratory when resources are available. Completing the laboratory work requires student teamwork under the instructor's supervision. Tasks should be assigned to individual students to ensure successful completion and guarantee each student's participation. Multiple runs may be necessary to promote participation and reduce the chances of human error. Specific tasks include timing with a stopwatch, monitoring water levels in tanks and piezometer tubes or manometers, and collecting water volumes for discharge rate estimation. In experiments where the water level changes rapidly, it is recommended to use a marker to indicate the water level, either directly on the instrument (such as a sand tube for experiment 5) or on an attached paper strip (e.g., next to a manometer tube for experiments 8 and 10) for later recording.

10.2 EXPERIMENT 1: FLOW OVER A WEIR

As discussed in Chapter 5, Section 5.2.2.2, weirs are structures built across small to medium-sized streams, typically spanning a few meters, to measure flow. They also provide a stable backwater for supporting power or irrigation practices. Accurate measurements require that all the stream's water flows over the weir. Various types of weirs, including rectangular and V-notch shapes, can be used. The water level above the weir's crest is measured, and an appropriate equation is applied to estimate the flow rate.

DOI: 10.1201/9781003587149-10

10.2.1 Equipment and Procedure

The experiment requires equipment that is readily available from commercial vendors, such as Armfield (2020) and Edibon (2023). As outlined in the Applied Fluid Mechanics Lab Manual (2023), the photo in Figure 10.1 depicts a flume equipped with an inlet nozzle and a stilling baffle designed to ensure smooth flow conditions. The nozzle, not shown in the figure, is positioned at the upstream end of the flume. The system should also include a point gauge to measure the water depth above the base of the notch, located at the near edge of the flume. Various types of weir notches can be installed, including the V-notch shown in the figure, as well as other options like a rectangular notch.

The experimental procedure, closely following the guidelines outlined in the Applied Fluid Mechanics Lab Manual (2023), is carried out as follows:

1. Start the pump to fill the channel upstream of the weir with water. Then stop the pump once water begins to flow over the weir. Wait a few minutes to allow the water to stabilize.
2. Measure the water level in the channel (h_o) by adjusting the point gauge to match the stabilized level.
3. Move the sliding mast upwards by a selected increment above h_o and record this precise reading as h. The size of the increment, usually between 1 and 5 cm, depends on the type of weir and the flow rate.

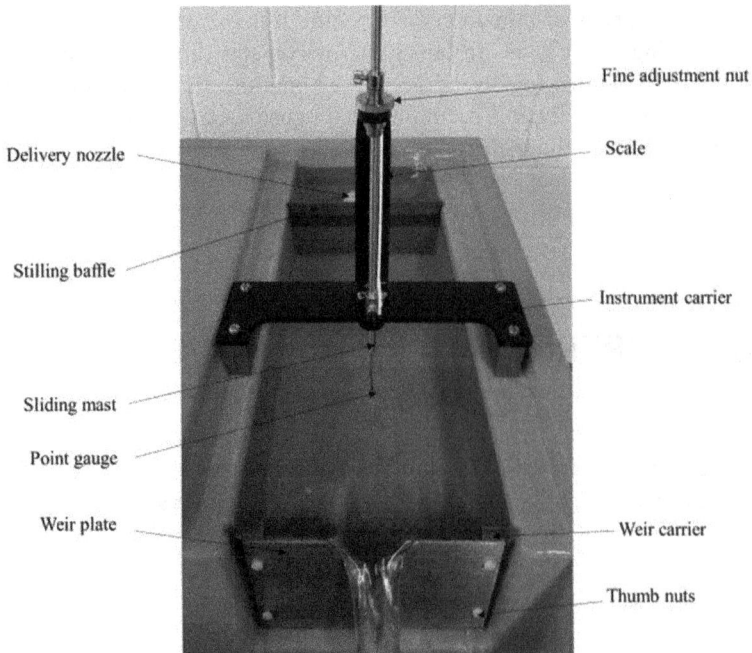

FIGURE 10.1   Weir instrument.

4. Turn on the pump and adjust the flow until the water level aligns with the point gauge. Confirm that the water level has stabilized before taking readings.
5. Measure the water volume and record the collection time to estimate the flow rate (equals the collected volume divided by the time of collection).
6. Repeat steps 3 through 5 by gradually increasing h by the selected increment as the flow increases.
7. Complete the calculations in a table format similar to Table 10.1.
8. The theoretical discharge can be estimated using the appropriate equation based on the type of weir (refer to Section 5.2.2.2). For convenience, the equations are repeated here.

   For a rectangular weir, the equation reads

$$Q = 1.84(L - 0.2H)H^{2/3} \tag{10.1}$$

   In equation (10.1), Q is the discharge rate in cubic meters per second, L is the length of the weir crest in meters, and H is the height of the water above the crest also in meters. For the V-notch weir, equation (10.2) applies

$$Q = 1.379H^{2.5} \tag{10.2}$$

   which is valid for a 90° notch.
9. For each test step, determine the coefficient of discharge by dividing the measured discharge by the theoretical value.

10.2.2 Example Data Set and Reporting Instructions

Table 10.1 presents data from a rectangular weir experiment. Using h_o of 0.1 m and L of 0.04 m, complete the calculations in the table to estimate the coefficient of discharge. Graphically illustrate the relationship between H and both the measured and theoretical discharges, as well as the coefficient of discharge. Provide commentary on the results. The report should also include a detailed outline of the calculation steps.

10.3 EXPERIMENT 2: ESTIMATING MANNING COEFFICIENT FROM A FLUME EXPERIMENT

As described in Section 5.2.2.3, Manning's equation is used to calculate water velocity in open channels as a step toward estimating the flow rate. The equation is expressed as

$$V = k_m \frac{R^{2/3}S^{1/2}}{n} \tag{10.3}$$

TABLE 10.1 Record of measured and calculated data for a weir experiment

TEST	MEASURED VOLUME (m³)	TIME OF COLLECTION (S)	h (M)	H=h−h₀ (M)	MEASURED DISCHARGE (m³/S)	THEORETICAL DISCHARGE (m³/S)	COEFFICIENT OF DISCHARGE
1	0.06	26	0.1136				
2	0.06	23	0.1182				
3	0.06	20	0.1209				
4	0.06	18	0.1027				
5	0.06	24	0.1161				
6	0.06	16	0.1355				

In equation (10.3), V is the average velocity, S is the water slope (rise over run; which can be approximated by the channel slope because the flow is under gravity), and R is the hydraulic radius, defined by equation (10.4):

$$R = \frac{A}{p} \tag{10.4}$$

where A and p are the cross-sectional area and wetted perimeter, respectively, and n is the Manning roughness coefficient. The factor K_m equals unity for meter-second units and 1.49 for foot-second units. For example, for V in meters per seconds, the area A should be in square meters, R and p in meters, the slope S is dimensionless, and n is dimensionless (unitless).

10.3.1 Equipment and Procedure

Various types of flumes are discussed in LibreTexts (2023). A complete flume setup is available from commercial vendors, including GUNT (2023) and EME (2018). The essential features of such setups include the ability to adjust the flume slope, options for controlling and measuring discharge, and the capacity to measure water depth at various points along the channel.

The following steps are adopted for the experimental procedure:

1. Follow the flume's instructions to run the apparatus and wait for the water in the flume to stabilize at a certain flow.
2. Set and record the slope of the flume.
3. Record the discharge reading.
4. Measure the depth at three different locations for the measured discharge.
5. For the same slope of the flume, repeat steps 3 and 4 for multiple runs, and record the discharge and water depths for each.
6. Complete the calculations in a table similar to Table 10.2 to estimate Manning's coefficient.

10.3.2 Example Data Set and Reporting Instructions

Table 10.2 provides data for estimating Manning's coefficient n, with the first row completed as a guide. In the table, d_{ave} is the average of the water depths d_1, d_2, and d_3, and n is calculated using the Manning equation with the corresponding velocity (V), hydraulic radius (R), and slope (S). The flume width is $b = 30\,\text{cm}$. The area is estimated by multiplying d_{ave} by b, and the wetted perimeter (p) is calculated as $b + 2d_{ave}$ for a rectangular cross section. Complete the calculations in the centimeter-second system as listed in the table. Just before calculating n, convert V to meters per second and R to meters by dividing both by 100, as the Manning equation requires meter-second units. Alternatively, you can convert all the data in the table to meter-second units first, although this approach is more tedious.

TABLE 10.2 Example measured flume sata

RUN	CHANNEL SLOPE S (DIMENSIONLESS)	DISCHARGE Q (cm³/s)	d_1 (cm)	d_2 (cm)	d_3 (cm)	d_{AVE} (cm)	AREA A (cm²)	WETTED PERIMETER P (cm)	R=A/p (cm)	V=Q/A cm/s	MANNING COEFFICIENT n
1	0.002	7,998	3.03	4.75	5.10	4.29	128.8	38.59	3.34	62.10	0.007466
2	0.002	9,795	3.93	5.37	5.70						
3	0.002	11,311	4.23	5.96	6.10						
4	0.002	12,646	4.61	5.94	6.47						
5	0.002	13,853	4.96	7.00	7.04						
6	0.002	15,996	5.43	6.66	7.28						

Source: From Scribd (2009).

The experiment report should include a sketch of the apparatus, based on the sources cited in the experimental procedure, along with calculation steps and a discussion of the results. Additionally, plot Manning's coefficient against velocity and document any observations regarding the findings.

10.4 EXPERIMENT 3: POROSITY OF GRANULAR MATERIAL

As discussed in Chapter 6 (Section 6.2), porosity is the ratio of the volume of void spaces in a rock or sediment to the total volume of the rock or sediment. The porosity of unconsolidated material depends on factors such as grain size distribution, grain shape, and packing.

10.4.1 Equipment and Procedure

Figure 10.2 illustrates some of the essential supplies for this and other porous media experiments, including glass beakers, graduated cylinders, and plastic tubes. A variety of these items in different sizes should be readily available.

In this specific experiment, metal shots are used as surrogates for granular material (see Figure 10.3). This choice is primarily driven by safety considerations, as they can be dried in an oven without significant risks. Additionally, their relatively large size makes it feasible to estimate the volume of each shot, which would be impractical with real soils. To conduct this experiment, you will need the following equipment: an oven, a scale, cylinders, and glass beakers. You will also require two types of sands with contrasting textures.

The main objectives of this experiment are as follows:

1. Determine the porosity of metal shots using three different methods: oven-drying, volumetric calculation, and water displacement.
2. Determine the porosity of two types of sand using the water displacement method.

A critical step in these measurements is estimating the volume of voids. For shots, each of the methods employs a distinct approach, as detailed below.

a. **Oven-Drying**:

In this method, the volume of void space is determined by calculating the difference between the weight of the beaker with saturated shots and the weight of the beaker with dry shots. This weight difference represents the volume of water that was removed, which is equal to the volume of the voids

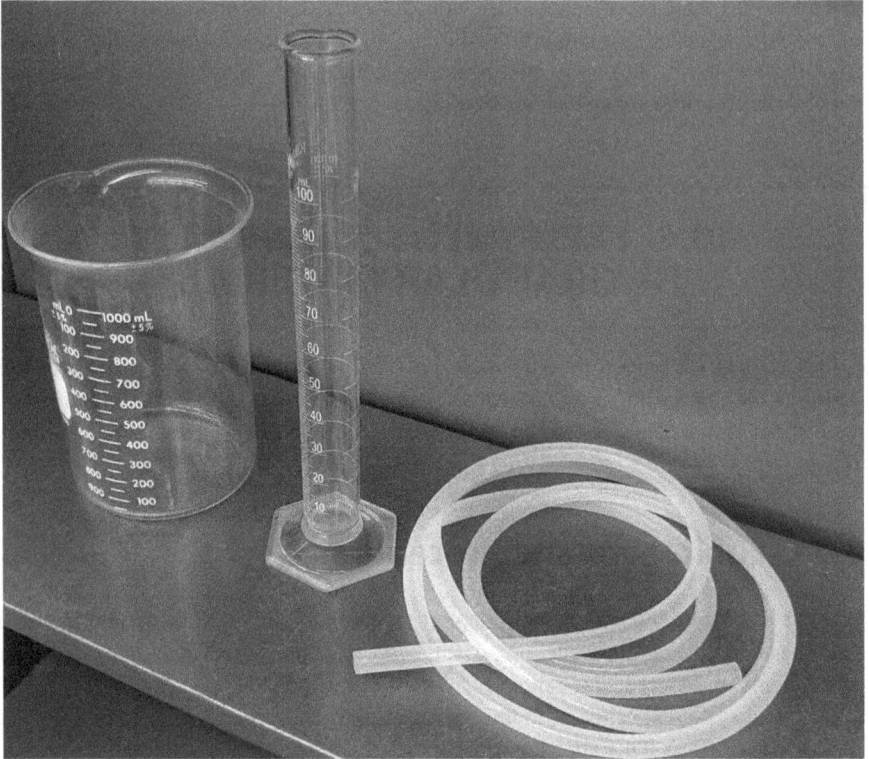

FIGURE 10.2 Some essential laboratory supplies, specifically a glass beaker, a cylinder, and a plastic tube. (Photograph Courtesy of Hong Zhang and Xiaolong Geng.)

previously occupied by the water. Porosity is then calculated by dividing the volume of voids by the total sample volume. The steps are listed below, starting with a glass beaker containing metal shots, as those shown in Figure 10.3, completely saturated with water.

1. Record the volume of saturated shots.
2. Weigh the beaker with saturated shots.
3. Use an oven to fully dry the shots (about 1 hour or until completely dry) (Figure 10.4). In true experiments, the temperature should not exceed 150° Fahrenheit to avoid damaging the structure of the porous material.
4. Weigh the oven-dried beaker with shots.

b. **Volumetric Calculations**:

This technique is used to illustrate the concepts of porosity evaluation and is not suitable for applications to natural materials. In this method, the volume of voids is determined by subtracting the volume occupied by the metal shots (solids) from the total volume. The solid volume is calculated

FIGURE 10.3   Glass beaker with dry metal shots. (Photograph by the author.)

based on the volume of a single shot, assumed to be spherical with a measured diameter (d), using the formula in equation (10.5).

$$V = \frac{4}{3}\pi\left(\frac{d}{2}\right)^3 \tag{10.5}$$

Porosity is calculated by dividing the volume of voids by the total volume. The specific steps are as follows.

1. Measure the average diameter of several shots.
2. Measure the weight of 30–40 shots and record the exact number (use this to calculate the weight of a single shot by dividing the weight by the number of shots.)
3. Pour 40–50 mL of shots into the cylinder and record both the actual weight and the volume of shots. The total number of shots in the sample equals the actual weight divided by the weight of a single shot (the number should be in the hundreds of grams for 40–50 mL of shots).
4. Calculate the total volume of shots by multiplying the number estimated in step 3 by the volume of one shot.

FIGURE 10.4   Oven-drying the sample. (Photograph by the author.)

5. Porosity is calculated by dividing the volume of voids (total volume minus volume of shots) by the total volume.

c. **Water Displacement**:

This technique is essentially the reverse of the oven-drying method. Here, porosity is estimated by dividing the volume of water required to fully saturate the sample by the total sample volume. The following steps are used in this method, starting from a fully dry shot.

1. Record the actual weight and the volume of shots.
2. Take an empty cylinder similar to that in Figure 10.2, fill it with water, and record the volume of water.
3. Pour water into the beaker containing the shots until the water just reaches the surface of the shots; record the volume of water left in the cylinder.

The water displacement steps for sand samples are similar to those for shots, specifically steps 1 to 3 above.

10.4.2 Data Set and Reporting Instructions

Example experimental data are provided in Table 10.3. The report should include

1. Calculating the porosity of the shot using three independent methods.
2. Comparing the results from the three determinations and discussing reasons for differences in the results and possible sources of error in each method.
3. Calculating the porosity of each sand sample using the water displacement method. How do the porosities of sands and shots compare? Explain any differences.

10.5 EXPERIMENT 4: SPECIFIC YIELD

Total porosity is a measurement of the entire volume of pore space in relation to the total soil volume. However, not all of this pore space is available for water transport. The porosity that can be utilized for storage and transportation of water is termed effective porosity. Specific yield is a metric used to quantify the amount of water that can be drained from a saturated sample of water-bearing material due to gravity. It is typically expressed as a percentage of the total volume of the material that was drained. In theory, it takes an infinite amount of time for complete drainage to occur; however, in

TABLE 10.3   Example measured data for porosity experiments

a. Oven-drying (shots)	Weight of saturated shots (g)	367
	Weight of oven-dried shots (g)	341
	Sample volume (mL[a])	70
b. Water displacement (shots)	Initial volume of water in cylinder (mL)	100
	Water left in cylinder (mL)	84
	Sample volume (mL)	40
c. Volumetric calculations (shots)	Weight of 30 shots (g)	10
	Diameter of one shot (cm)	0.4
	Weight of shot sample (g)	239.2
	Volume of shot sample (mL)	40
Water displacement (sands)	Sand 1: initial water volume in cylinder (mL)	100
	Sand 2: initial water volume in cylinder (mL)	100
	Sand 1: water left in cylinder (mL)	53
	Sand 2: water left in cylinder (mL)	43
	Sand 1: sample volume (mL)	100
	Sand 2: sample volume (mL)	125

[a] mL refers to milliliter, which is a cubic centimeter.

practical applications, most of the water drains within 24 hours. As a result, the values obtained in this experiment will generally be lower than the specific yield.

The term "specific retention" refers to the fractional volume that remains in a porous material after gravity drainage, and it is calculated as the ratio of the retained volume to the total sample volume. The relationship among effective porosity (n_e), specific yield, and specific retention can be expressed as follows in equation (6.6):

$$n_e = S_y + S_r \tag{6.6}$$

where S_y and S_r are the specific yield and specific retention, respectively.

10.5.1 Equipment and Procedure

This experiment is conducted using a sand-filled tube, similar to the one shown in Figure 10.5. The following steps are used to estimate the sample's specific yield.

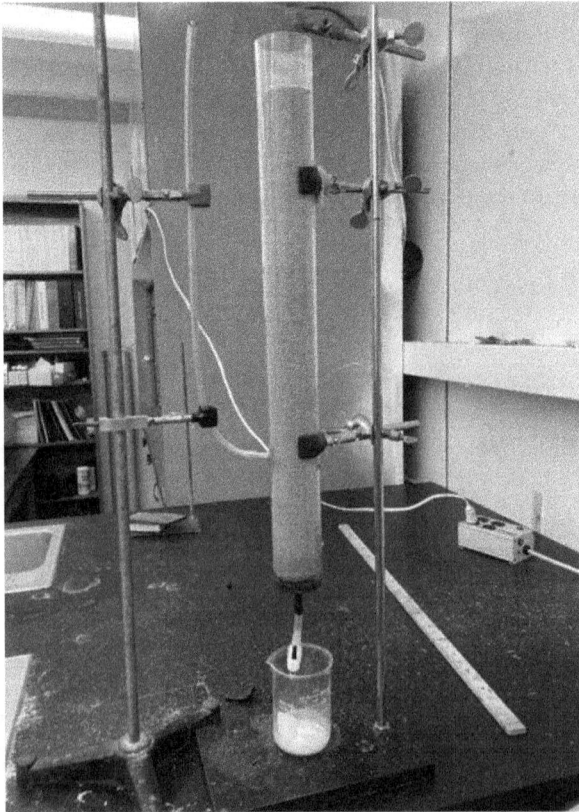

FIGURE 10.5 The sand tube used in the specific yield experiment. (Photograph by the author.)

1. Measure the internal diameter and length of the tube containing the sand.
2. Saturate the sand tube and record the volume of water added (V).
3. Continuously drain water from the sand column. Record the cumulative volume drained (V_w) over time, taking one reading every 2 seconds for the first 30 seconds, then after 1, 10, 20, and 30 minutes.

10.5.2 Example Data Set and Reporting Instructions

Example data for this experiment include a sand-filled tube with a diameter of 5.5 cm and a soil column length of 30 cm. The measured data for the specific yield experiment are recorded in Table 10.4. Your report should include the following:

1. Data sheet and diagram of the apparatus.
2. Graph of the volume of water drained and retained as a function of time.
3. Estimate effective sand porosity (equals volume of water V divided by total sample volume), which equals the area of the tube multiplied by the column length.
4. Estimate the specific yield and specific retention of the sand.
5. Compare your estimations to literature values and discuss errors.

TABLE 10.4   Data for a specific yield experiment

TIME (SECONDS)	V_w (cm³)	$S_y = V_w/V$	$S_r = N_e - S_y$
2	1.5		
4	11		
6	17		
8	21		
10	26		
12	31		
14	37		
16	45		
18	54		
20	60		
22	67		
24	75		
26	81		
28	87		
30	93		
90	182		
660	223		
1,860	233		
3,660	234		

10.6 EXPERIMENT 5: CAPILLARY RISE

As discussed in Chapter 3 (Section 3.6.1), capillary effects cause the movement of water through soil due to the forces of adhesion, cohesion, and surface tension. Water moves upward or laterally through the tiny spaces between soil particles. The smaller the soil pores, the higher and farther the water can move.

10.6.1 Equipment and Procedure

In this experiment, a tube is filled with dry sand and immersed in dyed water (Figure 10.6). The rate of capillary rise and the final capillary head in the sand will then be observed. The procedure begins by immersing the bottom of the tube approximately

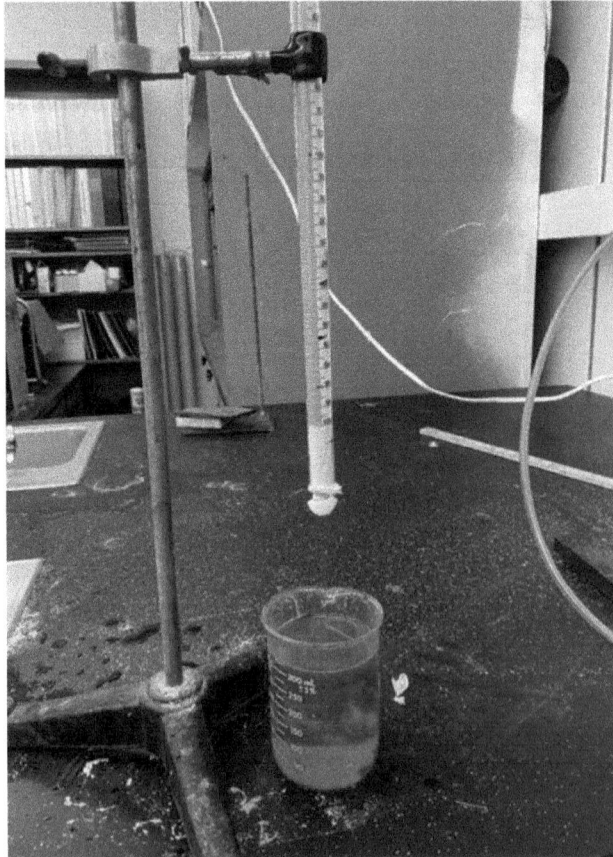

FIGURE 10.6 Capillary experiment rise setup. (Photograph by the author.)

1 inch into a beaker of dyed water. The height of the capillary fringe inside the tube should be measured above the water level in the beaker at intervals of 1, 2, 5, 10, 20, and 60 minutes, or longer if the capillary fringe continues to rise. Use a marker to record the water level on the tube at each interval for later measurement. Since the capillary fringe may exhibit irregularities, both the highest and lowest points should be recorded, and the average of these values used for calculations. Be sure to account for any reduction in the water level within the beaker when recording your measurements.

10.6.2 Example Data Set and Reporting Instructions

An example of measured data for a capillary rise experiment is shown in Table 10.5.

The report should include a graph of capillary rise versus time for the sand. According to theory, the capillary head h_c (the final rise) is related to the throat radius r by the approximate equation $h_c=0.15/r$, where both h_c and r are in centimeters. The capillary rise is used to calculate r, the radius of the average pore space. This idealized scenario is illustrated in Figure 10.7. As shown in the figure, the grain radius R is approximately related to r by equation (10.6) derived from the right-angled triangle formula, which

$$R = \frac{r}{0.4} \tag{10.6}$$

TABLE 10.5   Example experimental measurements for the capillary rise experiment

TIME (min)	DISTANCE (cm)
1	3.7
2	6.4
5	10.8
10	14.9
20	17.2
60	18.3

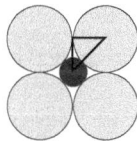

R = grain radius
r = void radius

$(R + r)^2 = R^2 + R^2$
$(R + r) = \sqrt{2} . R$
$\sim 1.4\,R$
$R = r/0.4$

FIGURE 10.7   Calculating grain size from capillary rise.

10.7 EXPERIMENT 6: DARCY'S LAW AND CONSTANT HEAD HYDRAULIC CONDUCTIVITY TEST

This experiment is designed to illustrate the validity of Darcy's law according to equation (10.7):

$$Q = KAS \tag{10.7}$$

where Q is the discharge (units of volume per unit time, e.g., cubic centimeters per second), A is the area (in length squared, e.g., square centimeters), K is the hydraulic conductivity, and S is the hydraulic head slope [dimensionless (has no units)]. In the experiment, the area is that of a circle, and the slope S is given by equation (10.8):

$$S = \frac{h_1 - h_2}{L} = \frac{h}{L} \tag{10.8}$$

where L is the distance between the two manometers, and h_1 and h_2 represent the water levels separated by this distance L, all measured in centimeters.

10.7.1 Equipment and Procedure

The experimental setup requires a permeameter available from commercial vendors (e.g., CERTIFIED MTP, 2023; see Section 6.5.1 in Chapter 6 for detailed information about this test). The system is equipped with two manometers, as shown in Figure 10.8, to measure water levels at points 3 and 4, which are spaced a specific distance apart. The procedure is outlined below.

1. For a specific tank location, adjust the faucet flow to the tank and fully open the outflow valve at point 2. Ensure that water does not overflow from the constant head tank.
2. Verify that the tubes are free of air bubbles.
3. Once the flow stabilizes, measure the difference in head levels between the two manometers (h).
4. Estimate the flow rate, Q, several times by collecting a volume of water and measuring the time taken for collection. Use the average Q from these trials in your calculations. Ensure that h remains constant.
5. Carefully adjust the height of the constant head tank and repeat steps 2 and 4 for approximately 10 intervals. Close both the inflow and outflow valves before adjusting the height. Afterward, reopen the valves and wait for the flow to stabilize before measuring Q and h.
6. Check if the ratio Q (average)/h remains roughly constant, as hypothesized by Darcy's law.

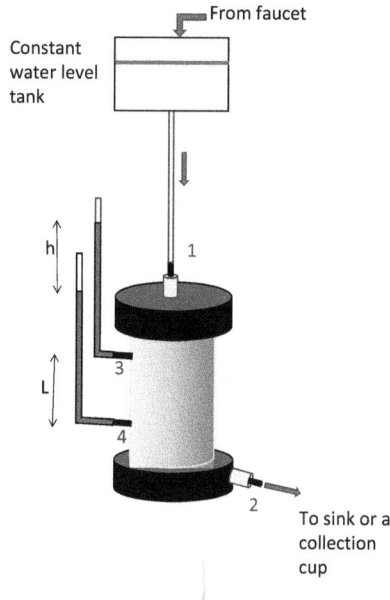

FIGURE 10.8   A schematic illustration of the constant head experiment setup.

10.7.2 Example Data Set and Reporting Instructions

Example experimental results are presented in Table 10.6. The separation distance is $L=6.35$ cm, and the chamber area is $A=31.67$ cm². Ten tests were conducted, with volumes collected and corresponding collection times recorded three times for each test. Note that the average discharge, based on the three collected volumes and their corresponding collection times, is used in the calculations, as demonstrated in the table for the first tank position.

The report should include the following:

1. Complete the table, including using Darcy's law to estimate the hydraulic conductivity for each test by using equation (10.9):

$$K = \frac{Q}{AS} = \frac{QL}{Ah} \tag{10.9}$$

2. Estimate the average hydraulic conductivity value and the standard deviation based on all tests, and discuss possible reasons for any variability in measured values.

3. Create a plot of $y=Q/h$ versus $x=h/L$. Is Darcy's law valid? Explain why or why not.

4. Use the plot to calculate the hydraulic conductivity. If using a trendline, ensure that the line passes through the origin, where $Q=0$ when $h=0$. How does this compare to the average value estimated in step 2?

TABLE 10.6   Data for a constant head experiment

POSITION	t_1 (s)	V_1 (cm³)	t_2 (s)	V_2 (cm³)	t_3 (s)	V_3 (cm³)	Q (cm³/s)	h (cm)	K (cm/s)	h/L	Q/A (cm/s)
1	15.7	180	10.3	185.0	15.5	182.0	13.74	10.0	0.276	1.57	0.43
2	15.4	181	5.3	100.5	8.5	115.0		10.5			
3	10.5	118	5.5	128.0	10.6	124.0		10.9			
4	10.6	130	5.5	126.0	10.6	120.0		10.8			
5	10.5	122	5.3	126.0	11.0	128.0		10.8			
6	10.6	122	5.0	122.0	10.2	126.0		11.2			
7	10.3	124	5.3	128.0	10.5	128.0		11.4			
8	10.3	128	5.2	128.0	10.4	132.0		11.7			
9	10.5	128	5.3	128.0	10.5	128.0		11.5			
10	10.3	132	5.2	132.0	10.5	126.0		12.0			

10.8 EXPERIMENT 7: VARIABLE HEAD HYDRAULIC CONDUCTIVITY TEST

10.8.1 Equipment and Procedure

The setup for this experiment is shown in Figure 10.9. It uses the same permeameter from Experiment 10.7 but replaces the constant water-level section with a tube, as illustrated in the figure. In this case, the manometers are not required. Details of the experiment are provided in Section 6.5.2. The following steps outline the procedure.

1. Measure the diameter of the tube and the length of the sample (between the two porous plates) as well as the chamber diameter.
2. Select several starting points (seven in the example data set in Section 10.8.2) on the tube, closer to the top end, and measure their heights above the outflow level.
3. Select several ending points (three in the example data set) on the lower part of the tube.
4. Establish a steady flow by measuring the discharge every few minutes until it becomes nearly constant and the water height remains unchanged at the desired value (h_o). Check the water level in the tube and adjust the mark and re-measure the height if necessary.
5. Measure the time required for the water level to fall from the top marked level to each of the ending marks using a stopwatch, and record the respective values.
6. Repeat steps 4 and 5 for the other starting points.

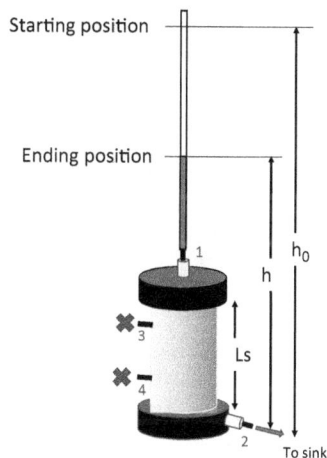

FIGURE 10.9 A schematic illustration of the variable head permeability test.

7. Equation (10.10) can be used to calculate the hydraulic conductivity based on the variable head apparatus:

$$K = \frac{2.3 d_t^2 Ls}{d_c^2 t} \log\left(\frac{h_o}{h}\right) \tag{10.10}$$

where d_t and d_c are the diameters of the tube and chamber, respectively, Ls is the length of the sand column, t is time, and h_o and h are the starting and ending water heights, respectively. Note that Ls represents the *total* soil column length, which differs from L, the distance between the two manometers.

10.8.2 Example Data Set and Reporting Instructions

Example measured data are shown below and in Table 10.7 for a total of seven starting points and three ending positions for each, resulting in 21 runs (7×3). The diameters d_t and d_c are 1.5 and 6.35 cm, respectively.

TABLE 10.7 Data for a falling head experiment listing the initial water levels, respective dropped levels, and the times to reach the latter

STARTING POINT	h_o (cm)	ENDING POINT	h (cm)	t (s)	K (cm/s)
1	66.00	a	49.53	4.41	
	66.00	b	43.94	5.84	
	66.00	c	36.83	7.89	
2	77.47	a	49.53	3.96	
	77.47	b	43.94	5.54	
	77.47	c	36.83	7.55	
3	69.85	a	49.53	4.13	
	69.85	b	43.94	5.67	
	69.85	c	36.83	7.93	
4	67.31	a	49.53	3.76	
	67.31	b	43.94	5.27	
	67.31	c	36.83	7.70	
5	72.39	a	49.53	4.73	
	72.39	b	43.94	6.23	
	72.39	c	36.83	8.86	
6	65.41	a	49.53	3.65	
	65.41	b	43.94	5.23	
	65.41	c	36.83	7.73	
7	62.86	a	49.53	3.33	
	62.86	b	43.94	4.91	
	62.86	c	36.83	7.46	

In the report, include the following.

1. Calculate hydraulic conductivity (cm/s) and estimate various means (arithmetic, geometric, and harmonic) and the standard deviation (see Section 2.3.3 in Chapter 2).
2. Draw a duration curve on probability paper and estimate the 10% and 90% expected values (see Section 5.2.4 in Chapter 5).
3. Discuss any sources of error.

10.9 EXPERIMENT 8: TRANSIENT AND STEADY-STATE AQUIFER PUMPING TEST

The experiments in Sections 10.9 through 10.11 are designed to use aquifer pumping tests to estimate the hydraulic properties of aquifers. Section 10.9 covers both steady-state and transient pumping tests, while Section 10.11 focuses on transient well recovery tests. Section 10.10 specifically evaluates the accuracy of approximate formulas in predicting an aquifer's response to pumping. The techniques for analyzing aquifer pumping tests are discussed in detail in Section 6.11 of Chapter 6.

10.9.1 Equipment and Procedure

The pumping test experiments require a manufactured Plexiglass tank and a peristaltic pump. As shown in Figure 10.10, to minimize construction costs, the sand tank represents only a small segment of the area enclosed by a circle. The flow within this segment is assumed to be representative of the entire area due to the symmetrical radial flow

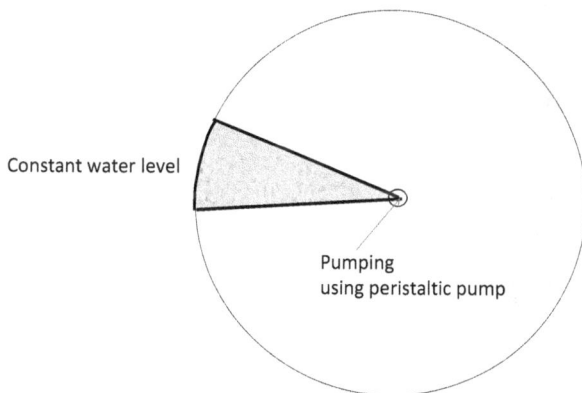

FIGURE 10.10 A schematic of the sand tank represented by the shaded area, which is only a small segment of the area bounded by a circle.

around the well, with the assumption that friction along the tank walls is negligible. Flow near the narrow edge of the tank may not be accurate due to the greater influence of the glass walls. Nevertheless, the test offers significant educational value.

Figure 10.11 provides a perspective view of the tank, which is equipped with five manometers for measuring the water levels within. The gray line in the figure represents the water level, as indicated by the manometer readings. These levels will vary over time during transient conditions but remain constant in steady-state situations. In all runs, the water level at the wider end of the tank should remain fixed at a specified value by providing faucet water through a tube.

The objective of these experiments is to estimate the hydraulic conductivity and storage coefficient of an aquifer model. Under a prescribed well flow rate, the water levels within the tank will be continuously monitored and used to make these estimations under both unsteady (transient)- and steady-state conditions. Note that the water level will change rapidly in the manometers during the transient experiment, especially in those closer to the pumping well. Therefore, the levels should be recorded using a marker on a paper strip placed next to the manometer tube for later measurement.

Given that the equations applied pertain to a 360° radial flow field, it is imperative to determine the correct discharge through the model. The flow rate used in the calculations should be the measured well value multiplied by the number of wedges that can fit within the full cylinder. Various tank dimensions are listed in Tables 10.8 and 10.9.

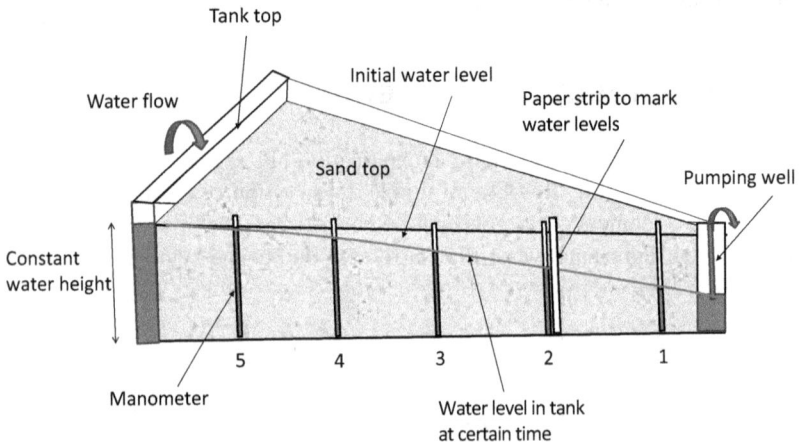

FIGURE 10.11 The sand tank with five manometers. A paper strip should be attached next to each manometer to mark water levels for later measurement. For illustration, the strip is shown next to manometer 2.

TABLE 10.8 Sand tank dimensions

Aquifer (sand) thickness (b)	26.8 cm
Angle of wedge (degrees)	15
Number of wedges in a full circle	24
Length of the tank	237.5 cm

TABLE 10.9   Manometers' location distances

	r_1	r_2	r_3	r_4	r_5
Distance from well (cm)	19.8	50	99.8	149.5	197

10.9.1.1 Unsteady (transient) flow experiment

In the transient experiment, the water levels in the manometers will be recorded as functions of time. The following measurements need to be taken:

1. Mark water levels in all the manometers before the pump is turned on. The markings should be placed on a paper strip attached next to the manometers as shown in Figure 10.11.
2. After the pump is activated, mark water levels in all the manometers at decreasing frequencies as time progresses: For the initial minute, mark levels every 10 seconds, and for the subsequent 2 minutes, mark levels every 30 seconds. Following this, mark once every minute until the water levels in the manometers reach a stable state. (Note that less frequent measurements are acceptable for manometers farther from the pump source.)
3. Measure discharge at least 2–3 times during the test by collecting a volume of water and noting the respective time of collection.

Hydraulic conductivity can be estimated by the Jacob equation, which approximates the relationship between drawdown s (decline in water level) and the logarithm of the time (t) as a straight line; then conductivity K can be estimated from equation (10.11):

$$K = \frac{2.3Q}{4\pi\Delta sb} \tag{10.11}$$

where b is the aquifer thickness, Δs is the drawdown for a log cycle, and Q is the discharge.

Aquifer storativity S is estimated by equation (10.12):

$$S = \frac{2.25Tt_o}{r^2} \tag{10.12}$$

where r=distance from the center of the pumped well; T is transmissivity; and t_o=time intercept where $s=0$.

10.9.1.2 Steady-state experiment

The procedure begins by establishing a suitable pumping rate that ensures the well does not completely run dry. This process is conducted for a sufficient duration to achieve a steady state, ensuring that the water level in the narrower end of the tank remains 5 cm above the bottom. No water level readings are taken at this stage. This process is repeated multiple times, starting from an initially uniform water level, until a pumping rate is determined for use throughout the subsequent experiments. The pump is then turned off to attain a flat water level before commencing the actual experiments.

The steady-state pumping test is conducted at the end of the transient test, when water levels in the manometers have stabilized. Both discharge and water levels in all the manometers are recorded one final time.

For steady state, the following Thiem equation (10.13) expresses the hydraulic conductivity of an unconfined aquifer:

$$K = \frac{2.3Q\log\left(\dfrac{r_2}{r_1}\right)}{\pi\left(h_2^2 - h_1^2\right)} \tag{10.13}$$

where h_1 and h_2 are the elevations of the water table above the base of the aquifer at distances r_1 and r_2, respectively.

10.9.2 Example Data Set and Reporting Instructions

Example experimental data are listed in Tables 10.10 and 10.11.
  Three discharge values for the transient case (in mL/s): 7.1, 4.5, and 5.7

TABLE 10.10   Water levels for the unsteady case as functions of time

TIME (s)	h_1 (cm)	h_2 (cm)	h_3 (cm)	h_4 (cm)	h_5 (cm)
0	26	26	27.4	27.3	27
10	23.2	25.9	27.4	27.2	27
20	21.5	25.5	27.4	27.1	27
30	20.5	25.1	27.4	27.1	27
40	19.7	24.9	27.4	27	27
50	19	24.7	27.4	27	27
60	17	24.1	27.4	27	27
90	13.6	21.9	27.35	27	27
120	12.9	20.6	27.3	27	27
150	12.5	19.9	27.1	27	27
180	12.3	19.4	26.75	27	26.9
240	12.3	19.2	26.5	26.75	26.9
300	11.9	18.9	25.95	26.75	26.8
360	12	18.8	25.6	26.7	26.8

TABLE 10.11   Initial and final water levels for the steady-state case

	h_1 (cm)	h_2 (cm)	h_3 (cm)	h_4 (cm)	h_5 (cm)
Initial height	25.8	26.8	27.3	26.9	27
Final height	11.9	17.9	23.4	25.7	26.6

Final discharge value for the steady-state case: 5.24 mL/s
The laboratory report should include the following.

Unsteady (Transient) Flow Experiment:
1. Plot the water table profiles at different times.
2. Plot drawdown versus the logarithm of time for the two manometers nearest to the pumping well.

Parameter Estimation:
1. For the transient test, use the plot of drawdown versus log(t) for the three manometers close to the pumping well to calculate K (in cm/s) and S values using the Jacob solution.
2. Using the Thiem equation, calculate K in cm/s using three pairs of manometers. If your answers differ, discuss possible reasons for the discrepancies.

10.10 EXPERIMENT 9: APPLYING THE THEIS AND THIEM SOLUTIONS TO AQUIFER TEST EXPERIMENTS

The calculations completed here follow up on Experiment 10.8, and no new data are collected. They assess the suitability of the mathematical Theis and Thiem expressions applied for estimating aquifer response to pumping.

10.10.1 The Theis Solution for Unsteady Aquifer Response

Use the Theis solution in equation (10.14) to simulate the drawdown at various times for the two manometers nearest to the pumping well (see Section 6.9.3). Assume reasonable values for hydraulic conductivity (based on previous experiments) and storage coefficient ($S \sim$ specific yield). Use the average water height as aquifer thickness.

$$h = h_o - \frac{Q}{4\pi T} W \qquad (10.14)$$

where u is given by equation (10.15):

$$u = \frac{Sr^2}{4Tt} \qquad (10.15)$$

Graphically compare these values with those measured in item 4 above for a number of K values.

10.10.2 The Thiem Solution for Steady-State Aquifer Response

Use the Thiem equation (10.16) for unconfined aquifers to estimate the steady-state water table profile (see Section 6.9.2):

$$h_2^2 = \frac{Q}{\pi K} \ln\left(\frac{r_2}{r_1}\right) + h_1^2 \tag{10.16}$$

where h_1 and h_2 are the water table heights at any two wells at distances r_1 and r_2 from the center of the pumping well. For all calculations, use h_1 as the water table height in the manometer nearest to the pumping well and r_1 as the distance from the center of the pumping well to that manometer. Compare results by plotting the estimated versus measured values.

10.11 EXPERIMENT 10: AQUIFER TRANSIENT RECOVERY TEST

10.11.1 Equipment and Procedure

This experiment utilizes the same sand tank described in Section 10.9. The recovery process involves pumping a well to achieve a steady-state condition and then shutting off the pump. The water level subsequently increases until it fully (or nearly) recovers to its pre-pumping condition. (For a detailed description of the well recovery procedure, refer to Section 6.9.3.3 and Figure 6.21.)

The first step is to attain a steady-state condition by operating the well until the water levels in the manometers have stabilized. There is no need to document the well levels during this phase. Take multiple discharge measurements. After achieving a steady state, perform one final discharge measurement and mark the water levels in all the manometers along with the time it took to reach the steady state (t_o).

After achieving the aforementioned steady state, initiate a transient recovery test. To do this, turn off the pump and start marking the water levels in the manometers closest to the pumping well over time since pumping ceased (t'). These measurements are referred to as residual drawdowns (s'). During the first minute, mark levels every 10 seconds; during the subsequent 2 minutes, mark levels every 30 seconds. Afterward, mark levels at 1-minute intervals until the water levels in the manometers stabilize.

10.11.2 Example Data Set and Reporting Instructions

Example measurements are listed below and in Table 10.12.
    Steady-State data

TABLE 10.12   Recovery data at the three manometers nearest to the pumping well

t' (s)	s_1' (cm)	s_2' (cm)	s_3' (cm)
10	14.5	8.9	3.5
20	12.7	8.6	3.3
30	11	8.1	3.1
40	9.5	7.3	2.9
50	8.1	6.3	2.5
60	6.8	5.3	2.2
90	5.9	4.7	1.8
120	3.3	3.8	1
150	2.1	2.2	0.5
180	1.2	1.2	0.2
240	0.5	0.6	0
300	0.2	0.3	0

$$t_o = 720 \text{ s}$$

$$Q \text{ average} = 111.12 \text{ cm}^3/\text{s}$$

$$b = 25.5 \text{ cm}$$

For the laboratory report, create a semi-logarithmic plot of s' against t/t' (where $t=t_o+t'$) for the three manometers nearest to the pumping well. Discuss any potential deviations from a linear relationship and provide reasons for such variations. Calculate hydraulic conductivity (K) values using the recovery equation (10.17):

$$K = \frac{Q}{4\pi bi} \tag{10.17}$$

where Q is the well pumping rate, b is aquifer thickness, and i is the slope of the straight-line fit of s' versus t/t'.

10.12 EXPERIMENT 11: TRACER TEST

A tracer is a substance introduced into water to monitor its movement, including direction and velocity. Groundwater tracers are typically soluble chemicals. The ideal tracer does not react with its environment, is easily detectable, non-toxic, and moves at the same velocity as the water particles. Common table salt (sodium chloride) is often preferred for many applications, though it may not be suitable for near-shore field studies due to interference from ocean saltwater.

The primary objective of this experiment is to demonstrate how tracers can be used to determine water velocity and illustrate the dispersion effects when freshwater displaces brackish water or vice versa. The study also estimates the time for the first arrival of a solute and explores the relationship between maximum and average water velocities.

10.12.1 Equipment and Procedure

The tracer test is conducted within a sand tube (see Figure 10.5) under steady-state water flow conditions. The complete setup is shown in Figure 10.12. Water exiting the tube is analyzed for electrical conductivity using a measuring device within the smaller tube in the figure. Electrical conductivity, measured in siemens per meter (S/m) according to SI units, serves as a proxy for salinity. The goal is to estimate the relative change in salinity, rather than its exact value. Commercially available electrical conductivity meters, such as those from Hanna Instruments, can be used.

A brief introduction of the tracer occurs after a period of steady-state "clean" water flow followed by clean water. The conductivity meter display updates automatically as salinity levels change. Fluorescein dye is used to track the progress of the "contaminant."

The experiment procedure comprises the following steps.

1. Measure the sand tube diameter and length of the sand column.
2. Measure sand porosity in the beaker using the water displacement method.
3. Measure conductivity of the brackish water (C_b) in the beaker.
4. Create a steady-state flow in the sand tube by opening the output valve fully and carefully adjusting the flow from the tap. Ensure the water level remains as close as possible to the marked level in the tube.
5. Measure the discharge rate a few times and check for consistency as an indicator of steady-state condition.

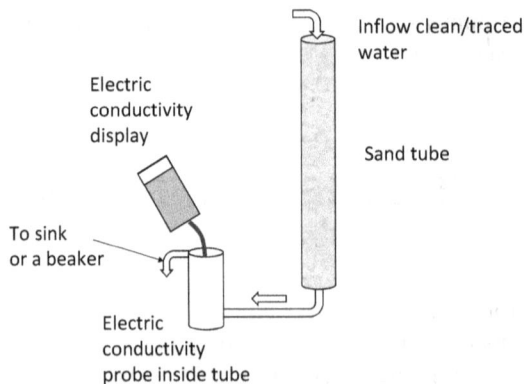

FIGURE 10.12   Setup for the tracer test experiment.

6. Monitor the conductivity of the flowing water using the probe and wait for it to reach a near-constant value.
7. Measure the discharge rate one more time.
8. Once steady state is reached:
   a. Shut off the tap water.
   b. Manually introduce brackish water at the top of the tube and start the timer at the beginning of this process. Allow the timer to run for the entire duration of the experiment. Gradually pour the water into the tube over approximately 30 seconds, ensuring that the water level aligns with the mark on the tube.
   c. Promptly return the tap water flow to the top of the tube after ending the tracer introduction and monitor for a steady-state condition.
9. Record the conductivity of the pumped water (C) and the respective time. Increase the frequency when you notice a rapid change in conductivity.
10. Continue to record data until the conductivity roughly returns to the background value.

10.12.2 Example Data Set and Reporting Instructions

Example experimental data are listed below and in Table 10.13:

Sand tube diameter=6 cm (Area=28.26 cm²)
Sand tube length=57 cm
Sand porosity=0.446
C_b=33.2 S/m
C_f=0.409 S/m
Measured discharges=3, 3.4, 3, and 3.1 cm³/s

TABLE 10.13   Measured electric conductivity vs. time for the tracer test

t (s)	C (s/m)	t (s)	C (s/m)	t (s)	C (s/m)	t (s)	C (s/m)
30	0.409	210	32.62	330	0.97	480	0.52
60	0.409	220	28.99	340	0.86	490	0.52
90	0.417	230	22.05	350	0.79	500	0.5
120	0.427	240	16.4	360	0.68	510	0.49
130	1.29	250	12.3	370	0.64		
140	1.29	260	8.92	380	0.62		
150	11.5	270	6.77	390	0.62		
160	17.78	280	4.69	400	0.59		
170	23.23	290	3.69	410	0.61		
180	29.57	300	2.58	420	0.57		
190	31.56	310	2.14	430	0.57		
200	33.2	320	1.67	440	0.55		

The laboratory report should include the following.

1. Compile the collected data.
2. Graph $C^* = (C - C_f)/(C_b - C_f)$ versus time. C^* represents a normalized value between zero and one (dimensionless). This graph is known as the breakthrough curve.
3. Use the breakthrough curve to determine the time for the first arrival of the tracer, the time to reach 50% concentration, and the time to reach the peak of the breakthrough curve.
4. Estimate the average pore (seepage) velocity by using the average discharge and the measured tube area and porosity.
5. Estimate the dispersivity of the sand by using the relationship in equation (10.18), which is valid for one-dimensional columns,

$$\alpha = \frac{V^2}{4\pi S^2 L} \qquad (10.18)$$

where V is the seepage velocity (cm/s), S is the slope of breakthrough at $C^* = 50\%$ (1/s), and L is the column length (cm).

REFERENCES

Applied Fluid Mechanics Lab Manual. (2023). Flow over weirs. https://uta.pressbooks.pub/appliedfluidmechanics/chapter/experiment-9/

Armfield. (2020). F1–13-MKII / F1–13a flow over weir. https://armfield.co.uk/product/f1-13-flow-over-weir/

CERTIFIED MTP. (2023). Permeameters. https://certifiedmtp.com/astm-aashto-permeameter-2-5-diameter/

Edibon. (2023). FME02 flow over weirs. https://www.edibon.com/en/flow-over-weirs

EME. (2018). Fluid mechanics / hydraulic lab equipment. https://www.enggmod.com/tilting-flume_computerised.html

GUNT. (2023). GUNT experimental flumes. Retrieved from https://www.gunt.de/images/download/flumes_english.pdf

LibreTexts. (2023). 8.6: Flumes. Retrieved from https://geo.libretexts.org/Bookshelves/Sedimentology/Book%3A_Introduction_to_Fluid_Motions_and_Sediment_Transport_(Southard)/08%3A_Sediments%2C_Variables%2C_Flumes/8.06%3A_Flumes

Scribd. (2009). Experiment # 1. Retrieved from https://www.scribd.com/doc/116285777/Determine-Mannings-Roughness-Coefficient-and-Chezy-Roughness-Coefficient-in-a-Labortary-Flume#

Hydrological and Climate Modeling

11

11.1 INTRODUCTION

Hydrologic conceptual models are simplified representations or simulations of parts of the real-world hydrologic cycle. These models encompass processes such as surface runoff, subsurface flow, evapotranspiration, and channel flow, as discussed in Chapter 3. Their classification as conceptual reflects the incorporation of various concepts that mirror our understanding of reality. These models can vary in complexity, contingent on the level of detail and the number of processes considered. Mathematical expressions are employed in these models to represent various processes and their interactions. These formulations are implemented within solution schemes by using simulation software, also known as simulation models. These simulation models serve the purpose of understanding hydrologic processes, making hydrologic predictions, and estimating information related to hydrologic systems. The results generated by these models are essential for managing water resources encompassing developing and utilizing water sources and designing and implementing cleanup strategies. Based on the processes represented, simulation models can be categorized into watershed, surface, or subsurface models.

It is important to understand that conceptual models are not precise or unique representations of reality; they are a reflection of our current understanding of hydrologic systems. Therefore, the results provided by simulation models are contingent on the specific applications and the quality of information utilized for the simulations. Nevertheless, these models are valuable tools, especially when combined with field assessment techniques.

In Chapter 8, Section 8.2.3 and Example 8.2 present elements of a simplified modeling approach used to assess contamination distribution in an aquifer. Initially, certain concepts are developed, and simplified assumptions are made to derive equations for predicting contaminant concentrations at different locations and times. Subsequently, these equations are applied using calculation methods, either through scientific calculators or spreadsheets, for prediction purposes. These two steps respectively represent the conceptual and simulation phases. In this context, the conceptual model is a simplified representation of reality, and its use should be limited to preliminary analyses.

DOI: 10.1201/9781003587149-11

257

More advanced conceptual models and the resulting simulation counterparts are required for in-depth assessments, which are more suitable for field applications. In such cases, the mathematical techniques known as numerical methods, involving increased complexity and extensive data requirements, are used, and spreadsheets may not suffice. As we will discuss later, many simulation models are available and can be selected and adopted based on specific needs. However, in some instances, there might be a need to develop a new simulation model or modify an existing one.

One example of modeling involves addressing a water quality issue, such as aquifer contamination. Initially, the understanding of the problem's severity and sources may be limited. The first step would be to create a conceptual model for the aquifer, taking into account all relevant information, including the aquifer's extent, type (confined or unconfined), sources of aquifer recharge, and potential contamination sources. This information is then incorporated into a simulation model suitable for assessing flow and transport conditions. Typically, simple or basic models can be used to quickly estimate the extent and severity of the problem. More advanced efforts can be employed to predict future contamination extents under various potential cleanup scenarios. In this context, models are valuable for extrapolating from current conditions to potential future conditions, which are obviously not known beforehand.

Another example of modeling involves addressing questions related to the environmental consequences of factors like climate change and increased population within a watershed, including changes in land use and land cover. Various scenarios must be explored to safeguard the watershed from adverse changes, and modeling serves as a valuable tool for identifying effective management options in this context.

Additionally, models can be used to compare the impacts of different methods for reducing pollution from various known sources. For instance, sediment in a river or bay can originate from different land uses, such as agriculture or conservation lands. In such cases, effective schemes for reducing sediment loads need to be designed. Models can also identify the most cost-effective best management practices for controlling chemical pollutant loads from urban versus agricultural areas.

As demonstrated by the above examples, models can address both water quantity, represented by water levels and discharges, and water quality, represented by chemical concentrations. Water quality issues are complicated by the characteristics of the contaminants of interest, including potential chemical degradation or phase changes, as well as persistence or bioaccumulation in the environment. Additionally, practical and sociological factors, such as government regulations, stakeholder interests, funding availability, staff time, historic data, and user expertise, can influence the applications and usefulness of model results.

Before using a model for predictions, several important steps are necessary to gain confidence in the model's reliability. These steps include verification, which assesses the model's accuracy in solving governing equations for various hydrologic processes. This often involves simulating simplified cases and comparing results with known solutions or those from other published models. Calibration is another critical step that ensures the model can replicate measured field data. Calibration entails estimating various model-controlling parameters, such as aquifer hydraulic conductivity and soil properties, to achieve the best match between model results and measured data, such

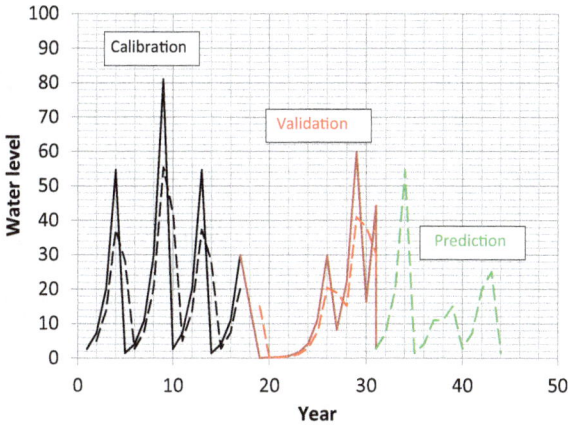

FIGURE 11.1 Example results for model calibration, validation, and prediction processes. Solid lines represent field-measured data; dashed lines represent model results.

as water levels and specific chemical concentrations. The fitting process may involve manual or automatic trial-and-error iterations until the best match is achieved. Finally, a validation process is conducted, requiring a best fit without adjusting the parameters established during calibration.

Figure 11.1 illustrates the steps for calibrating and validating a model, as well as applying the model to predict an aquifer's behavior based on 31 years of water-level field data. The water level fluctuations result from various aquifer inflows and outflows, including recharge, pumping, and surface water leakage. The first 17 years of data are used for calibration, while the subsequent 14 years serve for validation. Assuming a reasonable fit between field and modeled data, the model can be acceptably employed for predictive purposes over a 13-year period in this case. However, perfect match between the model and observed data is typically not achievable, so it is crucial to evaluate expected errors (deviations) shown in the figure.

11.2 WATERSHED MODELING

Watershed models are commonly employed in the simulation of various components of the water cycle. Like all models, their primary purposes are hydrologic prediction and understanding of controlling processes, which are detailed in Chapter 3. These processes encompass precipitation, evaporation, transpiration, infiltration, runoff, and groundwater. Some models offer more intricate elements, such as interflow (the flow of water within the surface soil layers), routing (representing surface water movement details), pond and reservoir storage, crop growth, and irrigation. Differences arise regarding the extent to which various components are simulated and the level of detail included. Variations also exist in terms of spatial representation, the time domain, and the incorporation of management-related aspects.

Precipitation and weather data are main input requirements by playing a pivotal role in driving water storage and flow within the watersheds. Certain models even provide the option to generate synthetic rainfall data based on statistical rainfall information. Other weather-related input data include temperature, solar radiation, relative humidity, and wind speed.

In general, the outcomes of a watershed simulation model primarily document a water budget for the watershed, encompassing elements like evapotranspiration, groundwater recharge, and stream flows, alongside water-flow hydrographs in streams. Of these, groundwater recharge is vital for groundwater assessment and modeling. In addition to water quantity and flows, most of these models can also be used for water quality studies, enabling the identification of the origin and fate of various contaminants within a watershed, including sediments, nutrients, and other chemicals.

List of available simulation models are listed in Texas A&M University (2023), USGS (2023a), and the Minnesota Pollution Control Agency (2023). The latter includes flooding simulation software that can utilize water flow hydrographs obtained from watershed models to estimate the extent of flooding. Many of these models are available in the public domain, meaning they are accessible at no cost. One popular model among these is SWAT (Soil and Water Assessment Tool; FAO, 2023), designed for large, complex watersheds. SWAT simulates various elements of the hydrologic cycle, including weather, surface runoff, irrigation return flow, evapotranspiration, stream transmission losses, pond and reservoir storage, crop growth, irrigation, groundwater flow, routing, and nutrient and pesticide loading. Its specific utilization is to predict the long-term impacts of agricultural practices on the environment, thus aiding in assessing the environmental efficiency of best management practices and alternative policies for land management. The model is also valuable for simulating water and nutrient cycles within the watershed.

Another example is Arc/EGMO, a multilayer model for predicting components of the water budget in small and large watersheds without a water quality component (Becker et al., 2002). Models combining water flow and quality include the Hydrological Simulation Program FORTRAN (HSPF) (EPA, 2002) and the USGS Water, Energy, and Biogeochemical Model (WEBMOD) (USGS, 2018; Webb and Parkhurst, 2017). Watershed Modeling System (WMS; AQUAVEO, 2023a) is a proprietary software (requires a license) that serves as an interface for several watershed and hydraulic (surface-water) models.

Case Study 11.1: Watershed Modeling: Oahu Island, Hawaii, USA

Example watershed modeling results are shown in Figure 11.2, which illustrates the distribution of aquifer recharge based on a water budget for the Island of Oahu, Hawaii, as detailed in Engott et al. (2017). This water budget accounts for contributions from rainfall, surface-water runoff, potential evapotranspiration, and irrigation water return. The recharge data were employed in a groundwater modeling simulation that will be discussed in Section 11.4.

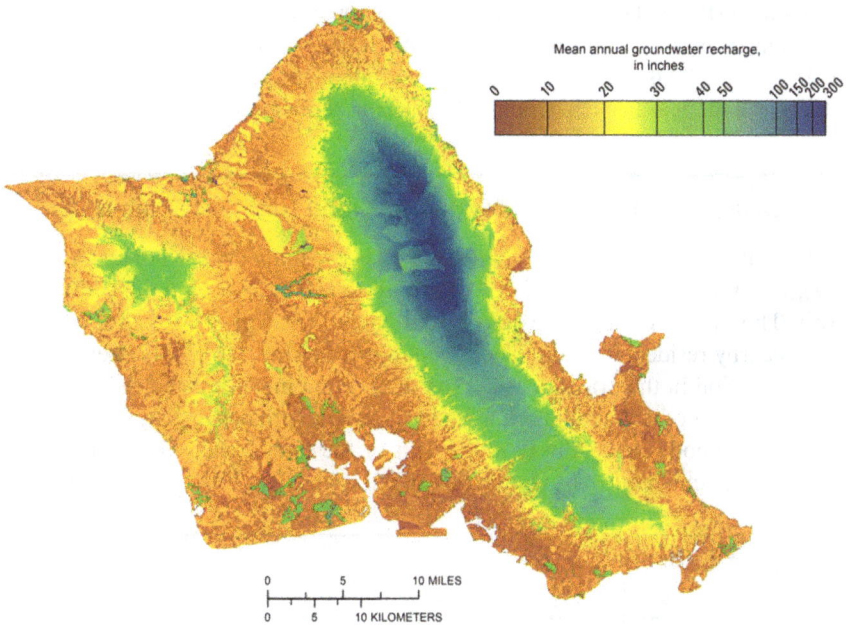

FIGURE 11.2   The distribution of mean annual recharge for average climate conditions (1978–2007) in inches for the Island of Oahu, Hawaii (Engott et al., 2017).

11.3 SURFACE WATER OR HYDRAULIC MODELING

As discussed in Section 11.2, watershed models encompass various components of the water budget, including elements such as rainfall, evapotranspiration, and both surface and subsurface flows. These models are equipped to analyze the influence of land surfaces and soil conditions on the timing and volume of runoff. Hydraulic models then leverage this information in conjunction with channel characteristics to simulate river channel velocity and stage in response to specific river discharges. These models are frequently employed in studies pertaining to flood hazards and environmental analyses.

The United States Geological Survey (USGS) has developed a range of hydraulic models, as listed in USGS (2023a). Notably, the HEC-RAS system, developed by the U.S. Army Corps of Engineers, encompasses multiple components for river analysis, including flow simulations, sediment transport computations, and water quality analysis. River geometric and hydraulic data are essential inputs for these computation schemes. Accurately estimating the volume of runoff resulting from rainfall is crucial for precise river response calculations. Moreover, assessments of water quantities are indispensable for evaluating water storage in reservoirs and determining the risk of flooding. Flood inundation is estimated through a combination of topographical features and volumetric calculations. Proprietary software, such as the Surface Water Modeling System (SMS; AQUAVEO, 2023b) is capable of simulating a wide range of processes, including riverine analysis and the transport of contaminants and sediments.

Case Study 11.2: Flood Modeling: Oahu Island, Hawaii, USA

An illustrative case of flooding-related research is the study by El-Kadi and Yamashita (2007), which focused on a flood occurrence in October 2004 on the Island of Oahu, Hawaii. This flood resulted in extensive damage to the University of Hawaii's campus and nearby residential areas. The modeling study aimed to assess streamflow and flood delineation in the affected area. The delineation model treated the flood as if it were due to a hypothetical dam break and was used to predict various characteristics, including the floodwater pathway, the extent of the flood zone, the maximum flood depth, and the time required to reach that depth. Figure 11.3 illustrates the simulated

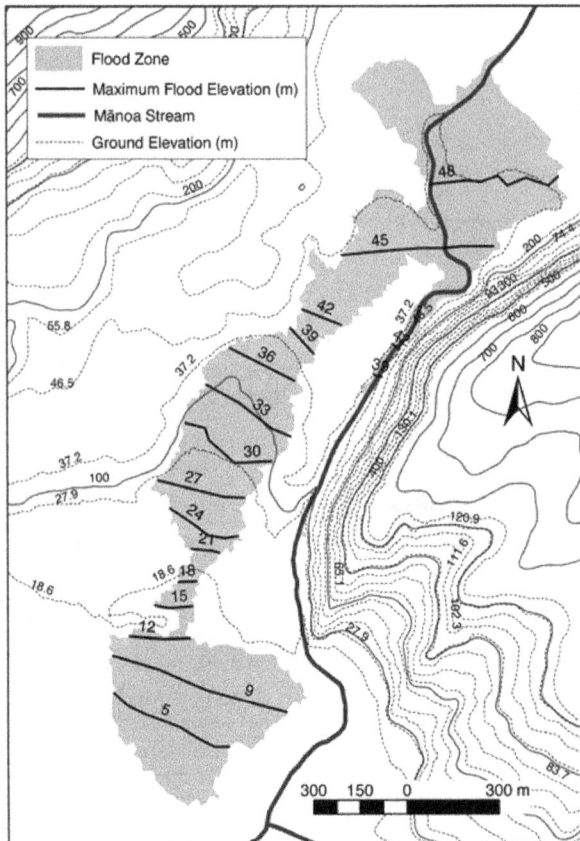

FIGURE 11.3 Simulated flood zone and the related maximum flood elevations within the area (El-Kadi and Yamashita, 2007).

flood zone and the related maximum flood elevations within the area. It is worth noting that the maximum flood elevations shown in the figure do not occur simultaneously, but rather at times estimated based on the advancement of the flood wave toward the downstream side.

In addition to modeling the effects of flooding, the study provided practical recommendations such as mitigating flood risks through streamflow diversion, stream dredging, and the realignment and lining of stream channels. It emphasized that future modeling scenarios should define a range of potential options before the collection of relevant data, thereby optimizing both data collection and the decision-making process. Among the recommendations was the creation of a new management watershed, the West Manoa Watershed, currently a part of the Manoa-Palolo Watershed, which would encompass the University of Hawaii campus. Currently, the campus is part of the larger Ala Wai Watershed, which is mostly urbanized. The study suggested that flooding in the proposed West Manoa Watershed would likely result from streamflow leakage originating in the Manoa-Palolo Watershed rather than from direct rainfall within the new watershed boundary.

11.4 SUBSURFACE WATER MODELING

While watershed models do incorporate a subsurface component, the spatial and temporal scales of the watershed processes make it challenging to conduct a detailed simulation of this component. The mathematical formulation for the subsurface zone is intricate, and its solution necessitates dividing the zone into small three-dimensional sections, a process that renders the integration of the zone within a watershed impractical, particularly for large-scale watersheds. Additionally, the time scale differs between surface and subsurface processes due to groundwater's much slower movement compared to surface water. To address this issue, a common approach is to establish a soft linkage by sequentially running the watershed model and the subsurface model, using the former's output as input for the latter. The primary link is established through surface water recharge, estimated by the watershed model, and serves as the main source of water in the subsurface zone. Flow in open channels also serves as a linking component, reflecting the interaction between the two models.

The same modeling principles that apply to watersheds and surface water can also be applied to subsurface modeling. The subsurface is generally divided into a deeper, fully saturated zone where water resources are developed for human use and a shallower, partially water-saturated zone where agricultural and near-surface practices dominate. Depending on the specific focus, a conceptual model can be designed to address each zone separately or as an integrated view of the two zones.

As with other models, after designing a conceptual model for the aquifer, a simulation model is chosen to address either water quantity alone or a combination of water quantity and quality. Creating the conceptual model generally requires simplifying assumptions about the aquifer and the physical processes governing groundwater and contaminant flow. The conceptual model encompasses aquifer type and data, including the spatial extent and conditions at the aquifer's boundary, as well as the nature of the fluids and contaminants.

The USGS (2023b) provides a list of groundwater models and related software developed by the agency. One example is the widely-used public-domain modular finite-difference groundwater flow model, MODFLOW, which simulates water flow in groundwater systems (Harbaugh et al., 2000). MODFLOW also accounts for various hydrologic elements such as rivers, streams, drains, springs, reservoirs, wells, evapotranspiration, and recharge from precipitation and irrigation. Over time, the program has evolved with the development of numerous new packages and related programs for groundwater studies, including contaminant transport and saltwater-freshwater interaction. Proprietary software options include the Groundwater Modeling System (GMS; AQUAVEO, 2023c) and FEFLOW (MIKE, 2023).

Case Study 11.3: Subsurface Modeling: Oahu Island, Hawaii, USA

Examples of model calibration are presented in numerous studies, including Whittier et al. (2010) and Gingerich et al. (2024). An illustrative result from calibrating the groundwater flow model MODFLOW (Harbaugh et al., 2000) is shown in Figure 11.4 for the Island of Oahu, Hawaii (R. Whittier, personal communication, October 15, 2024). The simulation follows the approach of Whittier et al. (2010), where the island is divided into several aquifers, each characterized by varying hydraulic properties based on its specific geological features, and treated as a homogeneous, anisotropic system.

The island includes a high-level groundwater zone with vertical flow barriers, such as volcanic dikes, which dominate the rift zones of the volcanoes. Other zones feature freshwater-lens aquifers formed in highly conductive flank lava flows, embedded with coarser clinker beds. Brackish water mixing zones separate the freshwater lens from the underlying ocean water. The hydraulic head of the freshwater, referred to as low-level water, is generally a few meters above sea level with a low hydraulic gradient. However, it can fluctuate locally due to influences such as discharging springs, heavily pumped wells, and barriers like valley-fill deposits. The flank lavas are overlain by coastal sediments, commonly referred to as caprock, which restricts discharge to the ocean and acts as a semi-confining layer. Additionally, the island has alluvial valley-fill deposits that hinder lateral groundwater movement between stream valleys.

The main assumptions involve approximating the various geological features in the rift zone as porous media rather than treating the dikes as distinct bodies with significantly different hydraulic properties from the interbedded lava flows. A more realistic approach, which would involve accurately characterizing the three-dimensional nature and hydraulic features of the dikes, is practically challenging. Another key assumption involves disregarding the saltwater interaction with the freshwater aquifer and the resulting brackish zone. For simplification, only freshwater flow is considered, with the bottom of the freshwater lens estimated using the Ghyben-Herzberg equation based on the assumption of a sharp interface (see Section 8.2.5 in Chapter 8). While the approach described here may yield less accurate results near the coast, it is useful farther inland where freshwater management is the primary concern.

The recharge data used in the analysis were based on a watershed model, as shown in Figure 11.2. Simulation accuracy in Figure 11.4 was evaluated by comparing simulated versus measured water-levels at various aquifer wells. The accuracy is considered reasonable if the observed and simulated data points fall near a 45° line on an *x-y* plot (Figure 11.5). It is important to note that perfect accuracy is unattainable, and some scatter around the 45° line is expected, as shown in the figure. The model generally underestimates values in the high-water level range (with symbols falling below the 45° line) and overestimates

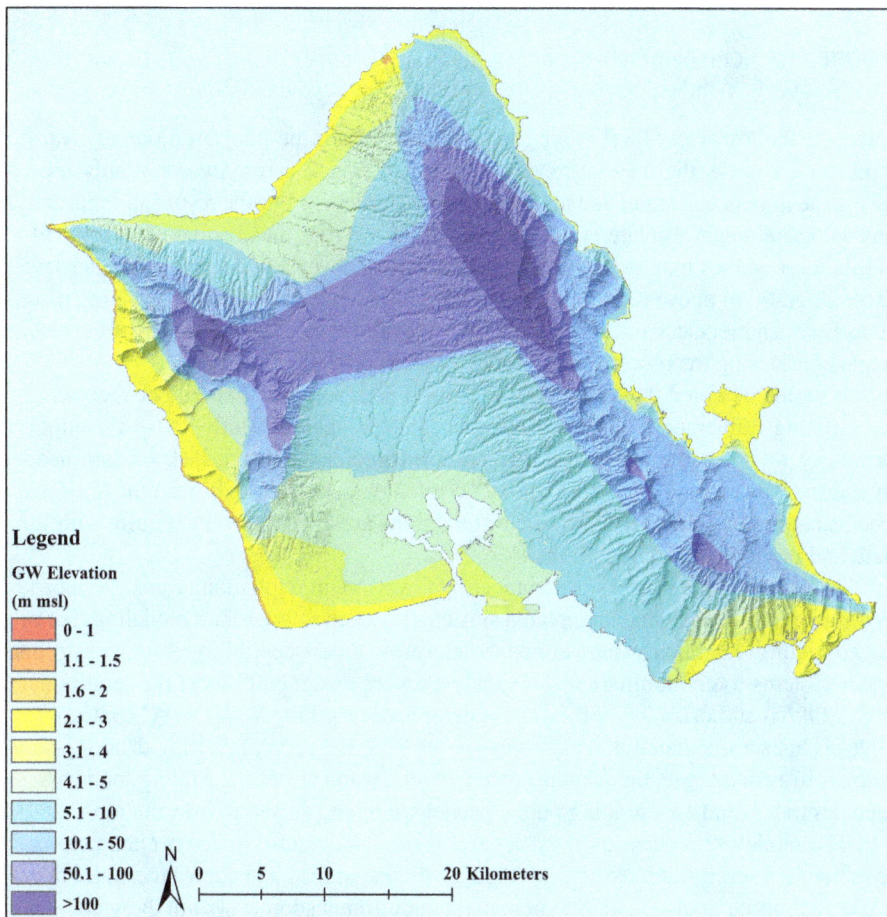

FIGURE 11.4 Simulated groundwater elevations in meters above sea level. (Source: R. Whittier, personal communication, October 15, 2024.)

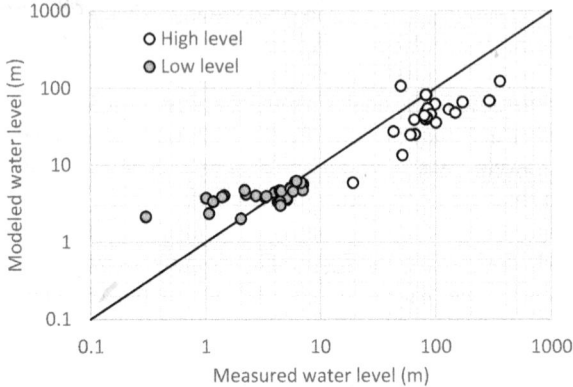

FIGURE 11.5   Comparison between measured and estimated water levels against the 45° line. (Source: R. Whittier, personal communication, October 15, 2024.)

values in the low-water level range (with symbols above the line). Additionally, various indicators, such as the mean error and root mean square error, are commonly used to assess the fit between simulated and measured water levels. For the results in Figures 11.4 and 11.5, the mean absolute error is 13 m, while the root mean squared error is 39 m. While these values may seem high, the calculations cover a wide range of water levels, exceeding 300 m above sea level. Further calibration may be required, involving discussions between modelers and decision-makers to evaluate the reasonableness of the results, while considering the objectives and scope of the modeling efforts.

It should be noted that the modeling simplified assumptions should be reconsidered if modeling refinements become necessary. In this specific study, these assumptions primarily relate to the island's highly complex geology and the lack of data needed to accurately describe such formations. While other models that account for density effects caused by saltwater-freshwater interactions are available, they require additional data and more costly computational resources.

An example of a model's use in meeting government regulations was introduced by Whittier et al. (2010), which adopted a similar approach to the island modeling that produced Figure 11.4. Such a study aimed to determine the susceptibility of various groundwater systems to contamination. The analyses were designed to meet the requirements of the 1996 reauthorization of the U.S. Federal Safe Drinking Water Act (U.S. Congress, 1996). This act mandated that each state in the U.S. focus on protecting drinking-water sources by providing technical assistance to public water systems, utilizing sound assessment methods, and encouraging public participation and access to information.

The modeling process used water-level results and aquifer transport characteristics to delineate source-water assessment areas, which are areas that contribute water flow to a specific source. It also identified potential contaminant sources within these areas. The final product included delineated capture zones for Oahu's groundwater sources and susceptibility scores based on potential contaminating activities. Areas of highest susceptibility were found, as expected, in regions dominated by high-density population centers, agricultural activities, and industrial and military operations (Whittier et al., 2010).

Case Study 11.4: Sustainability Modeling: Jeju Island, Korea

An example of the application of models in aquifer management for sustainability was introduced by El-Kadi et al. (2013) in a study focused on Jeju Island, a volcanic island located approximately 140 km south of the coast of South Korea. According to Sohn (1996), the island's geology primarily consists of permeable basalt at the surface, underlain by layers of low-permeability volcanic and unconsolidated sediments. El-Kadi et al.'s study aimed to evaluate the sustainability of groundwater use under projected scenarios of climate change, land use, and increased pumping.

The study employed a calculation scheme to assess the groundwater system's response to varying scenarios, incorporating different recharge values based on selected climate conditions. Both climate change impacts and prolonged drought periods, lasting up to ten years, were considered. Results for each level of water withdrawal were used to predict potential negative effects on water levels, spring flows, and salinity. The analysis integrated multiple hydrological processes controlling water flow, using available data on the island's water resources.

At the time of the study, the sustainable yield of the island was estimated at 1.77 million m³/day. As a precautionary measure, authorities set the recommended maximum extraction limit at 64% of this value, or approximately 1.5 million m³/day. The study concluded that even the recommended extraction limit could result in some long-term stress on groundwater resources, regardless of drought or climate change. Under a 10-year drought scenario, the situation could worsen, with spring flows projected to decrease by an average of 20%, and 12% of springs potentially drying up. Additionally, island-wide water levels were expected to drop by an average of 3 m. In scenarios involving full sustainable yield usage, salinity levels could rise in some areas to as much as 9,000 mg/L.

The study recommended implementing water-use policies in the western part of the island, where water-level declines could be particularly severe under various scenarios. Additionally, restrictive management practices were advised in coastal areas to prevent further declines in water levels and seawater intrusion.

11.5 CLIMATE MODELING

Just like other classes, climate models are essential for understanding related processes, identifying necessary information, planning field data collection studies, and predicting future conditions. The modeling process starts with developing a conceptual model, followed by utilizing a numerical scheme to solve the model's governing mathematical equations. For reliable predictions, the model requires calibration and validation.

Climate conceptual models incorporate equations that describe the atmosphere, oceans, land surface, and ice. The atmospheric component concerns air temperature, humidity, wind patterns, and atmospheric pressure. The ocean component covers sea surface temperatures, currents, salinity, and ocean heat content. The land surface component includes soil moisture, vegetation, and surface temperature. The ice component accounts for the dynamics of glaciers, ice sheets, and sea ice.

Climate models are crucial for addressing climate change by enhancing the understanding of its mechanisms and predicting future scenarios based on different greenhouse gas emission trajectories. The results are used to assess the potential impacts of climate change on ecosystems, human health, agriculture, and infrastructure, providing valuable data for policymakers to develop strategies for mitigation and adaptation.

Types of Climate Models are listed as follows.

1. **Global Climate Models** (e.g., Washington and Parkinson, 1986; IPCC, 2013): Simulate the climate on a global scale by dividing the Earth into a three-dimensional grid. These models are used for long-term climate projections.
2. **Regional Climate Models** (e.g., Giorgi et al., 1993; Jacob et al., 2014): Study specific regions, providing higher resolution data for more detailed climate impact assessments.
3. **Earth System Models** (e.g., Taylor et al., 2012; Eyring et al. 2016): Integrate additional components, such as the carbon cycle and vegetation, offering a more detailed assessment of interactions within the climate system.

Challenges in climate modeling stem from the climate system's complexity, with numerous interacting components that are difficult to accurately describe. Uncertainties in model projections arise from incomplete understanding of climate processes and their interactions, as well as data limitations. Future human activities, an essential element of model conceptualization, also contribute to uncertainties. Additionally, higher resolution modeling, which enhances accuracy, requires significant computational power and resources.

However, recent advances are promising. Improved data from satellite observations and ground-based measurements are contributing to enhanced model calibration and validation. Advances in supercomputing allow for higher resolution and more detailed simulations, enabling further process integration. Finally, global interest has driven collaborative projects that improve models by comparing different approaches and sharing data and results.

REFERENCES

AQUAVEO. (2023a). Watershed modeling system. https://www.aquaveo.com/software/wms-watershed-modeling-system-introduction

AQUAVEO. (2023b). Surface water modeling system. https://www.aquaveo.com/software/sms-surface-water-modeling-system-introduction

AQUAVEO. (2023c). Groundwater modeling system. https://www.aquaveo.com/software/gms-groundwater-modeling-system-introduction

Becker, A., Klöcking, B., Lahmer, W., & Pfützner, B. (2002). The hydrological modelling system ARC/EGMO. In: Singh V. P., Frevert, D. (eds), *Mathematical Models of Large Watershed Hydrology* (pp. 321–384). Water Resources Publications. ISBN 1887201343, 9781887201346.

El-Kadi, A. I., & Yamashita, E. (2007). Modeling streamflows and flood delineation of the 2004 flood disaster, Mânoa, Oʻahu, Hawaiʻi. *Pacific Science, 61*(2), 235–255. https://www2.hawaii.edu/~ericyama/61.2el-kadi.pdf

El-Kadi, A. I., Tillery, S., Whittier, R. B., Hagedorn, B., Mair, A., Ha, K., & Koh, G.-W. (2013). Assessing sustainability of groundwater resources on Jeju Island, South Korea, under climate change, drought, and increased usage. *Hydrogeology Journal, 22*(3), 625–642. https://doi.org/10.1007/s10040-013-1084-y

EPA. (2002). Hydrological Simulation Program - Fortran (HSPF). https://19january2021snapshot.epa.gov/ceam/hydrological-simulation-program-fortran-hspf_.html

Engott, J. A., Johnson, A. G., Bassiouni, M., Izuka, S. K., & Rotzoll, K. (2017). Spatially distributed groundwater recharge for 2010 land cover estimated using a water-budget model for the Island of Oʻahu, Hawaiʻi (ver. 2.0, December 2017). U.S. Geological Survey Scientific Investigations Report 2015–5010, 49 p. https://doi.org/10.3133/sir20155010

Eyring, V., Bony, S., Meehl, G. A., Senior, C. A., Stevens, B., Stouffer, R. J., & Taylor, K. E. (2016). Overview of the Coupled Model Intercomparison Project Phase 6 (CMIP6) Experimental Design and Organization. *Geoscientific Model Development, 9*(5), 1937–1958. https://doi.org/10.5194/gmd-9-1937-2016

FAO. (2023). Soil and Water Assessment Tool (SWAT). https://www.fao.org/land-water/land/land-governance/land-resources-planning-toolbox/category/details/en/c/1111246/

Gingerich, S. B., Boschmann, D. E., Grondin, G. H., & Schibel, H. J. (2024). Groundwater model of the Harney Basin, southeastern Oregon (U.S. Geological Survey Scientific Investigations Report 2024–5017, 104 p.). U.S. Geological Survey. https://doi.org/10.3133/sir20245017

Giorgi, F., Marinucci, M. R., Bates, G. T., & De Canio, G. (1993). Development of a Second-Generation Regional Climate Model (RegCM2). Part I: Boundary-Layer and Radiative Transfer Processes. *Monthly Weather Review, 121*(10), 2794–2813. https://doi.org/10.1175/1520-0493(1993)121<2794:DOASGR>2.0.CO;2

Harbaugh, A. W., Banta, E. R., Hill, M. C., & McDonald, M. G. (2000). MODFLOW-2000, the U.S. Geological Survey modular ground-water model; user guide to modularization concepts and the ground-water flow process. U.S. Geological Survey Open-File Report 00–92, 121 pp. https://doi.org/10.3133/ofr200092

IPCC. (2013). Climate change 2013: The physical science basis. Contribution of working group I to the fifth assessment report of the Intergovernmental Panel on Climate Change [Stocker, T. F., Qin, D., Plattner, G.- K., Tignor, M., Allen, S. K., Boschung, J., ... & Midgley, P. M. (eds)]. Cambridge University Press. https://www.ipcc.ch/report/ar5/wg1/

Jacob, D., Petersen, J., Eggert, B., Alias, A., Christensen, O. B., Bouwer, L. M., ... & Teichmann, C. (2014). EURO-CORDEX: New high-resolution climate change projections for European impact research. *Regional Environmental Change, 14*(2), 563–578. https://link.springer.com/article/10.1007/s10113-013-0499-2

MIKE. (2023). All-in-one groundwater modelling solution. https://www.mikepoweredbydhi.com/products/feflow

Minnesota Pollution Control Agency. (2023). Available stormwater models and selecting a model. https://stormwater.pca.state.mn.us/index.php/Available_stormwater_models_and_selecting_a_model

Sohn, Y. (1996). Hydrovolcanic processes forming basaltic tuff rings and cones on Cheju Island, Korea. *Geological Society of America Bulletin, 108*(10), 1199–1211. https://pubs.geoscienceworld.org/gsa/gsabulletin/article-abstract/108/10/1199/183078/Hydrovolcanic-processes-forming-basaltic-tuff?redirectedFrom=fulltext

Taylor, K. E., Stouffer, R. J., & Meehl, G. A. (2012). An overview of CMIP5 and the experiment design. *Bulletin of the American Meteorological Society, 93*(4), 485–498. https://doi.org/10.1175/BAMS-D-11-00094.1

Texas A&M University. (2023). Watershed models. https://hydrologicmodels.tamu.edu/watershed-models/

U.S. Congress. (1996). Safe Drinking Water Act Amendments of 1996 (S.1316). *104th Congress (1995–1996).* https://www.congress.gov/bill/104th-congress/senate-bill/1316

U.S. Geological Survey. (2018). Water, Energy, and Biogeochemical Model (WEBMOD). https://www.usgs.gov/software/water-energy-and-biogeochemical-model-webmod

U.S. Geological Survey. (2023a). Water resources surface water software. https://water.usgs.gov/software/lists/surface_water

U.S. Geological Survey. (2023b). Water resources groundwater software. https://water.usgs.gov/software/lists/groundwater

Washington, W. M., & Parkinson, C. L. (1986). *An Introduction to Three-Dimensional Climate Modeling.* University Science Books. https://www.researchgate.net/publication/31845793_An_Introduction_to_Three-Dimensional_Climate_Modeling_WM_Washington_CL_Parkinson#read

Webb, R. M. T., & Parkhurst, D. L. (2017). Water, energy, and biogeochemical model (WEBMOD), user's manual, version 1 (U.S. Geological Survey Techniques and Methods, book 6, chap. B35, 171 p.). https://doi.org/10.3133/tm6B35

Whittier, R., Rotzoll, R., Dhal, S., El-Kadi, A. I., Ray, C., & Chang, D. (2010). Groundwater source assessment program for the state of Hawaii, USA: Methodology and example application. *Journal of Hydrogeology, 18,* 711–723. https://link.springer.com/article/10.1007/s10040-009-0548-6

Advanced Data Collection Techniques 12

12.1 INTRODUCTION

Hydrologic investigations primarily rely on a blend of experimental and computational approaches, and significant advancements in both techniques have occurred throughout the 21st century. Experimental approaches involve collecting data through field and laboratory work to support computational analyses conducted in office settings. Fieldwork is crucial for site assessment, and the data obtained are invaluable for theoretical investigations. The objective of these investigations is to attain a deeper understanding of the underlying physical, chemical, and biological processes as well as to forecast future conditions related to water flow and chemical transport. Although laboratory experiments typically emphasize process comprehension, they often do not fully account for the broader-scale scenarios encountered in the field.

12.2 WATER QUALITY ANALYSIS

Assessing the quality of groundwater and surface water is crucial for both human health and environmental sustainability. Direct and indirect techniques are employed in this process. Water contaminants may include turbidity and concentrations of substances such as heavy metals, nutrients, emerging contaminants (pharmaceuticals, PFAS, etc.), and pathogens (bacteria and viruses). Direct water quality measurements involve collecting water samples and analyzing them to determine these parameters, providing an immediate assessment of the water source's quality. Indirect methods, on the other hand, use proxies or indicators to evaluate water quality. The presence or absence of

DOI: 10.1201/9781003587149-12

271

these species reflects the overall health of the water ecosystem. One such example is bioindicators—organisms sensitive to changes in water conditions. These include certain species of fish, aquatic plants, algae, and microorganisms.

Surface water quality techniques are generally non-intrusive and can also measure river flow. In contrast, groundwater field techniques primarily involve collecting aquifer data through well drilling, an intrusive method that can disturb the aquifer and often incurs significant costs. Wells can be used to measure water levels and assess contamination through discrete or continuous sampling. Although techniques exist for onsite water quality measurements, it is common practice to utilize water laboratories for chemical analyses, which use a range of specialized equipment to measure diverse physical, chemical, and biological parameters (e.g., Hauser, 2001). The overarching goals of these analyses are to determine whether water is suitable for consumption, environmental well-being, or industrial processes. To conduct water quality tests, various meters analyze pH, indicating the acidity or alkalinity of water; conductivity (measuring electrical current, which correlates with ion concentrations in the water, such as salt); dissolved oxygen; turbidity; and water flow. Depending on the target chemicals, equipment may include spectrophotometers, mass spectrometers, and atomic absorption spectrophotometers. Chromatography is employed for the separation and quantification of chemical compounds in water samples, while ion chromatography is specifically used for the analysis of ions and anions in water, such as sodium, nitrate, sulfate, and chloride. Other instruments encompass microbiological testing equipment, biochemical oxygen demand incubators, chemical oxygen demand reactors, ammonia analyzers, and total organic carbon analyzers. Various analyses are supported by water sampling equipment, autosamplers, data loggers, and data analysis software.

12.3 DOPPLER RADAR

Doppler radar technology represents a specialized form of radar system designed to detect the velocity of objects such as aircraft, ships, water, or weather phenomena by measuring the Doppler shift in the frequency of radar signals (e.g., Bringi and Chandrasekar, 2001; Rinehart, 1996). Doppler radar systems function by emitting electromagnetic waves, typically in the microwave or radio frequency range. When these radar waves encounter a moving object, some of the waves are reflected back toward the radar antenna, with their frequency changing due to the Doppler effect. If the object is moving toward the radar system, the frequency of the returning waves increases (known as blue shift); if the object is moving away, the frequency decreases (known as red shift). The radar system analyzes the frequency shift of the returned signals to calculate the velocity of the objects. By comparing the transmitted frequency with the received frequency, the radar can determine the speed and direction of the moving target.

This technology finds applications in diverse fields, including air traffic control, military use, traffic law enforcement, and various industrial applications, such as measuring vehicle speed in automotive contexts. It is also employed in manufacturing and quality control to detect motion and assess production processes.

Doppler radar can be utilized in several hydrology-related applications, such as monitoring and studying precipitation, wind patterns, and storm systems (Brown and Johnson, 2020; National Weather Service, n.d.). These radar systems offer real-time information regarding the location, intensity, and movement of precipitation events and can estimate the amount of rainfall over a specific area (e.g., Martinez, 2016). They are also used to monitor snowfall in regions where snowmelt contributes to water resources. In areas where snowpack is a significant source of water, Doppler radar technology can estimate the water content in the snow. Understanding how precipitation patterns are changing over time is crucial for future water resource planning and management.

Additionally, Doppler radar technology is valuable for tracking and predicting severe weather events, including thunderstorms, hurricanes, and tornadoes. Through continuous monitoring, radar technology aids authorities in issuing timely flood alerts and taking necessary measures to mitigate the impact of floods. Furthermore, Doppler radar data collected over extended periods can contribute to studies on the long-term climate change impacts on water resources.

An important application of ADCP technology is measuring stream flows, serving as an alternative to various classic approaches described in Section 5.2.2 (Chapter 5). This technology has extensive applications in both oceanography and hydrology (e.g., Woods Hole Oceanographic Institution, 2023; Côté et al., 2011; Appell et al., 1991). The process begins with the emission of high-frequency sound waves into the water. These waves travel through the water and are either dissipated or reflected back toward the ADCP when they encounter suspended particles or objects, such as sediment, plankton, or even water molecules. The ADCP then measures the frequency of the returning sound waves, which is altered due to the relative motion between the ADCP and the water particles. By analyzing the frequency shift from multiple beams at different angles and depths, the ADCP can calculate the three-dimensional velocity profile of the water column.

ADCPs come in various configurations, including downward-looking, upward-looking, and side-looking instruments, each tailored for specific applications. They can be installed on moorings or buoys, or deployed on research vessels to collect data over varying time scales and depths. Mueller et al. (2013) detail the approach for measuring discharge with an ADCP from a moving boat.

One notable advantage of ADCPs is their higher efficiency and accuracy compared to traditional current meters. They are well-suited for measuring larger rivers and small-scale currents and are effective for water columns up to 1,000 m in length. Importantly, unlike other technologies, ADCPs measure the absolute speed of the water, not just the relative speed of one water mass in relation to another. However, there are several disadvantages associated with ADCPs. For instance, depending on the type of sound waves used, there is often a compromise between measurement distance and precision. Additionally, ADCPs have a relatively rapid battery consumption rate. In clear waters, such as those found in tropical regions, the sound waves may not encounter enough particles to produce reliable data. Lastly, the presence of bubbles in turbulent water or swimming marine life can lead to miscalculations of velocities.

12.4 TELEMETRY

Telemetry plays a crucial role in sophisticated systems designed for real-time data acquisition, transmission, and monitoring of various parameters related to water resources (Anderson and Burt, 2010). Continuously recorded and transmitted data are commonly presented through graphical interfaces and charts, facilitating interpretation and supporting decision-making processes. Ultimately, these systems empower hydrologists and water resource managers by providing comprehensive access to real-time information for analysis. In addition to offering real-time updated data, the detailed data collected significantly improve the understanding of hydrological processes, thereby improving the effectiveness of water resource management.

Applications of automated data-related technologies span various domains, including weather and climate monitoring (Trenberth et al., 2006; Longman et al., 2024), dam and reservoir monitoring (U.S. Army Corps of Engineers, 2013), groundwater monitoring (Butler and Healey, 2008), and flood warning systems (Horritt and Bates, 2002).

Within these advanced systems, telemetry involves transmitting data from monitoring stations to central databases, which is crucial for remote data collection. This process utilizes communication technologies such as satellite communication, which is widely used in remote areas to ensure reliable and continuous information flow. Alternatively, cellular networks are suitable for more accessible regions due to their cost-effectiveness and extensive coverage. In contrast, radio frequency communication is deployed in areas with limited cellular coverage, particularly beneficial in rugged terrains where other communication methods may be challenging.

12.5 STABLE AND RADIOACTIVE ISOTOPES

Further advanced approaches involve the application of stable and radioactive isotopes to study and trace water within hydrological systems (e.g., LLNL, 2023; ScienceDirect, 2023; Clark and Fritz, 1997). Isotopes are variations of a chemical element characterized by an identical number of protons but a differing number of neutrons in their atomic nuclei. They are categorized as stable or radioactive based on their nuclear stability. Stable isotopes remain unchanged over time, as they do not undergo spontaneous radioactive decay. In contrast, radioactive isotopes, also known as radioisotopes, possess an unstable nucleus and undergo radioactive decay, emitting radiation in the form of alpha or beta particles or gamma rays to achieve a more stable state.

Both stable and radioactive isotopes play crucial roles in hydrology, serving as valuable tools for tracing water movement, identifying sources of water or contaminants, understanding residence times, and studying various hydrological processes.

This information is essential for water resource management and environmental studies.

Stable isotopes, such as oxygen-18 and deuterium (hydrogen-2), are frequently employed as tracers to investigate water movement across various hydrologic compartments, including rivers, lakes, and groundwater (e.g., Gat, 1996). The ratios of oxygen-18 to oxygen-16 ($\delta^{18}O$) and deuterium to hydrogen (δD) in water help identify water sources and understand processes like evaporation and condensation. Carbon-13, another stable isotope, is used to study the carbon cycle in aquatic ecosystems (e.g., NOAA Global Monitoring Laboratory, 2024).

On the other hand, radioactive isotopes like tritium (hydrogen-3) and iodine-131 serve as tracers to determine flow rates and water paths in hydrological systems (e.g., Mahlangu et al., 2020). The movement of naturally occurring tritium, for instance, can be tracked over time, providing valuable insights into water transport and mixing. Sabarathinam et al. (2023) provide a literature review of carbon isotopes in groundwater and rainwater. As an application, a combination of carbon-13 and carbon-14 isotopes is used by Bhandary et al. (2015) to determine the age, flow direction, flow velocity, and recharge area of groundwater aquifers. Carbon-14 and chlorine-36 are employed for dating relatively older groundwater, as the decay of these isotopes allows scientists to estimate the time since the water entered the aquifer.

12.6 REMOTE SENSING

Advanced hydrology techniques encompass a wide array of tools, including remote sensing technologies such as satellite imagery, aerial photography, and light detection and ranging (Jensen, 2007). These technologies play a pivotal role in monitoring various aspects of hydrology, including land use, precipitation, snow cover, and changes in water bodies. They are particularly vital in flood forecasting and early warning systems, where real-time data and weather forecasts are integrated into hydrological models to predict potential flood events.

12.7 GEOPHYSICAL APPROACHES

Process-based groundwater studies greatly benefit from non-intrusive aquifer sensing techniques that do not significantly disturb the natural environment, as opposed to intrusive methods such as drilling wells. Geophysical techniques are primarily used to investigate and characterize subsurface water resources and assess the properties of the subsurface geology. Integrating these methods with traditional hydrological investigations can lead to a more comprehensive understanding of the hydrological system in a given area. Key geophysical techniques used in hydrology include the following.

1. Ground-penetrating radar is a technique that uses radar pulses to image the subsurface (e.g., Jol, 2009). The technique is especially useful for mapping the water table, identifying features such as buried channels, and locating pipes in water distribution systems and other linear subsurface structures.

2. Electrical resistivity imaging measures subsurface electrical resistivity, which can indicate the presence of water (e.g., Binley and Kemna, 2005). The technique relies on the fact that wet materials have lower resistivity than dry materials, and salty water has lower resistivity than freshwater. Thus, it is commonly employed for groundwater exploration and mapping the extent of aquifers.

3. Seismic refraction determines the depth and characteristics of subsurface layers based on the travel times of seismic waves (Sheriff and Geldart, 1995). It is effective for mapping the water table, subsurface geology, and aquifer properties, including specific features such as faults and folds.

4. Borehole geophysical tools gather information about the subsurface at specific locations by lowering equipment into boreholes (EPA, 2024a). Techniques include borehole logging and borehole seismic methods, which assess the properties of geological formations.

5. Gravity surveys measure variations in the Earth's gravitational field, which can identify subsurface features and density variations, including the presence of groundwater and other liquids (e.g., Kennedy et al., 2021).

6. Time-domain electromagnetic surveys induce electromagnetic fields in the subsurface and measure the response, thereby mapping electrical conductivity, which can provide information about groundwater presence. Available resources include EPA (2024b).

7. Hydromagnetic surveys measure the electromagnetic response of subsurface materials and are often used to study saline water intrusion into freshwater aquifers (Ward and Hohmann, 1988).

8. Thermal methods, including thermal logging and thermal tomography, use variations in subsurface temperature to infer properties such as the presence and dynamics of groundwater (Kurylyk et al., 2019).

9. As discussed earlier, remote sensing and light detection and ranging technology are valuable techniques for monitoring various aspects of hydrology.

12.8 SOIL MOISTURE MEASUREMENT

Numerous methods are available for measuring soil moisture at a specific point (e.g., Robinson et al., 2008). However, point measurements are not representative of the surrounding area due to the high spatial variability of soil moisture across domains of interest (e.g., Famiglietti et al., 2008). As an alternative, wireless and non-intrusive techniques are preferred. Recent developments in wireless soil moisture sensing have been reported by Bogena et al. (2022). Among the recent advances, the cosmic-ray soil moisture observing system (COSMOS) is a non-intrusive remote sensing technology that employs cosmic-ray neutrons to assess soil moisture on a large scale (Zreda et al., 2012). Cosmic-ray neutrons are high-energy neutrons originating from space, primarily from outside the solar system. They are a component of cosmic rays, which consist of energetic particles constantly bombarding the Earth from various sources in the universe. The COSMOS measurement system comprises an array of detectors that assess the flux of cosmic-ray neutrons. When these particles interact with the Earth's surface, they are moderated by the hydrogen atoms in the soil. The intensity of the moderated neutrons is inversely proportional to the soil moisture content. Therefore, by measuring the flux of these neutrons, soil moisture can be inferred.

COSMOS has an advantage over intrusive and satellite-based technologies by not requiring direct contact with the soil or being constrained by the high costs of satellite data. It provides soil moisture estimates over a large area and offers estimates at multiple depths in the soil profile, typically reaching depths of approximately 1 m. Thus, COSMOS arrays can cover extensive geographic areas, making them well-suited for regional and even global-scale soil moisture monitoring. A recent application of the technology is presented by Heistermann et al. (2023), who compiled 3 years of soil moisture observations at an agricultural research site in northeast Germany.

12.9 SATELLITE DATA COLLECTION

The Gravity Recovery and Climate Experiment (GRACE)[1] satellite mission (Tapley et al., 2004; JPL, 2023) utilized a non-intrusive technique to collect remotely sensed water resource data. The mission was a joint project between NASA and the German Aerospace Center and consisted of two satellites, GRACE-1 and GRACE-2, from its launch in March 2002 to the end of its science mission in October 2017, followed by GRACE Follow-On in 2018 (NASA Jet Propulsion Laboratory, 2024). These satellites were designed to precisely measure variations in Earth's gravitational field, providing valuable data for understanding Earth's climate, water resources, and geophysical processes (e.g., JPL, 2023; Swenson et al., 2006). The GRACE satellites used a unique technique that involved measuring the distance between the two satellites, which changed as they passed over areas with different gravitational pulls (JPL, 2023). These data were used to study changes in Earth's water distribution, such as ice melt, groundwater depletion, and ocean currents. The data allowed scientists to track changes in water distribution, such as the movement of groundwater, sea level rise, and ocean currents. Example studies in water resources include Famiglietti et al. (2011), Scanlon et al. (2018), and Richey et al. (2015). The latter study used GRACE data between 2003 and 2013 to conclude that 21 of the world's 37 largest aquifers have exceeded sustainability tipping points and are being depleted, with 13 of them considered significantly distressed. Hydrology-related data are useful for understanding and managing water resources, especially in regions prone to droughts, floods, and sea level rise.

NOTE

1 An acronym for Gravity Recovery and Climate Experiment.

REFERENCES

Anderson, M. G., & Burt, T. P. (eds). (2010). *Hydrological Measurements and Telemetry*, pp. 1–344. Chichester: John Wiley and Sons.
Appell, G. F., Bass, P. D., & Metcalf, M. A. (1991). Acoustic Doppler current profiler performance in near surface and bottom boundaries. *IEEE Journal of Oceanic Engineering*, 16(4), 390–396. https://doi.org/10.1109/48.90903
Binley, A., & Kemna, A. (2005). DC resistivity and induced polarization methods. In: Rubin, Y., Hubbard, S. S. (eds) *Hydrogeophysics*. Water Science and Technology Library, vol. 50. Springer, Dordrecht. https://doi.org/10.1007/1-4020-3102-5_5
Bhandary, H., Al-Senafy, M., & Marzouk, F. (2015). Usage of carbon isotopes in characterizing groundwater age, flow direction, flow velocity, and recharge area. *Procedia Environmental Sciences*, 25, 28–35. https://doi.org/10.1016/j.proenv.2015.04.005
Bogena, H. R., Weuthen, A., & Huisman, J. A. (2022). Recent developments in wireless soil moisture sensing to support scientific research and agricultural management. *Sensors*, 22(24), 9792. https://doi.org/10.3390/s22249792
Bringi, V. N., & Chandrasekar, V. (2001). *Polarimetric Doppler Weather Radar: Principles and Applications*, pp. 1–636. Cambridge: Cambridge University Press.
Brown, P. A., & Johnson, R. W. (2020). *Doppler Techniques in Water Resources: Principles and Applications*, pp. 1–412. Cham: Springer.
Butler Jr., J. J., & Healey, J. M. (2008). *Well Hydraulics and Telemetry for Environmental Monitoring*, pp. 1–256. Berlin: Springer.
Clark, I. D., & Fritz, P. (1997). *Environmental Isotopes in Hydrogeology*, pp. 1–328. Boca Raton, FL: CRC Press.
Côté, J. M., Hotchkiss, F. S., Martini, M., & Denham, C. R. (2011). Acoustic Doppler Current Profiler (ADCP) data processing system manual. U.S. Geological Survey Open File Report 00–458, vol. 4, 51 p. https://pubs.usgs.gov/of/2000/of00-458/
EPA. (2024a). Borehole geophysical methods. https://www.epa.gov/environmental-geophysics/borehole-geophysical-methods
EPA. (2024b). Electromagnetic methods: Time domain. Retrieved October 9, 2024, from https://www.epa.gov/environmental-geophysics/electromagnetic-methods-time-domain
Famiglietti, J. S., Ryu, D., Berg, A. A., Rodell, M., & Jackson, T. J. (2008). Field observations of soil moisture variability across scales. *Water Resources Research*, 44, W01423. https://doi.org/10.1029/2006WR005804
Famiglietti, J. S., Lo, M., Ho, S. L., Bethune, J., Anderson, K. J., Syed, T. H., & Rodell, M. (2011). Satellites measure recent rates of groundwater depletion in California's Central Valley. *Geophysical Research Letters*, 38(3), L03403. https://doi.org/10.1029/2010GL046442
Gat, J. R. (1996). Oxygen and hydrogen isotopes in the hydrologic cycle. *Annual Review of Earth and Planetary Sciences*, 24(1), 225–262. https://doi.org/10.1146/annurev.earth.24.1.225
Hauser, B. A. (2001). *Drinking Water Chemistry: A Laboratory Manual* (2nd ed.). CRC Press. https://doi.org/10.1201/9781420032451

Heistermann, M., Francke, T., Scheiffele, L., Dimitrova Petrova, K., Budach, C., Schrön, M., Trost, B., Rasche, D., Güntner, A., Döpper, V., Förster, M., Köhli, M., Angermann, L., Antonoglou, N., Zude-Sasse, M., & Oswald, S. E. (2023). Three years of soil moisture observations by a dense cosmic-ray neutron sensing cluster at an agricultural research site in north-east Germany. *Earth System Science Data, 15*, 3243–3262. https://doi.org/10.5194/essd-15-3243-2023

Horritt, M. S., & Bates, P. D. (2002). Evaluation of 1D and 2D numerical models for predicting river flood inundation. *Journal of Hydrology*, 268(1–4), 87–99. https://doi.org/10.1016/S0022-1694(02)00121-X

Jensen, J. R. (2007). *Remote Sensing of the Environment: An Earth Resource Perspective*, pp. 1–592. Upper Saddle River, NJ: Pearson Education.

Jol, H. M. (2009). *Ground Penetrating Radar: Theory and Applications*, pp. 1–544. Amsterdam: Elsevier.

JPL. (2023). GRACE. Retrieved from https://grace.jpl.nasa.gov/mission/grace/

Kennedy, J. R., Pool, D. R., & Carruth, R. L. (2021). Procedures for field data collection, processing, quality assurance and quality control, and archiving of relative- and absolute-gravity surveys (Techniques and Methods 2-D4). U.S. Geological Survey. https://doi.org/10.3133/tm2D4

Kurylyk, B. L., Irvine, D. J., & Bense, V. F. (2019). Theory, tools, and multidisciplinary applications for tracing groundwater fluxes from temperature profiles. *Wiley Interdisciplinary Reviews: Water*, 6(1), e1329. https://doi.org/10.1002/wat2.1329

LLNL. (2023). Isotope hydrology. Retrieved from https://water.llnl.gov/isotopes

Longman, R. J., Lucas, M. P., Mclean, J., Cleveland, S. B., Kodama, K., Frazier, A. G., Kamelamela, K., Schriber, A., Dodge II, M., Jacobs, G., & Giambelluca, T. W. (2024). The Hawai'i Climate Data Portal (HCDP). *Bulletin of the American Meteorological Society*, 105(7), E1074–E1083. https://doi.org/10.1175/BAMS-D-23-0188.1

Mahlangu, S., Lorentz, S., Diamond, R., & Dippenaar, M. (2020). Surface water-groundwater interaction using tritium and stable water isotopes: A case study of Middelburg, South Africa. *Journal of African Earth Sciences, 171*, 103886. https://doi.org/10.1016/j.jafrearsci.2020.103886

Martinez, S. M. (2016). Application of Doppler LiDAR for streamflow monitoring in mountain watersheds (Master's Thesis, University of Colorado Boulder). https://example.edu/thesis/doppler-lidar-streamflow

Mueller, D. S., Wagner, C. R., Rehmel, M. S., Oberg, K. A., & Rainville, F. (2013). Measuring discharge with acoustic Doppler current profilers from a moving boat (ver. 2.0, December 2013). U.S. Geological Survey Techniques and Methods, book 3, chap. A22, 95 p. https://dx.doi.org/10.3133/tm3A22

NASA Jet Propulsion Laboratory. (2024). GRACE follow-on mission overview. *GRACE-FO*. Retrieved October 9, 2024, from https://gracefo.jpl.nasa.gov/mission/overview/

National Weather Service. (n.d.). *Using and understanding Doppler radar*. National Oceanic and Atmospheric Administration, U.S. Department of Commerce. Retrieved July 3, 2025, from https://www.weather.gov/mkx/using-radar

NOAA Global Monitoring Laboratory. (2024). Carbon isotopes in atmospheric carbon dioxide. *NOAA*. https://gml.noaa.gov/ccgg/isotopes/chemistry.html

Richey, A. S., Thomas, B. F., Lo, M.-H., Reager, J. T., Famiglietti, J. S., Voss, K., Swenson, S., & Rodell, M. (2015). Quantifying renewable groundwater stress with GRACE. *Water Resources Research, 51*(7), 5217–5238. https://doi.org/10.1002/2015WR017349

Rinehart, R. E. (1996). *Radar for Meteorologists* (3rd ed.), pp. 1–428. Grand Forks, ND: Rinehart Publications.

Robinson, D. A., Campbell, C. S., Hopmans, J. W., Hornbuckle, B. K., Jones, S. B., Knight, R., Ogden, F., Selker, J., & Wendroth, O. (2008). Soil moisture measurement for ecological and hydrological watershed-scale observatories: A review. *Vadose Zone Journal*, 7(1), 358–389, https://doi.org/10.2136/vzj2007.0143

Sabarathinam, C., Al-Rashidi, A., Alsabti, B., Samayamanthula, D. R., & Kumar, U. S. (2023). A review of the publications on carbon isotopes in groundwater and rainwater. *Water,* 15(19), 3392. https://doi.org/10.3390/w15193392

Scanlon, B. R., Zhang, Z., Save, H., Wiese, D. N., Landerer, F. W., & Long, D. (2018). Global models underestimate large decadal declining and rising water storage trends relative to GRACE satellite data. *Proceedings of the National Academy of Sciences,* 115(6), E1080–E1089. https://doi.org/10.1073/pnas.1704665115

ScienceDirect. (2023). Water isotopes. Retrieved from https://www.sciencedirect.com/topics/earth-and-planetary-sciences/water-isotopes

Sheriff, R. E., & Geldart, L. P. (1995). *Exploration Sismology.* Cambridge University Press. https://doi.org/10.1017/CBO9781139168359

Swenson, S., Yeh, P. J.-F., Wahr, J., & Famiglietti, J. (2006). A comparison of terrestrial water storage variations from GRACE with in situ measurements from Illinois. Geophysical Research Letters, 33(16), L16401. https://doi.org/10.1029/2006GL026962

Tapley, B. D., Bettadpur, S., Watkins, M., & Reigber, C. (2004). The gravity recovery and climate experiment: Mission overview and early results. *Geophysical Research Letters,* 31(9), L09607. https://doi.org/10.1029/2004GL019920

Trenberth, K. E., Moore, B., Karl, T. R., & Nobre, C. (2006). Monitoring and prediction of the Earth's climate: A future perspective. *Journal of Climate*, 19(24), 5001–5008. https://doi.org/10.1175/JCLI3897.1

U.S. Army Corps of Engineers. (2013). *Instrumentation and Monitoring of Dams and Reservoirs* (Engineer Manual EM 1110-2-1906).

Ward, S. H., & Hohmann, G. W. (1988). Electromagnetic theory for geophysical applications. In: *Electromagnetic Methods in Applied Geophysics*, vol. 1, pp. 131–311. https://doi.org/10.1190/1.9781560802631.ch4

Woods Hole Oceanographic Institution. (2023). Acoustic Doppler Current Profiler (ADCP). https://www.whoi.edu/what-we-do/explore/instruments/instruments-sensors-samplers/acoustic-doppler-current-profiler-adcp/

Zreda, M., Shuttleworth, W. J., Zeng, X., Zweck, C., Desilets, D., Franz, T., & Rosolem, R. (2012). COSMOS: The COsmic-ray soil moisture observing system. *Hydrology and Earth System Sciences,* 16, 4079–4099. https://doi.org/10.5194/hess-16-4079-2012

Advanced Data Analysis Techniques

13

13.1 STATISTICAL HYDROLOGY

Hydrological studies are essential for understanding and managing water resources, predicting hydrological events, and assessing the impacts of human use and climate change on water systems. In general, studies adopt a physically based approach within a deterministic framework to examine water availability and quality within the water cycle. Simply stated, this approach involves conducting field studies to collect information about the hydrological system and using theoretical analyses to relate such data to issues of concern, such as water and contamination levels.

Alternatively, studies can adopt an approach that applies statistical methods to analyze and interpret hydrological data. Hydrological data can include measurements of precipitation (rainfall and snow), streamflow, groundwater levels and flow, and evaporation and transpiration. It should be emphasized that combining physical and statistical methods is expected to provide enhanced results by utilizing all available information and the best understanding of the controlling hydrological processes.

The following sections briefly discuss various statistical methods tailored to the book's audience. These methods can be roughly grouped into two categories: (1) data analysis and modeling and (2) simulation and forecasting. Appropriate training in statistical techniques is essential before delving into the details of these methods, which are beyond the scope of this textbook. Details on statistical hydrology can be accessed from various resources. Examples include Haan (1977), which covers statistical methods in hydrology, including probability distributions, regression analysis, and stochastic processes. Chow et al. (1988) discuss the principles of hydrology with a section dedicated to statistical methods. Beven and Freer (2001) explore statistical methods to quantify uncertainty and sensitivity in hydrological modeling. Viessman and Lewis (2003) provide an introduction to hydrology with a focus on statistical analysis and hydrological modeling.

DOI: 10.1201/9781003587149-13

13.1.1 Data Analysis and Modeling

Data analysis approaches include time series analysis, which examines patterns and trends in hydrological data over time, such as streamflow, precipitation, and groundwater levels. Approaches also include frequency analysis, which estimates the likelihood of extreme events (e.g., floods and droughts) by fitting probability distributions to historical data. Common distributions used in hydrology include the normal, log-normal, exponential, Gamma, and Gumbel distributions. These distributions are employed in modeling different types of hydrological processes, including predicting extreme events such as maximum rainfall or peak river flows.

Techniques also include trend analysis, which focuses on detecting long-term trends in hydrological data. For example, the objective might be to assess the expected future impacts of climate change or human activities on water resources. Using available data, statistical tests are conducted to identify temporal trends and account for seasonality. Techniques such as linear or nonlinear regression, autocorrelation, and time series analyses are employed to characterize relationships between variables. Geographic information systems (GISs, Section 13.2) are useful for visualizing spatial trends and correlations, while the kriging technique (Section 13.3) is utilized to estimate hydrologic variables at unmeasured locations. In some instances, hydrologic models (Chapter 11) are developed to simulate hydrologic processes as an important step in predicting future trends.

Data analyses also include risk assessment, which evaluates the risk and uncertainty associated with hydrological events, such as floods and droughts. This assessment is essential for developing effective water management strategies and designing infrastructure and other actions necessary for mitigating potential damage to life and property. The first step in risk assessment is to use historical data and statistical models to estimate the likelihood and severity of hydrological events. Next, models are employed to simulate the hydrological processes of concern, mapping and analyzing areas and populations that might be affected by hydrological hazards. The susceptibility of the exposed elements to the identified hazards is assessed based on land use, socio-economic conditions, and other factors. Strategies are then developed to reduce risk, such as flood protection plans and water conservation measures. Finally, plans for early warning systems and emergency responses are integrated with the monitoring of hydrological conditions, allowing for updates to risk assessments as new data become available.

13.1.2 Simulations and Forecasting

This class of techniques aims to understand, predict, and manage water resources processes through the development of stochastic simulation models. Such models include regression analysis, which is a statistical method used to model and analyze the relationships between hydrological variables. Examples include relationships between rainfall and runoff; streamflow and precipitation, evapotranspiration, and land use; pollutant concentrations and hydraulic head and porosity; and groundwater levels and precipitation and recharge rates. Section 2.3.4 discusses the application of the least squares method in regression analysis, introducing an example of linear regression and various expressions used in non-linear cases.

Through multivariate analysis, regression can be generalized to analyze multiple variables simultaneously to find patterns and correlations and understand how they relate to each other. This type of analysis is particularly suitable for cases involving complex data structures and relationships between variables. One such technique is multiple regression, which is used to predict the value of a variable based on the values of two or more other variables.

Stochastic models use statistical concepts to analyze hydrological processes and predict future conditions, incorporating the inherent randomness and variability of natural systems. Stochastic processes are defined by mathematical expressions that combine controlling variables representing the evolution of a system over time. For example, this analysis can describe temporal variations in stream flows, groundwater levels, and water quality indicators, such as chemical concentrations. In this regard, time series analysis employs statistical techniques to model and predict future values based on previously observed data. Common methods in this approach include autoregressive models, which use past values to predict future ones; moving average models, which use past forecast errors to predict future values; and autoregressive moving average models, which combine these two types of models.

The techniques also include the Markov chains model, which is represented by sequences of random events where the probability of each event depends only on the state of the previous event. Monte Carlo simulations are also used, involving the generation of a large number of random samples from a probabilistic model to assess the variability and uncertainty of hydrological processes. In general, one or more input variables are generated from their respective distributions and utilized as input to a suitable model that generates the outputs of interest. Such outputs can estimate the probability distributions of outcomes, such as flood risks or water availability.

Finally, stochastic differential equations can be used to describe the evolution of hydrological variables influenced by random fluctuations. This approach is suitable for complex systems where dynamics and stochasticity play a significant role, such as groundwater flow or chemical transport.

It is important to note that combining stochastic and deterministic models can enhance predictive capabilities and provide a more detailed understanding of hydrological processes. For example, a Monte Carlo application can utilize physically based governing equations, such as those controlling groundwater flow or chemical transport, to generate a large number of random samples through numerous simulations. The data input to the equations would be an equally probable distribution of parameters, such as hydraulic conductivity and dispersion coefficients. The output would be represented in statistical terms, such as averages and variability of water levels, or in probabilities of exceedance, such as potential risks to human health.

13.1.3 Probability Density Functions Commonly Used in Hydrology

Due to their inherent variability and uncertainty, variables such as rainfall, river discharge, and groundwater levels are often treated as random variables. Section 2.3.3 covered basic statistical terms and definitions, including probability density functions (PDFs).

A PDF describes the likelihood of a random variable taking on a particular value. For continuous variables, the PDF, $f(x)$, defines the probability of the variable falling within a specific range between a and b as given by the integral of the function over that range via equation (13.1):

$$P(a \geq X \leq b) = \int_a^b f(x)\,\mathrm{d}x \tag{13.1}$$

Among PDFs, the widely used normal (or Gaussian) distribution, discussed in Section 2.3.3, stands out. This and other distributions (Table 13.1) are employed in various statistical hydrology applications (Cheng et al., 2007; Hoskin and Wallis, 1997). Each expression contains the variable of interest, X, and parameters of the distribution. Consult the listed publications for definitions and additional information about each PDF.

13.1.4 Outline for Deterministic or Stochastic Approaches

The following outline is suitable for both deterministic and stochastic approaches. The main difference lies in the selection of the appropriate model (Step 4). The results of the analysis will be either deterministic or statistical in nature. For example, a deterministic model can predict a specific river stage value at a future time, while a stochastic model can estimate it as an expected value with a range of variability. The stochastic approach has the advantage of providing an estimate of uncertainty in predictions. As a result, the range of stochastic outcomes can greatly assist managers in developing optimal management strategies. For example, these strategies can include plans for worst-case scenarios, such as droughts caused by lower water levels or flooding due to high water levels.

1. **Data Collection:**
   - Collect historical hydrological data from various sources, such as weather stations, river gauges, groundwater levels, and satellite data.
2. **Data Preprocessing:**
   - Clean the data by addressing missing or erroneous entries.
   - Normalize the data to a common scale, for example, expressing water levels as a ratio relative to a baseline value.
   - Decompose the data time series into trend, seasonal, and residual components.
3. **Exploratory Data Analysis:**
   - Plot the data time series to visualize trends, cycles, and anomalies.
   - Estimate summary statistics, including the mean and standard deviation.
   - Assess autocorrelation[1] to understand relationships within the data.
4. **Model Selection:**
   - Select a deterministic (physically based) model that accounts for the underlying physical processes (e.g., a rainfall runoff or groundwater model).

- Alternatively, select a stochastic model that utilizes statistical methods to model the data.
5. **Model Fitting and Validation:**
   - Divide the data into training and testing sets.
   - Fit the model to the training data.
   - Validate the model using the testing set and assess performance by evaluating the errors.
6. **Forecasting:**
   - Use the model to make predictions about future hydrological events, such as those influenced by land use or climate change.
7. **Uncertainty Analysis:**
   - Assess the reliability of the forecasts and quantify their uncertainty.
8. **Visualization and Interpretation:**
   - Present the results in a suitable format for the intended audience.

13.1.5 Example Applications of the Monte Carlo Approach

13.1.5.1 Simple Flow Simulations

As discussed in Chapter 6, flow in groundwater aquifers can be represented by Darcy's law. For one-dimensional, steady-state flow in a confined aquifer, this equation can be easily integrated to yield

$$h = h_1 - \frac{Q}{Kb} \cdot x \tag{13.2}$$

In equation (13.2), Q, K, and b represent the aquifer discharge per unit width, hydraulic conductivity, and aquifer thickness, respectively. The variable h is the head value at any distance x, while h_1 is the head value at $x=0$. If all others are known, this equation can be used to estimate the head value at any distance x in the direction of the flow. For a deterministic case, assume the respective values of h_1, Q, K, b, and x are 100 m, 5 m³/day per unit meter, 10 m/day, 12 m, and 20 m. These values yield a value of h of 91.67 m.

However, the value of h can be subject to uncertainty due to the variability of hydraulic conductivity. Such uncertainty can be quantified by applying a stochastic approach. Assume that K is normally distributed, with a population mean of 10 and a standard deviation of 3 m/day. A Monte Carlo simulation can be completed by sampling a large number of values for K from its population distribution and calculating the respective h. The sample size should be iteratively set by nearly matching the statistics of the sample and the population, namely the mean, standard deviation, and the known PDF. However, a technique known as the Latin Hypercube (McKay et al., 1979) can be used to minimize the size of the sample through a systematic sampling scheme. Such an approach is beneficial considering the extensive calculations required for the Monte Carlo methodology, especially for cases requiring elaborate numerical solution techniques. However, due to the simplicity of the example discussed here, a conventional approach for Monte Carlo application will be used.

13.1.5.1.1 Generating K values

For this example, an Excel calculation scheme was fully utilized. The first step was to generate K values for a sample size of 250. The function =NORM.INV(RAND(), mean, sd)^2 is used to generate the values, where "mean" and "sd" represent the population mean and standard deviation of K, with values of 10 and 3 m/day, respectively, as listed in cells I7 and I8 in Figure 13.1. The generated values of K are displayed in column A of Figure 13.1. Note that in this and all Excel-related figures, only the top portion of the sheet is shown.

13.1.5.1.2 Validation of generated **K** values

The following steps are completed for validating K values. The process is done by calculating the general sample statistics and the distribution PDF. All cell and column identifications in the discussion refer to Figure 13.1.

1. Use the AVERAGE, STDEV, MIN, and MAX functions to calculate the sample's mean, standard deviation, minimum, and maximum, which are 10.37, 2.94, 1.72, and 17.87, respectively, as displayed in cells I12 through I15. The mean and standard deviation are close to the population values of 10 and 3. Increasing the sample size beyond 250 can improve accuracy. During Excel page manipulation, including saving the file, the generated K values may change despite using fixed values for the mean and standard deviation of K, due to the random process involved in generating the K values. Consequently, the h values would also change. These changes should not be of concern, as the distribution of the generated K values should converge to the population distribution, especially when the sample size is large.

	A	B	C	D	E	F	G	H	I
1	Generated	Integer		Sample	Sample	Population			
2	K	K	Frequency	PDF K	Normalized PDF K	PDF (normalized)		Frequncy Sum	6688
3	12.50	12	27	0.004037	0.69	0.71			
4	11.73	11	39	0.005831	1.00	0.85			
5	11.41	11	39	0.005831	1.00	0.90			
6	10.11	10	39	0.005831	1.00	1.00		Population	K
7	10.06	10	39	0.005831	1.00	1.00		Mean	10
8	10.60	10	39	0.005831	1.00	0.98		SD	3
9	12.19	12	27	0.004037	0.69	0.77			
10	12.15	12	27	0.004037	0.69	0.77			
11	11.87	11	39	0.005831	1.00	0.82		Sample	K
12	8.38	8	28	0.004187	0.72	0.86		Average	10.04
13	10.95	10	39	0.005831	1.00	0.95		SD	2.72
14	15.49	15	5	0.000748	0.13	0.19		Minimum	1.59
15	4.80	4	6	0.000897	0.15	0.22		Maximum	17.99
16	13.32	13	14	0.002093	0.36	0.54			
17	9.93	9	32	0.004785	0.82	1.00			

FIGURE 13.1 Generating values of hydraulic conductivity from a normal distribution and validating the generated values.

2. In cell B3, type the function =INT(A3) to convert the K values to whole (integer) values. Copy this to the rest of column B (titled "Integer K") for a total of 250 values.

3. In cell C3, use the COUNTIF function =COUNTIF($B$3:$B$253, B3) and copy it to the rest of column C (titled "Frequency") to estimate the frequency of each discrete K value.

4. In cell I2, use =SUM(C3:C253) to calculate the total number of frequencies.

5. In column D, calculate the PDF by dividing the frequency of each value in column C by the total number of frequencies in cell I2.

6. In column E, calculate the normalized PDF by dividing the values in column D by the maximum value in that column using the formula =D2/(MAX($D$2:$D$253)).

7. Calculate the population normalized PDF using the normal distribution equation from Table 13.1 by typing the expression =EXP(–(mean–C3)^2/2/sd^2) in cell F3 and copying it to the rest of the cells in column F.

Figure 13.2 compares the PDF of the sample with that of the population. A reasonable match can be observed, which is expected to improve with a larger sample size.

13.1.5.1.3 Estimation of hydraulic head h values

Figure 13.3 illustrates the various steps in calculating the head values h, their statistics, and their distribution PDF. In the figure, cells I4 and I5 contain the mean and standard deviation of the K population, while cells I13 through I16 contain the sample's mean, standard deviation, minimum, and maximum values. The generated K values in column A are used in equation (13.2) to provide 250 corresponding values for h (column B), which can be analyzed to assess the uncertainty in h. The h values are calculated by entering the formula =$I$7-$I$8/A3/$I$9*$I$13 in cell B3, then copying it to the rest of the cells in the column, resulting in 250 values. In this formula, cells I7 through I10 contain the respective values for the variables h_1, q, b, and x in equation (13.2), and cell A2 contains the generated value of K.

13.1.5.1.4 Statistical analysis of hydraulic head **h** values

Following the same steps used for K, statistical analysis of the calculated h values can be performed to estimate the mean, standard deviation, and PDF. The steps below outline the process for h. All cell and column references pertain to Figure 13.3.

1. Use the AVERAGE, STDEV, MIN, and MAX functions to calculate the mean, standard deviation, minimum, and maximum, which are 98.94, 4.84, 50, and 95 m, respectively. These values are listed in cells J13 through J16. As mentioned earlier, the stochastic approach provides valuable information compared to the single deterministic value of 91.67 m, which is subject to uncertainty. This deterministic value is calculated by applying equation (13.2), using the average K value of 10 m/day while keeping all other variables the same as in the stochastic solution.

TABLE 13.1   Probability distributions (PDFs) and their use in hydrology (Cheng et al., 2007; Hoskin and Wallis, 1997)

DISTRIBUTION	EXPRESSION	CHARACTERISTICS	USE IN HYDROLOGY
Normal (Gaussian) distribution	$f(X) = \dfrac{1}{\sqrt{2\pi\sigma^2}} e^{-\frac{(X-\mu)^2}{2\sigma^2}}$	Symmetrical, bell-shaped curve. Defined by mean (μ) and standard deviation (σ)	Hydrological variables that are symmetrically distributed around the mean, such as certain river flows under natural, unregulated conditions
Log-normal distribution	$f(X) = \dfrac{1}{X\sigma\sqrt{2\pi}}$ $e^{-\frac{(\ln X - \mu)^2}{2\sigma^2}}$ for $X \geq 0$	Distribution of a variable whose logarithm is normally distributed	Precipitation, streamflow, and groundwater data, as these variables often exhibit right skewness
Exponential distribution	$f(X) = \lambda e^{-\lambda X}$ for $X \geq 0$	Describes time between events in a Poisson process. Defined by rate parameter λ	Modeling time intervals between rainfall events and other similar processes
Beta distribution	$f(X) = \dfrac{X^{\alpha-1}(1-X)^{\beta-1}}{B(\alpha,\beta)}$ for $0 \leq X \leq 1$	Flexible shape depending on parameters α and β. Used in Bayesian statistics	Modeling soil moisture and proportions of land area affected by drought
Gamma distribution	$f(X) = \dfrac{\lambda^k X^{k-1} e^{-\lambda X}}{\Gamma(k)}$ for $X \geq 0$	Generalizes the exponential distribution. Defined by shape parameter k and rate parameter λ	Modeling rainfall amounts and streamflow, especially when the data are positively skewed
Weibull distribution	$f(X) = \dfrac{k}{\lambda}\left(\dfrac{X}{\lambda}\right)^{k-1}$ $e^{-\left(\frac{X}{\lambda}\right)^k}$ for $X \geq 0$	Used in reliability analysis and modeling life data. Defined by shape parameter k and scale parameter λ	Modeling wind speeds, flood frequency analysis, and rainfall durations
Pearson type III distribution	$f(X) = \dfrac{\beta^\alpha}{\Gamma(\alpha)}$ $(X-\gamma)^{\alpha-1} e^{-\beta(X-\gamma)}$	Distribution generalizes the gamma distribution and can be transformed into a normal distribution by appropriate choice of parameters	Widely used for flood frequency analysis and other skewed hydrological data

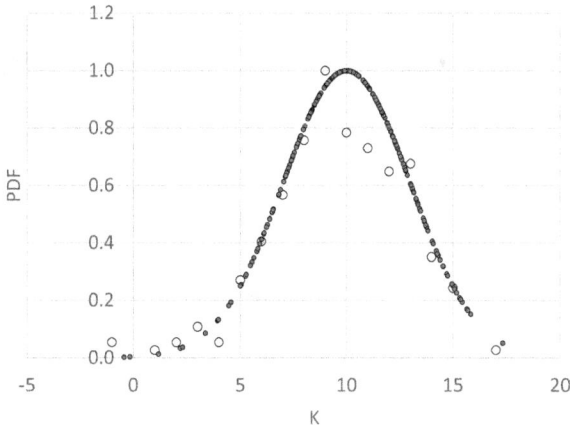

FIGURE 13.2   Comparison between the normalized PDFs for both population and sample of *K* (small circles and large circles, respectively).

L11		f_x								
	A	B	C	D	E	F	G	H	I	J
1	Generated	Calculated	Integer	Frequency	Calculated		Normalized		Frequency Count	8278
2	K	h	h	h	PDF		PDF			
3	8.20	89.84	89	22	0.002658		0.37		Population	K
4	8.16	89.79	89	22	0.002658		0.37		Mean	10
5	9.15	90.89	90	23	0.002778		0.39		SD	3
6	10.24	91.86	91	32	0.003866		0.54			
7	5.40	84.57	84	3	0.000362		0.05		h1	100
8	12.44	93.30	93	47	0.005678		0.80		Q	50
9	12.55	93.36	93	47	0.005678		0.80		b	12
10	13.25	93.71	93	47	0.005678		0.80		x	20
11	11.09	92.49	92	59	0.007127		1.00			
12	7.18	88.39	88	13	0.001570		0.22		Sample	K	h
13	7.20	88.42	88	13	0.001570		0.22		Mean	10.11	90.25
14	9.08	90.82	90	23	0.002778		0.39		SD	2.90	4.14
15	9.04	90.78	90	23	0.002778		0.39		Minimum	1.85	54.00
16	13.17	93.67	93	47	0.005678		0.80		Maximum	18.98	95.00
17	10.93	92.38	92	59	0.007127		1.00				

FIGURE 13.3   Calculating *h* values, the respective statistics, and the distribution PDF.

2. In cell C3, use the =INT(B3) function to convert the h values to whole (integer) values, then copy the formula to the rest of column C.
3. In cell D3, use the COUNTIF function =COUNTIF($B$3:$B$252, B3) to estimate the frequency of the discrete h value, then copy it to the rest of column D.
4. Use =SUM(D3:D252) in cell I1 to calculate the total frequency count.

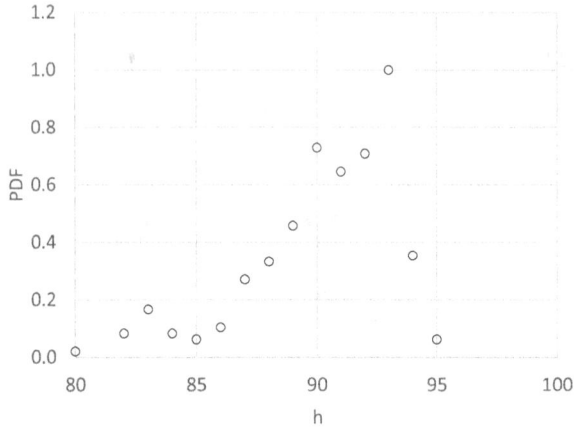

FIGURE 13.4   Graph of the normalized PDF for h.

5. In column E, calculate the PDF by dividing the frequency of each value in column D by the total number of frequencies in cell I1.
6. In column F, calculate the normalized PDF by dividing each value in column E by the maximum value in that column, using =E2/(MAX($E$3:$E$252)). These values are plotted in Figure 13.4. In some applications, it may be beneficial to fit the PDF to a theoretical distribution, such as one of those listed in Table 13.1, which could enhance the predictive capabilities of a relevant simulation model.

13.1.5.2 Generalized Flow Simulations

As discussed in Chapter 11, numerical simulations are essential for achieving a realistic representation of field conditions. The application of the Monte Carlo method for such cases will follow the same procedure described in the previous section, with the numerical model running numerous times based on input parameters generated from their assumed distributions. Similarly, the results are analyzed to assess their statistical features.

As an example, the MODFLOW model by Harbaugh et al. (2000) was applied within the GMS package (AQUAVEO, 2023) to a relatively simple aquifer bounded by known boundary conditions, with a fully penetrating single well operating at the center of the aquifer's plan view (Figure 13.5). The aquifer is confined and has dimensions of 1,000 m by 1,000 m in plan view, with a depth of 50 m. It is bounded by specified heads of 100 and 60 m on the left and right, respectively, while the top and bottom boundaries in the plan view are set as no-flow. The well's pumping rate was set at 10,000 m³/day. Figure 13.6 illustrates the deterministic solution, assuming a hydraulic conductivity (K) value of 10 m/day.

For the stochastic analysis, K was assumed to be normally distributed, with a population mean of 10 m/day and a population standard deviation of 1.0 m/day. The Latin Hypercube option (McKay et al., 1979) was used with 100 segments[3], resulting in K sample statistics that reasonably matched the population statistics.

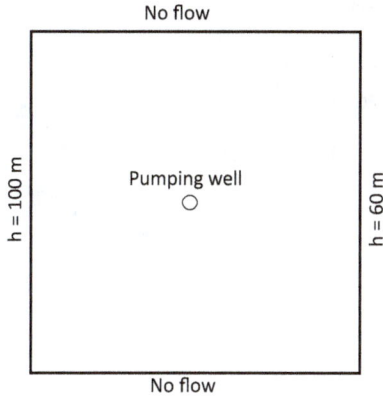

FIGURE 13.5   Sketch of the simulated aquifer site.

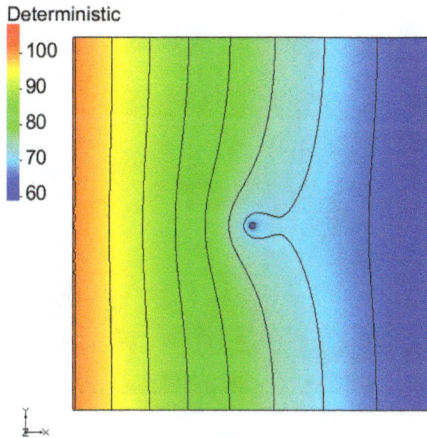

FIGURE 13.6   Deterministic results of the aquifer's hydraulic head obtained with a hydraulic conductivity of 10 m/day.

The simulation results for the hydraulic head h are illustrated in Figures 13.7 through 13.11. The mean hydraulic head does not match the deterministic solution (compare Figures 13.6 and 13.7). It can be concluded that the deterministic solution generally does not provide the most likely h outcome under uncertainty. Additionally, as mentioned earlier, the statistical results offer further advantages by providing information about the uncertainty of the results. As shown in Figure 13.10, uncertainties are highest near the well and non-existent at the boundary where the head is specified.

Risk analysis can also be completed to identify zones where water level and quality indicators exceed or fall below threshold values. For example, Figure 13.11 illustrates the probability of the hydraulic head being less than 75 m in different zones of the aquifer discussed above. The high probability primarily covers the zone around and

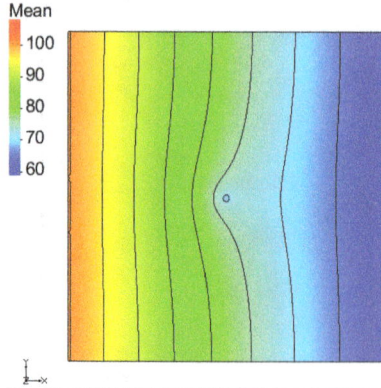

FIGURE 13.7   Stochastic results of the aquifer's mean hydraulic head.

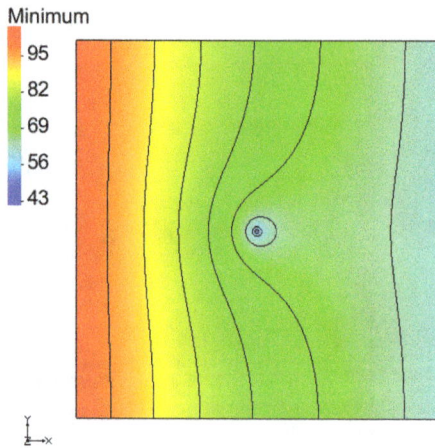

FIGURE 13.8   Stochastic results of the aquifer's minimum hydraulic head.

downstream of the well. In practical applications, actions such as restricting new water developments would be implemented in areas where high risk exists.

13.2 GEOGRAPHIC INFORMATION SYSTEM

A GIS is a powerful tool designed for the collection, management, and analysis of data associated with specific geographical locations. This system integrates diverse data layers, including maps, satellite imagery, aerial photographs, and tabular data. Examples of information include scientific and socio-economic data. Hydrological data can encompass river networks, aquifer spatial data (such as hydraulic conductivity), and

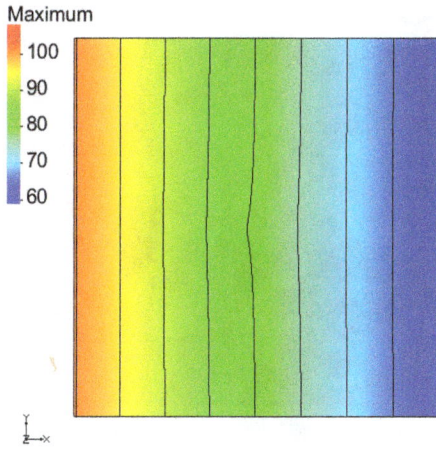

FIGURE 13.9   Stochastic results of the aquifer's maximum hydraulic head.

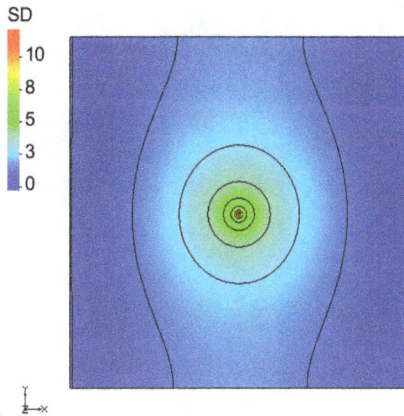

FIGURE 13.10   Stochastic results of the standard deviation of hydraulic head.

information about lakes and other surface water bodies. It can also include aquifer water levels and the distribution of water quality data. A GIS is highly effective in enhancing the comprehension of spatial information, as well as in analyzing and formatting data for modeling purposes. It is therefore a valuable tool for decision-making because it allows users to see relationships, patterns, and trends that might not be immediately apparent in tabular data or traditional maps.

In this context, establishing a connection between a GIS and a modeling framework significantly streamlines the modeling process, reducing the need for labor-intensive tasks. This is particularly crucial for complex hydrological models that rely on spatial data that conform to the model's requirements, often involving dividing the area of interest into a small-scale grid for numerical calculations.

Key components and concepts of a GIS include the information to be displayed and studied, which is manipulated and analyzed using specialized GIS software. Popular

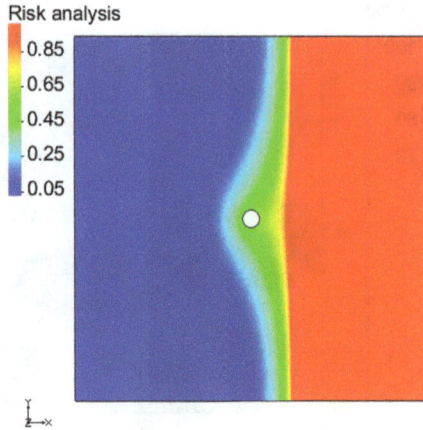

FIGURE 13.11   Spatial distribution of the probability that the hydraulic head is less than 75 m.

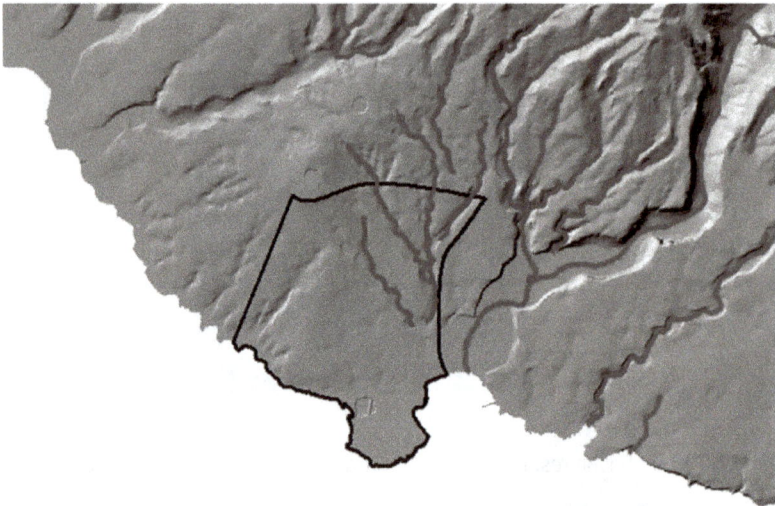

FIGURE 13.12   The modeling site on Kauai, Hawaii (shown by the black heavy line), super-imposed on GIS maps showing topography and water streams. ArcGIS (2023) was used to create this map as well as those in Figures 13.13 and 13.14 (Pap et al., 2023). Data were obtained from Hawaii OPSD (2023).

GIS software options include ArcGIS (ArcGIS, 2023), QGIS (QGIS, 2023), and Google Earth (Google Earth, 2023). GIS can be utilized on computers, global positioning systems devices, and remote sensing equipment for various data-related tasks. Skilled professionals are required to develop specific procedures and methodologies for data collection, analysis, and visualization when using GIS software.

Figures 13.12 through 13.15 exemplify a GIS application in groundwater modeling using the GMS user interface (AQUAVEO, 2023), which is designed to facilitate the

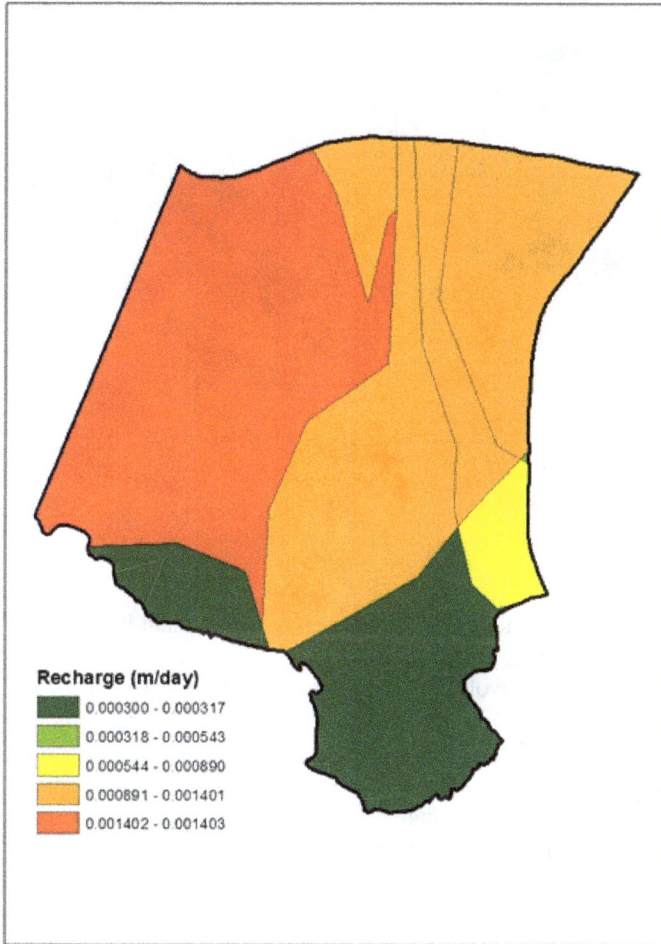

FIGURE 13.13   GIS Recharge map for the modeling site (based on Mair, 2024).

various steps described in this example. The primary aim of this analysis was to employ a groundwater model to simulate groundwater levels in the coastal area depicted in Figure 13.12, situated on the Island of Kauai, Hawaii. Among the objectives was an examination of the effects of sea level rise. Additionally, slow drainage issues were a concern, particularly for the Hanapēpē Salt Pond, where a key facility involved in salt production is negatively impacted by these problems.

Data essential for the model include the area outline, topography, and water streams (Figure 13.12), as well as groundwater recharge (Figure 13.13). Additionally, relevant information encompasses geological data and water-use wells. Hawaii-specific GIS spatial data are primarily accessible through Hawaii OPSD (2023) and extends island-wide. Further details about this study are available in Pap et al. (2023), which

The Salt Pond   The Salt Pond

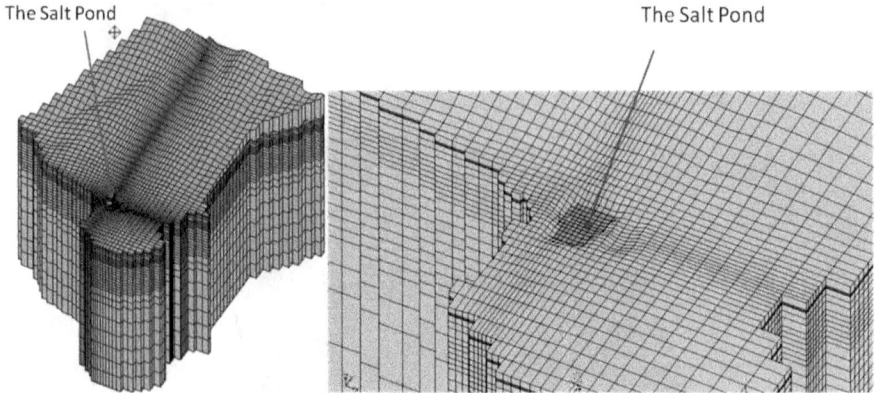

FIGURE 13.14   (a) A three-dimensional grid of the modeling site. The grid is refined where more detail is needed and (b) A zoomed-in view around the pond area illustrating the refined three-dimensional grid (Pap et al., 2023).

integrated groundwater modeling with comprehensive field-based geophysical and geochemical data collections.

The subsequent step involves the allocation of GIS data to the various cells. Data can exhibit vertical uniformity or variability. Typically, recharge data are assigned to the top layer, while soil and geological data may vary vertically according to their properties. Figure 13.13 illustrates a representation in which recharge data are clipped to fit within the modeled area. This clipping process may be necessary for display purposes but is not obligatory for data assignment to the cells.

To develop and apply the groundwater model, the study site was divided into smaller three-dimensional cells, as shown in Figure 13.14. The size of these cells can be either fixed or variable, with more intricate refinement in areas of particular interest. The final stages involve model calibration and execution to estimate water levels in the aquifer under various management scenarios. Among these simulations, the study modeled sea level rises of 0.6 and 1 m, which resulted in increases in the water level at the Salt Pond by approximately 0.8 and 0.5 m above the baseline, respectively (Figure 13.15). The figure also illustrates a flooded pond under these conditions.

13.3 GEOSTATISTICAL METHODS

Field data collection, particularly when involving subsurface intrusive techniques, is typically carried out at specific and often limited locations due to the associated high costs. Interpolation techniques are commonly used to assess the spatial distribution of data across the entire domain of interest. However, these conventional methods often

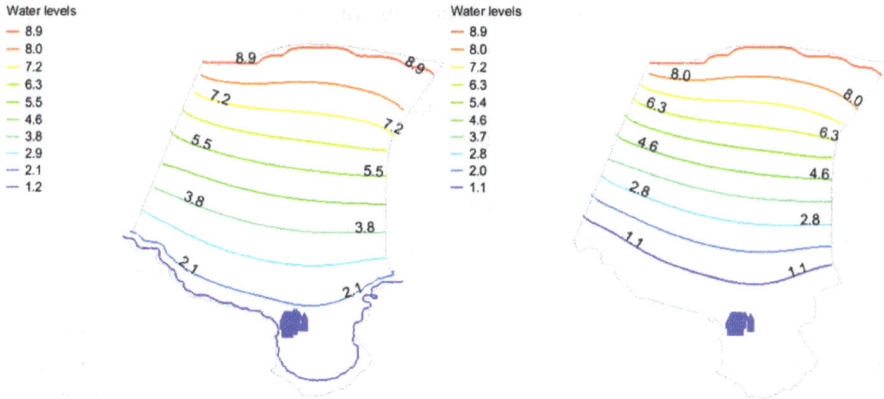

FIGURE 13.15 Water levels under sea level rise (a) 1 m and (b) 0.6 m. The blue area represents a flooded pond under such scenarios. The colored lines indicate groundwater levels above mean sea level (Pap et al., 2023).

fail to account for the spatial interdependence or correlation of measurements, which arises, for example, from the natural geological processes that shape the formation of aquifer materials. For instance, a high conductivity value at a specific location can suggest higher values in nearby areas that exhibit a correlation with the measured value. The same issues also arise for rainfall data, which are naturally spatially varied. This correlation tends to diminish as the distance from the measurement point increases. Consequently, simply averaging the measured values may lead to inaccuracies by neglecting this important spatial correlation structure.

Geostatistical analysis, on the other hand, offers a more precise approach. It is an advanced procedure that derives an estimated surface from a scattered set of point measurements, taking into consideration the spatial behavior of the phenomenon represented by the measured values (e.g., ESRI, 2023; Adobe Acrobat, 2023; GISGeography, 2023, Esri, 2021a). The technique is based on the principle that spatially correlated data points contain valuable insights into the spatial structure and variability of the measured parameter under investigation. This technique estimates values at unobserved locations by assigning varying weights to nearby sample points, guided by their spatial proximity. The approach consists of two major steps: (1) investigating the spatial behavior of the measured values and (2) estimating the output surface. These steps are further explained as follows.

The initial step in geostatistical analysis is crucial and involves the definition of a semi-variogram to quantify the spatial correlation between neighboring observations (also known as autocorrelation) by graphing the variance of all pairs of data based on their distances. This step is preceded by an examination of the data to identify its distribution, trends, directional components, and outliers. Figure 13.16 provides an illustration of a semi-variogram fitted to the semi-variogram calculations. The fitted line is typically represented by a polynomial that captures the overall large-scale trend. Examples of such lines include spherical, circular, exponential, Gaussian, or linear types, chosen to best represent the data. The semi-variogram reveals that closely situated data points exhibit stronger correlation and minimal semi-variance, while distant points show lower correlation and higher semi-variance. Autocorrelation becomes independent beyond a certain distance (range), marked by the point where the variation levels off, known as the "sill." At this juncture, there is no spatial correlation or relationship between the proximity of data points.

The ultimate goal of the next step is to provide an optimal estimation of the variable of interest at various locations by minimizing prediction errors, which is done through a technique called kriging. The optimal values are derived by utilizing weighted averages of available values from all data points, with the weights based on the estimated semi-variogram. Kriging not only yields an optimal prediction surface but also provides a confidence measure regarding the accuracy of the prediction. This confidence is quantified by estimating the standard error, which tends to be larger in situations with a limited number of data points. Figure 13.17 illustrates a map generated through kriging, accompanied by a corresponding map of the estimated standard error. The colored shades on the maps reflect the corresponding relative values for the variable of interest and the standard error. Several kriging techniques are suited to different applications. More details are available in other resources, such as Esri (2021b).

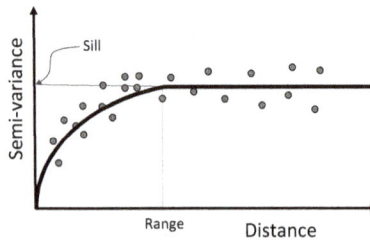

FIGURE 13.16   Example semi-variogram plot showing the calculated values (dots) and the fitted model (line).

FIGURE 13.17   Map of predicted surface and the estimated standard error (GISGeography, 2023). The map shows a hypothetical surface to elucidate the concepts of kriging.

13.4 MODELING AND OPEN-SOURCE PROGRAMMING

Advancements in computer software and hardware, as well as graphical representations, have greatly facilitated the study of large-scale sites with various interacting geological, physical, and chemical processes. Furthermore, methods have been refined to address the fate and transport of variable-density and variable-viscosity liquids, such as petroleum products. Research focusing on multiple flow domains is also making progress, which is especially pertinent in cases involving relatively large subsurface fractures within common porous materials.

Open-source techniques in hydrology have gained popularity in recent years due to their cost-effectiveness, accessibility, and collaborative nature. These techniques, combined with computational tools, empower hydrologists to conduct modeling-related research and share their findings with the broader scientific community, primarily without the cost and licensing constraints associated with proprietary software. Open-source techniques in hydrology are becoming increasingly popular for various hydrological studies and applications. These open-source tools and software provide flexibility, transparency, and affordability, making them valuable resources for researchers and professionals in the field of hydrology. Available open-source techniques and associated tools can be grouped into modeling, data sharing, analysis, visualization software, and education. Examples of available packages include OpenMI (Harpham et al., 2019; modeling software), QGIS (QGIS, 2023; GISs), MODIS (MODIS, 2023; remote sensing and data analysis), R language (R Core Team, 2023), Python and libraries (Python, 2023), HydroShare (HydroShare, 2023; data collection and sharing), and HydroLearn (HydroLearn, 2023; education and training).

Open-source efforts are facilitated by GitHub, a web-based platform designed for hosting and collaborating on code repositories (GitHub, 2023). It is utilized by various groups for a wide spectrum of software projects, ranging from small scripts to large, complex applications. The platform serves as a crucial tool for version control, enabling developers to efficiently track and manage changes to their code. This, in turn, allows multiple developers to work together on a project simultaneously, keeping a record of alterations and managing different versions of the code.

GitHub also equips developers with tools to host their code, making it easily shareable, facilitating collaborative work on projects, and granting access to the code from anywhere with an internet connection. Additionally, it offers options for reporting bugs, submitting feature requests, and managing general project tasks and communication related to them. These tasks encompass proposing changes to a project and enabling peer reviews of modifications before they are integrated into the main codebase. Furthermore, GitHub provides a wide array of integrations and extensions to enhance its functionality. It prioritizes secure development by offering features such as code scanning to identify vulnerabilities and access control measures to safeguard both the code and data.

An example of the application of the open-source methodology can be found in Shuler and Mariner (2020).

13.5 ASSIGNMENTS: STOCHASTIC ANALYSIS

1. As discussed in Section 6.4, flow in a one-dimensional unconfined aquifer can be described by Darcy's law, which can be integrated to yield equation (13.3):

$$h = \sqrt{h_1^2 - \frac{2Q}{K} \cdot x}$$ (13.3)

where Q and K represent the discharge per unit width and hydraulic conductivity. The variable h_1 is the head value at $x=0$. This expression gives the water level h at any distance x.

Using the method in Section 13.1.5.1, repeat the calculations to assess the statistical information about K and h. The specific data values are $h_1 = 100$ m, $Q=5$ m³/day per unit meter, and $x=200$ m. The hydraulic conductivity K is normally distributed with a mean of 10 m/day and a standard deviation of 1.0 m/day.

2. A confined aquifer's response to a single pumping well at a rate Q for a steady case condition is represented by the equation (see Section 6.9.1):

$$Q = 2\pi Kb \frac{(h_2 - h_1)}{\ln\left(\frac{r_2}{r_1}\right)}$$

where K and b are the hydraulic conductivity and aquifer thickness, and h_1 and h_2 are the hydraulic head values at distances r_1 and r_2 from the well, respectively. The specific data values are $b=2$ m, $h_1=99$ m, $h_2=100$ m, $r_1=0.5$ m, and $r_2=1,000$ m. The hydraulic conductivity K is normally distributed with a mean of 10 m/day and a standard deviation of 3 m/day. Follow the calculation scheme in Section 13.1.5.1 to calculate the statistical information for K and Q.

NOTES

1 Autocorrelation reflects the degree to which values of a certain variable are related. Points that are spatially close are expected to have a strong correlation, in contrast to points that are farther apart.

2 For columns A through F in Figure 3.1, the formula is entered in the top cell (shaded in yellow) and then copied to the remaining cells in the column, resulting in a total of 250 values.

3 In this technique, the probability distribution curve for K is divided into a number of segments of equal probability (equal areas under the curve). The parameter is then randomized until a value is found that lies within each probability segment. This approach ensures that possible combinations of parameter values are sampled as completely as possible with a limited number of model's runs.

REFERENCES

Adobe Acrobat. (2023). Kriging: An introduction to concepts and applications. https://www.esri.com/content/dam/esrisites/en-us/events/conferences/2020/federal-gis/kriging-an-intro-to-concepts.pdf

AQUAVEO. (2023). Groundwater modeling system. https://www.aquaveo.com/software/gms-groundwater-modeling-system-introduction

ArcGIS. (2023). ArcGIS Online. https://www.arcgis.com/index.html

Beven, K., & Freer, J. (2001). Uncertainty and sensitivity analysis of hydrological models. In: Anderson, M. G., McDonnell, J. J. (eds), *Encyclopedia of Hydrological Sciences*, pp. 1–17. Chichester, UK: Wiley. https://search.worldcat.org/title/Encyclopedia-of-hydrological-sciences/oclc/61247175

Cheng, K.-S., Chiang, J.-L., & Hsu, C.-W. (2007). Simulation of probability distributions commonly used in hydrological frequency analysis. *Hydrological Processes, 21*(1), 51–60. https://doi.org/10.1002/hyp.6176

Chow, V. T., Maidment, D. R., & Mays, L. W. (1988). Applied Hydrology. New York, NY: McGraw-Hill.

Esri. (2021a). Understanding geostatistical analysis. *ArcGIS Desktop*. https://desktop.arcgis.com/en/arcmap/latest/extensions/geostatistical-analyst/understanding-geostatistical-analysis.htm

Esri. (2021b). Kriging in geostatistical analyst. *ArcGIS Desktop*. Retrieved October 9, 2024, from https://desktop.arcgis.com/en/arcmap/latest/extensions/geostatistical-analyst/kriging-in-geostatistical-analyst.htm

ESRI. (2023). ArcGIS desktop: Release 10.8.1 [Software]. https://www.esri.com/en-us/arcgis/products/arcgis-desktop/overview

GISGeography. (2023). Kriging interpolation – The prediction is strong in this one. Retrieved from https://gisgeography.com/kriging-interpolation-prediction/

GitHub. (2023). Hydrology. Retrieved from https://github.com/topics/hydrology

Google Earth. (2023). Google Earth. Retrieved from https://earth.google.com/web/@0,-2.19919995,0a,22251752.77375655d,35y,0h,0t,0r/data=OgMKATA

Haan, C. T. (1977). Statistical *Methods* in *Hydrology*. Ames, IA: Iowa State University Press.

Harbaugh, A. W., Banta, E. R., Hill, M. C., & McDonald, M. G. (2000). MODFLOW-2000, the U.S. Geological Survey modular ground-water model; user guide to modularization concepts and the ground-water flow process. U.S. Geological Survey Open-File Report 00–92, 121 pp. https://doi.org/10.3133/ofr200092

Harpham, Q. K., Hughes, A., & Moore, R. V. (2019). Introductory overview: The OpenMI 2.0 standard for integrating numerical models. *Environmental Modelling and Software, 122*, 104549. https://doi.org/10.1016/j.envsoft.2019.104549

Hawaii OPSD. (2023). Hawaii statewide GIS program. Retrieved from https://planning.hawaii. gov/gis/download-gis-data-expanded/

Hosking, J. R. M., & Wallis, J. R. (1997). *Regional Frequency Analysis: An Approach Based on LMoments.* Cambridge, UK: Cambridge University Press. https://doi.org/10.1017/CBO9780511529443

HydroLearn. (2023). HydroLearn. Retrieved from https://www.hydrolearn.org/

HydroShare. (2023). HydroShare. Retrieved from https://www.hydroshare.org/

Mair, A. (2024). Mean annual groundwater recharge rates for Kauaʻi, Oʻahu, Molokaʻi, Maui, and the Island of Hawaiʻi, for a set of drought and land-cover conditions [Data set]. U.S. Geological Survey. https://doi.org/10.5066/P9DDP1C6

McKay, M. D., Beckman, R. J., & Conover, W. J. (1979). A comparison of three methods for selecting values of input variables in the analysis of output from a computer code. *Technometrics, 21*(2), 239–245. https://doi.org/10.1080/00401706.1979.10489755

MODIS. (2023). MODIS: Moderate resolution imaging spectroradiometer. https://modis.gsfc. nasa.gov/

Pap, R., Lerner, D., Nobrega-Olivera, M., Dulai, H., El-Kadi, A., Lautze, N., Wallin, E., de Bolós, X., Ferguson, C., Glazer, B., & Habel, S. (2023). Salt pond hydrogeologic investigation. Prepared by the University of Hawaiʻi Sea Grant College Program for the County of Kauaʻi Public Access, Open Space & Natural Resources Preservation Fund Commission. https://seagrant. soest.hawaii.edu/wp-content/uploads/2023/11/Salt-Pond-Hydrogeologic-Investigation-final. pdf

Python. (2023). Python. https://www.python.org/

QGIS. (2023). QGIS: A free and open source geographic information system. https://www.qgis. org/en/site/

R Core Team. (2023). R: A language and environment for statistical computing. R Foundation for Statistical Computing. https://www.R-project.org/

Shuler, C. K., & Mariner, K. E. (2020). Collaborative groundwater modeling: Open-source, cloud-based, applied science at a small-island water utility scale. *Environmental Modelling & Software, 127*. https://doi.org/10.1016/j.envsoft.2020.104693

Viessman, W., Jr., & Lewis, G. L. (2003). Introduction to *Hydrology* (5th ed.). Upper Saddle River, NJ: Pearson Education.

Appendix A
Example Spreadsheet Use

Spreadsheets are excellent tools for applying hydrology principles by solving equations and analyzing both experimental and computer-generated data, particularly when repeated calculations are necessary. Their wide range of mathematical and statistical functions allows for in-depth analyses. Additionally, spreadsheets provide various graphical tools that enhance the visual interpretation and examination of the results, facilitating better understanding and decision-making in hydrological studies.

As an example, this appendix demonstrates how to calculate stream velocity and discharge, as well as how to plot discharge against river stage using a simple equation. The Manning equation is applied to calculate the stream water velocity, V, through the following formula:

$$V = \frac{R^{2/3} S^{1/2}}{n} \tag{A.1}$$

The discharge is estimated using $Q = VA$, where A represents the stream's cross-sectional area. In equation (A.1), R, the hydraulic radius, is calculated by dividing the area A by the wetted perimeter p, which are illustrated in Figure A.1. The variables S and n represent the water surface slope and the Manning roughness coefficient, respectively. In this example, the calculations are based on the cross-sectional profile shown in Figure A.1, where the stream stage (water level height) determines the corresponding values of A and p. The water surface slope S is set to 0.01, and the roughness coefficient n is 0.04. Further details regarding this approach are elaborated in Chapter 5, Section 5.2.2.3.

The calculation scheme involves determining $R^{2/3}$ (hydraulic radius raised to the power of 2/3) and $S^{1/2}$ (slope raised to the power of 1/2), then multiplying these values

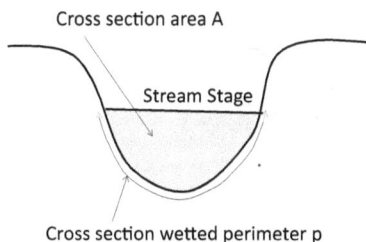

FIGURE A.1   A sketch of a stream cross section for the example spreadsheet application.

together and dividing by the Manning roughness coefficient, n. While it is possible to perform these calculations in one step, it is recommended to carry them out in sequential steps for clarity and to help identify potential errors. The calculation formulas are entered into the first row of the spreadsheet for the initial stream stage and subsequently repeated for each additional row (stage). The following steps are implemented using Microsoft Excel®©.

1. Figure A.2: In cells A1 through H1, enter the column titles. Then, input the data values for stage, area, and wetted perimeter in the corresponding cells in rows 2 through 7.
2. Figure A.3: To calculate R, select cell D2, type the equal sign "=", then select cell B2, type the division operator "/", and select cell C2.
3. Figure A.4: Press the return button, which will display the value of R (the area A divided by the wetted perimeter p) for the first stage.
4. Figure A.5: To calculate $R^{2/3}$, select cell E2, type the equal sign "=", then select cell D2, followed by the power operator "^" and (2/3), ensuring the parentheses are included.
5. Figure A.6: Press the return button, which will display the value of R raised to the power of 2/3 for the first stage.
6. Figure A.7: To calculate $S^{1/2}$, select cell F2, type the equal sign "=", followed by 0.01 (the value of the water slope), then type the power operator "^" and (1/2), ensuring the parentheses are included.

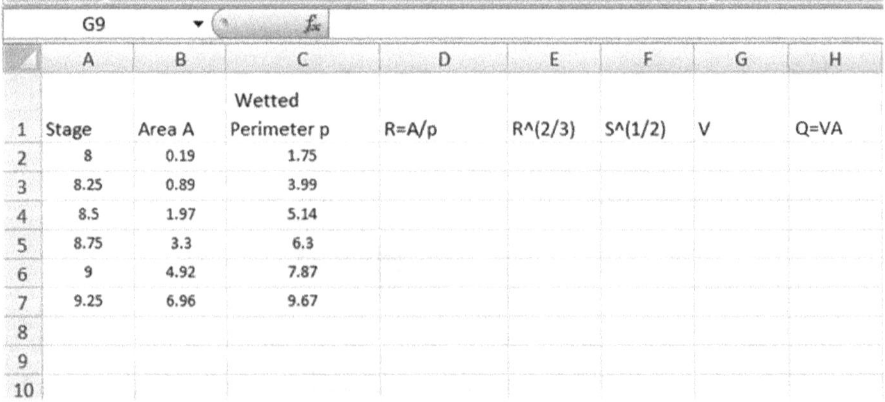

	A	B	C	D	E	F	G	H
1	Stage	Area A	Wetted Perimeter p	R=A/p	R^(2/3)	S^(1/2)	V	Q=VA
2	8	0.19	1.75					
3	8.25	0.89	3.99					
4	8.5	1.97	5.14					
5	8.75	3.3	6.3					
6	9	4.92	7.87					
7	9.25	6.96	9.67					
8								
9								
10								

FIGURE A.2   Adding column titles in cells A-1 through H-1.

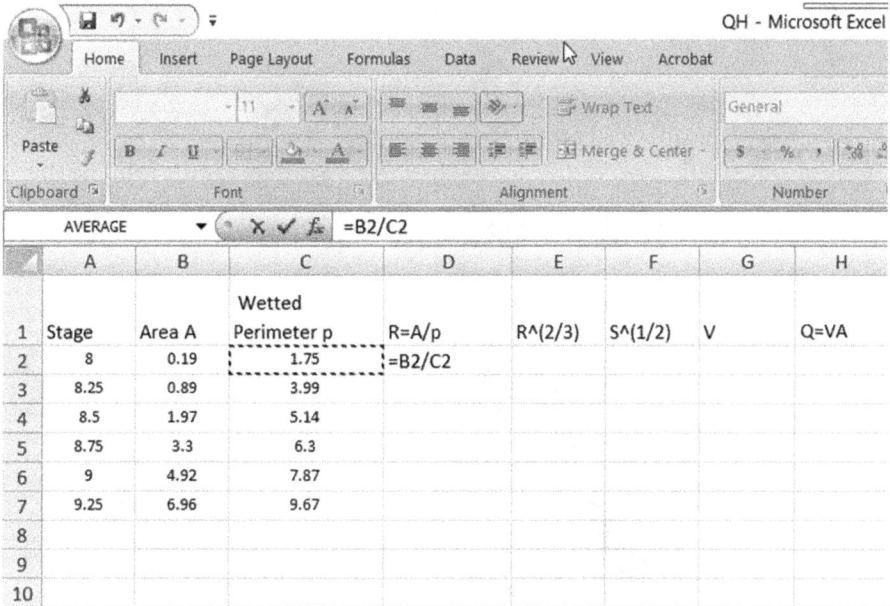

FIGURE A.3   Entering the formula for wetted perimeter *R* in cell D2.

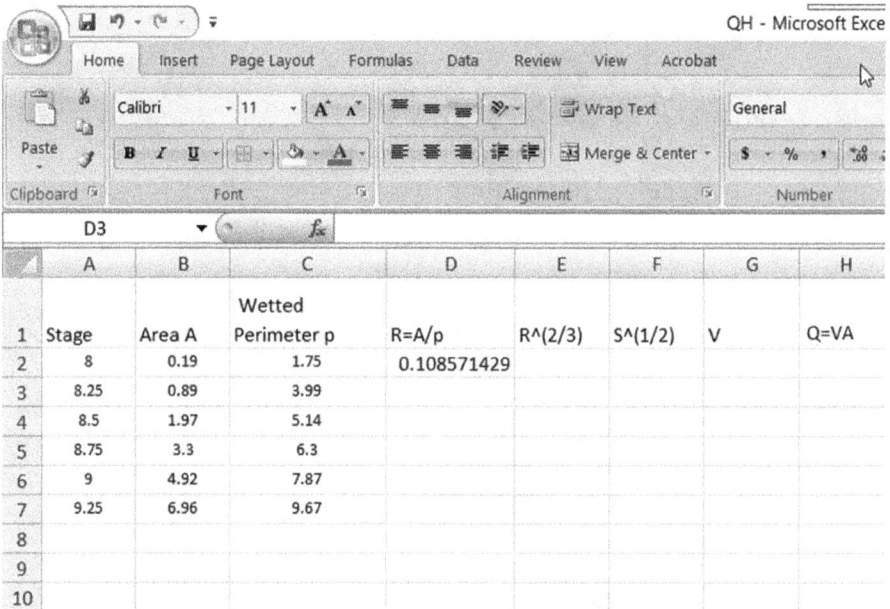

FIGURE A.4   Displaying the value of *R* for the first stage.

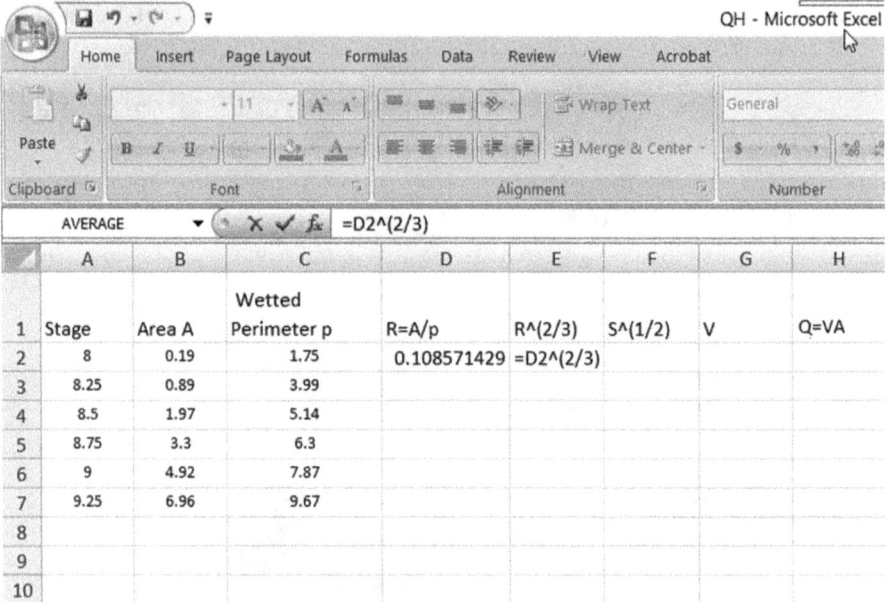

FIGURE A.5 Setting *R* to the power of 2/3 in cell E2.

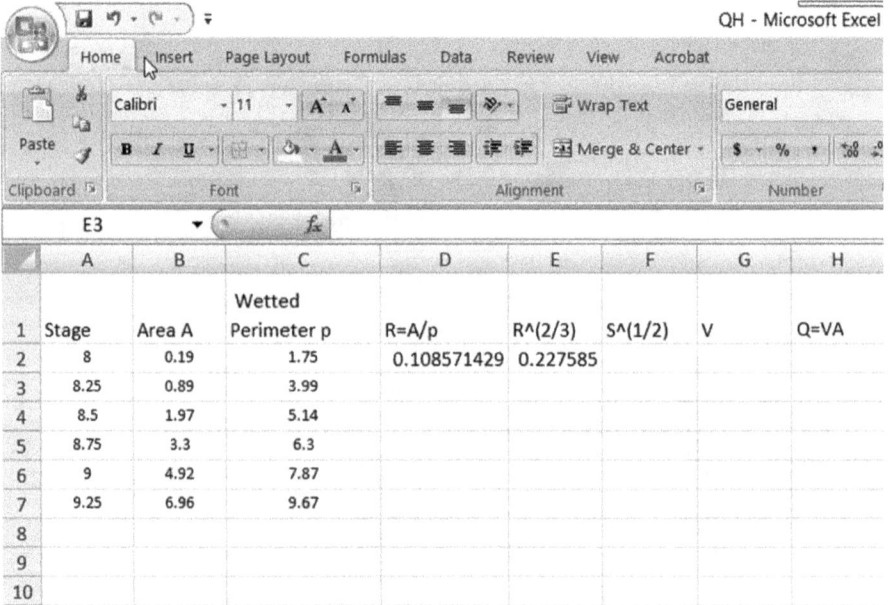

FIGURE A.6 Displaying the value of *R* to the power of 2/3 for the first stage.

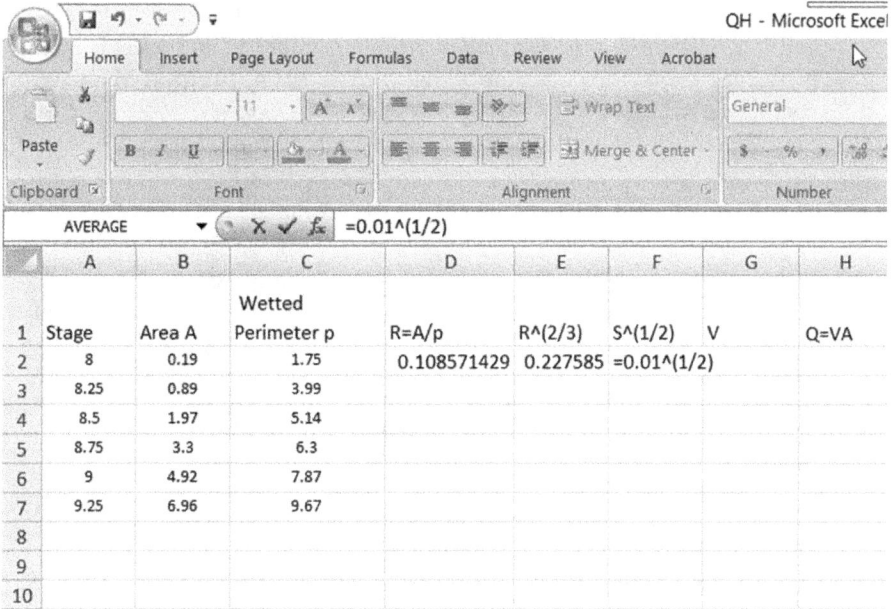

FIGURE A.7 Setting water slope *S* to the power of ½ in cell F2.

7. Figure A.8: Press the return button, which will display the value of S raised to the power of 1/2 for the first stage.

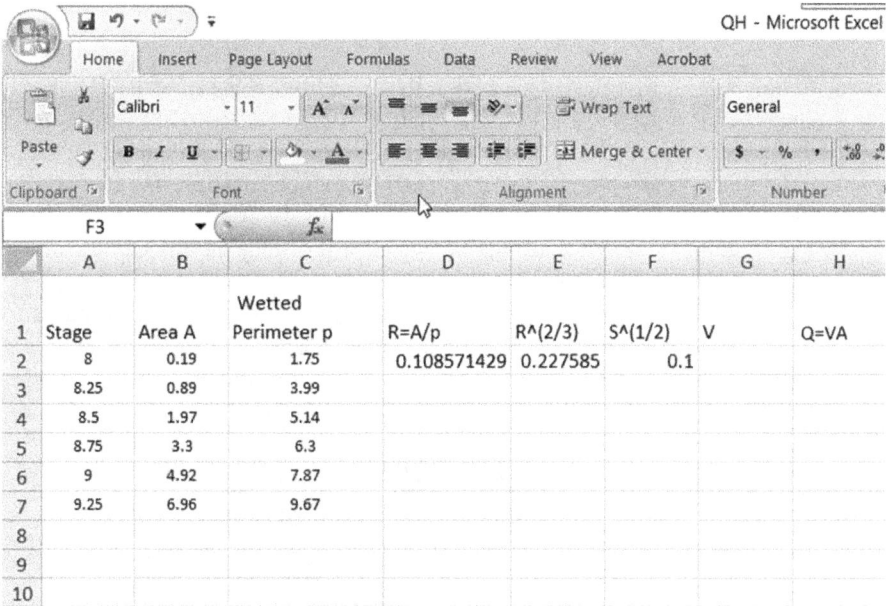

FIGURE A.8 Displaying *S* to the power of ½ for the first stage.

8. Figure A.9: To calculate V, select cell G2, type the equal sign "=", then select cell E2. Afterward, type the multiplication operator "*" and select cell F2. Next, type the division operator "/" followed by "0.04" (which is Manning's coefficient, n).

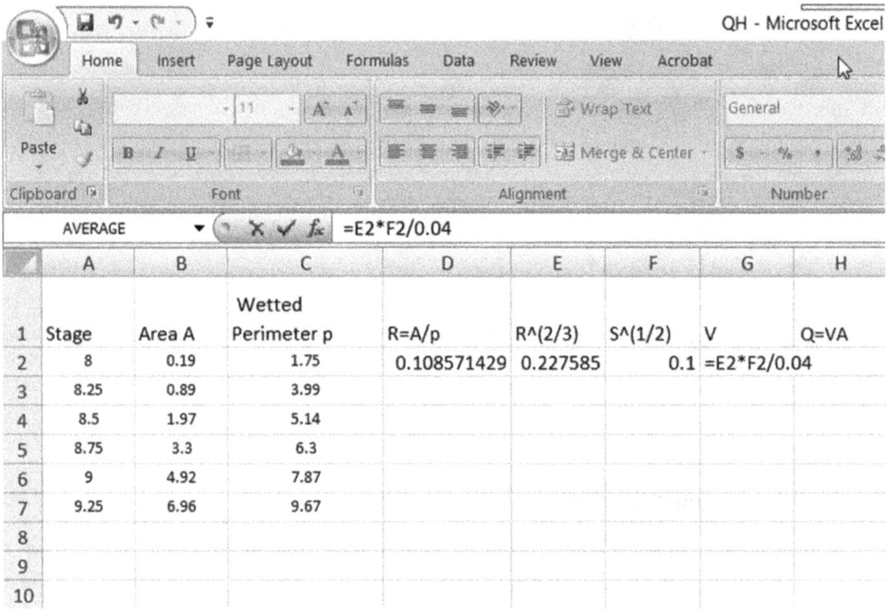

	A	B	C	D	E	F	G	H
1	Stage	Area A	Wetted Perimeter p	R=A/p	R^(2/3)	S^(1/2)	V	Q=VA
2	8	0.19	1.75	0.108571429	0.227585		0.1 =E2*F2/0.04	
3	8.25	0.89	3.99					
4	8.5	1.97	5.14					
5	8.75	3.3	6.3					
6	9	4.92	7.87					
7	9.25	6.96	9.67					
8								
9								
10								

FIGURE A.9   Entering the expression for velocity V in cell G2.

9. Figure A.10: Press the return button to display the value of V for the first stage. Next, move to cell H2 to calculate Q. Begin by typing the equal sign ("="), then select cell G2 (which contains the value of V). Afterward, type the multiplication operator "*" and select cell B2 (which contains the value of A, the cross-sectional area).

10. Figure A.11: Press the return button, which will display the value of the discharge Q for the first stage.

11. Figure A.12: To repeat all the calculations for the remaining river stages, begin by selecting cell D2. Hold down the Shift key, and then select cell H2 to highlight the range. Once selected, click the Copy button from the Excel menu bar to duplicate the formulas for use in the other stages.

12. Figure A.13: Select cell D-3, press the Shift key, then select cell H-7, and press the Paste button. The results will be displayed for all river stages.

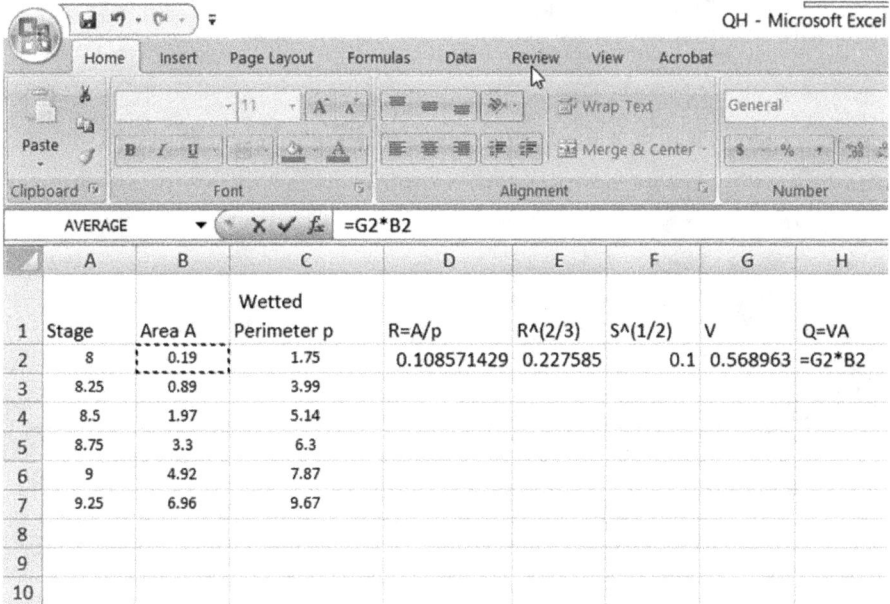

FIGURE A.10   Displaying the value of V and entering the expression for Q (equals V times A) for the first stage.

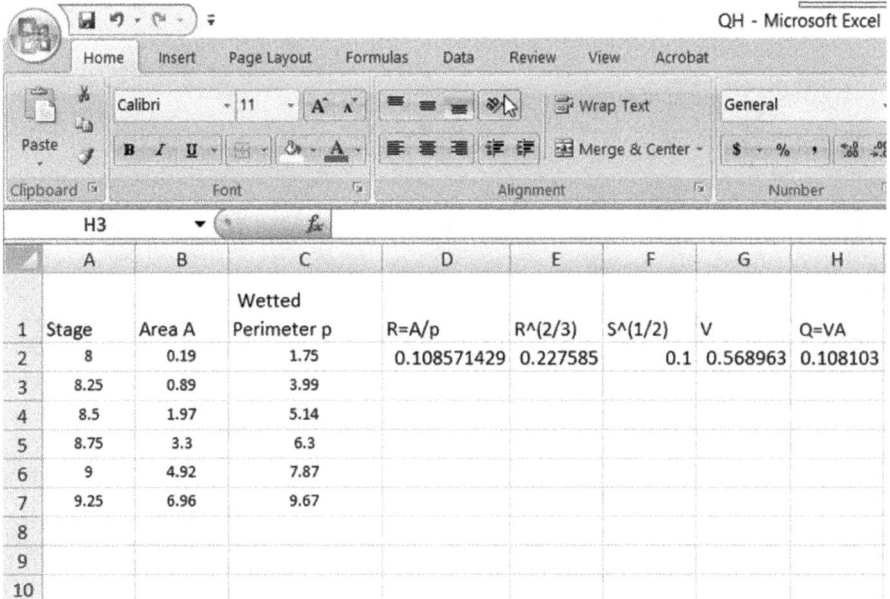

FIGURE A.11   Displaying the value of Q for the first stage.

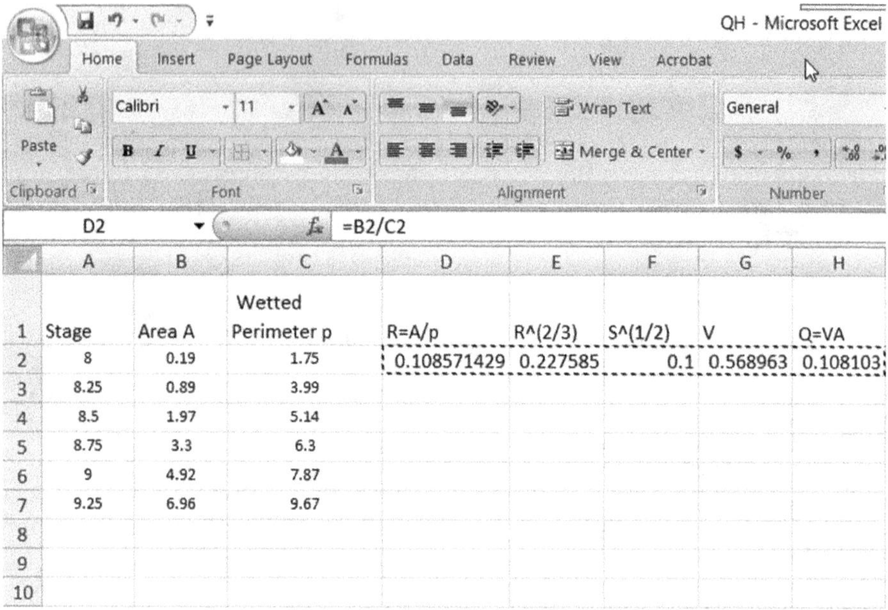

FIGURE A.12   Preparing to repeat the calculations for the remaining values of the stage.

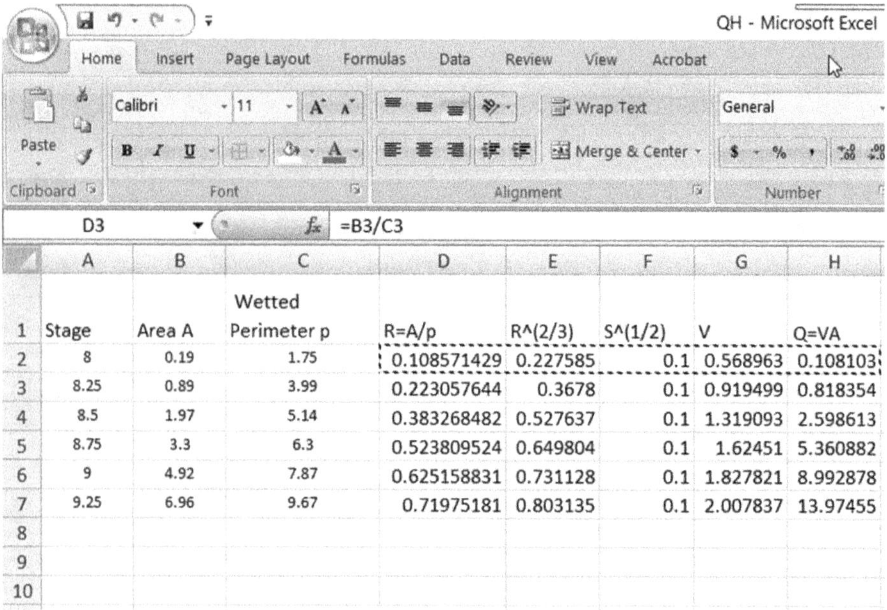

FIGURE A.13   Repeating the calculations for remaining rows.

13. Figure A.14: To plot the stage versus discharge, highlight the relevant data by selecting cell A-2 and holding down the Shift key while selecting cell A-7. Then, while holding the Ctrl key, select cell H-2 and, with the Shift key still held, select cell H-7. From the menu, choose "Scatter" under the Insert tab to create the plot.
14. Figure A.15: Choose the right-side plot option, which will produce the plot shown in the figure.
15. Figure A.16: The graph can be enhanced by adding axis titles using the "Layout" option. Additionally, to display the plot in the commonly used format, you can switch the axes by selecting the dataset and adjusting the series x- and y-values accordingly.

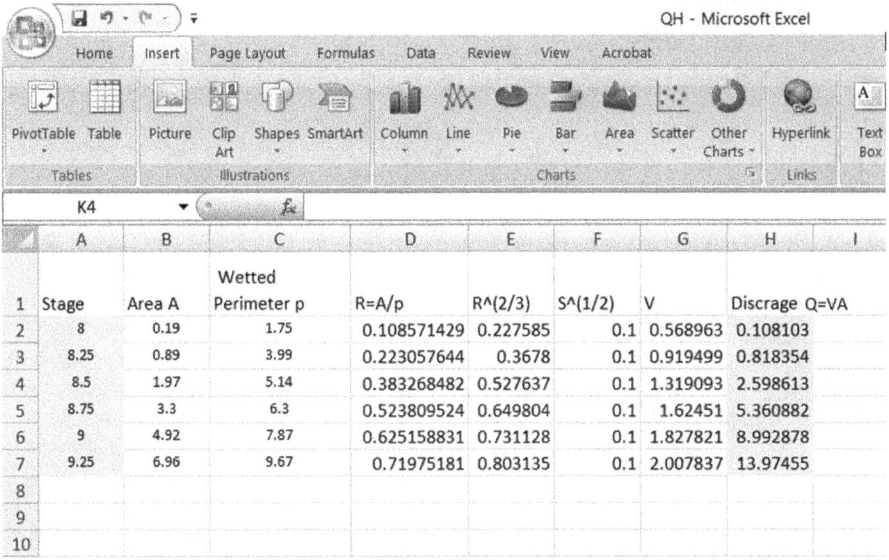

	A	B	C	D	E	F	G	H	I
1	Stage	Area A	Wetted Perimeter p	R=A/p	R^(2/3)	S^(1/2)	V	Discrage Q=VA	
2	8	0.19	1.75	0.108571429	0.227585	0.1	0.568963	0.108103	
3	8.25	0.89	3.99	0.223057644	0.3678	0.1	0.919499	0.818354	
4	8.5	1.97	5.14	0.383268482	0.527637	0.1	1.319093	2.598613	
5	8.75	3.3	6.3	0.523809524	0.649804	0.1	1.62451	5.360882	
6	9	4.92	7.87	0.625158831	0.731128	0.1	1.827821	8.992878	
7	9.25	6.96	9.67	0.71975181	0.803135	0.1	2.007837	13.97455	
8									
9									
10									

FIGURE A.14   Steps for plotting stage versus discharge.

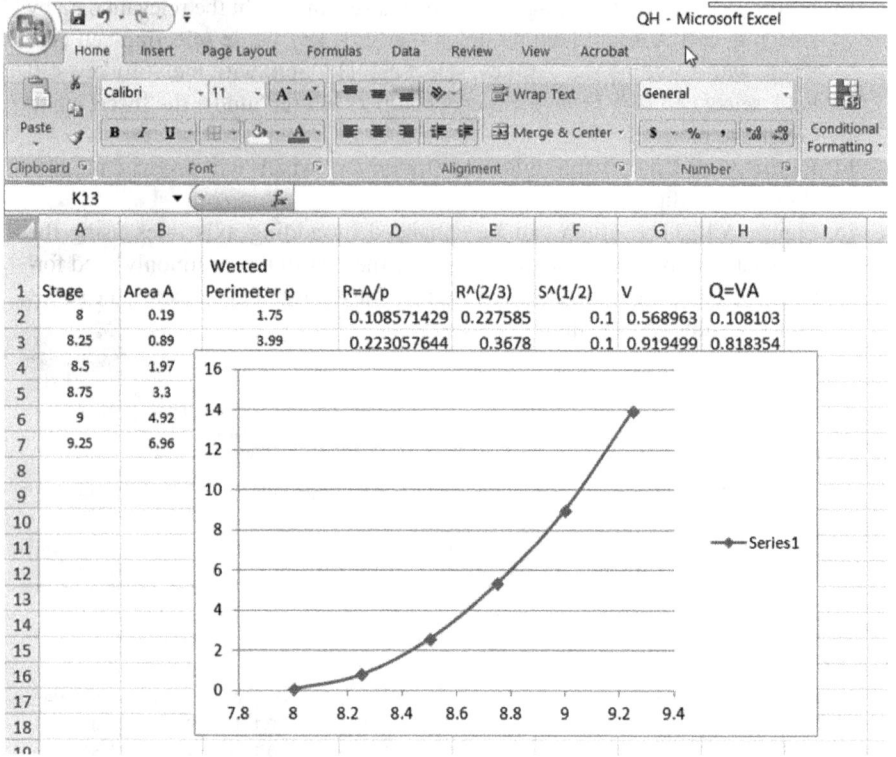

FIGURE A.15 Creating the plot.

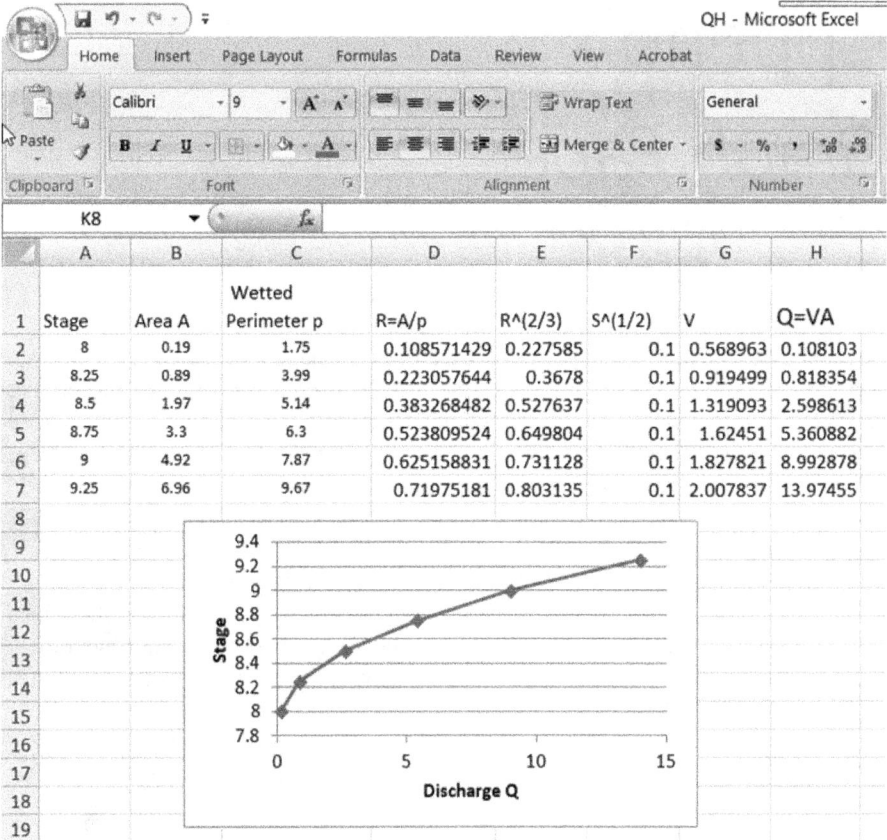

QH - Microsoft Excel

	A	B	C	D	E	F	G	H
1	Stage	Area A	Wetted Perimeter p	R=A/p	R^(2/3)	S^(1/2)	V	Q=VA
2	8	0.19	1.75	0.108571429	0.227585	0.1	0.568963	0.108103
3	8.25	0.89	3.99	0.223057644	0.3678	0.1	0.919499	0.818354
4	8.5	1.97	5.14	0.383268482	0.527637	0.1	1.319093	2.598613
5	8.75	3.3	6.3	0.523809524	0.649804	0.1	1.62451	5.360882
6	9	4.92	7.87	0.625158831	0.731128	0.1	1.827821	8.992878
7	9.25	6.96	9.67	0.71975181	0.803135	0.1	2.007837	13.97455

FIGURE A.16   Formatting the graph.

Appendix B
Graph Papers

This appendix includes various types of blank graph papers designed for use in specific applications discussed throughout the text. While many tasks that utilize these papers can be completed with specialized or general software, it is strongly encouraged that students hand-draw figures to gain a deeper grasp of the fundamental concepts. The appendix features the following types of graph papers.

1. **Square Paper (Figure B.1)**: This can be used for creating linear graphs, such as simple x–y plots representing two variables, including river discharge versus time or river stage. It can also be used to roughly estimate the relative watershed subareas required for calculating a watershed's average rainfall, also referred to as the rainfall effective uniform depth (see Chapter 4, Section 4.3).
2. **Log–Log Graph Paper (Figure B.2)**: This can be used in aquifer test analysis through the log–log curve matching method (see Chapter 6, Section 6.11.1).

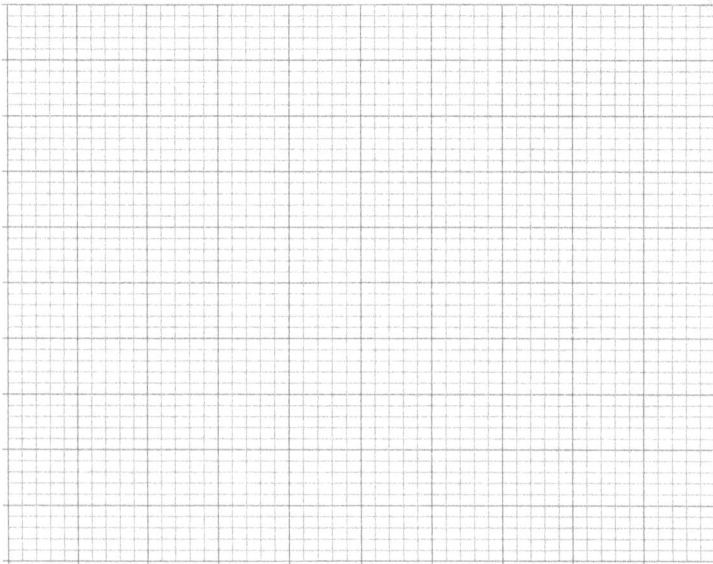

FIGURE B.1   Blank linear graph or square paper.

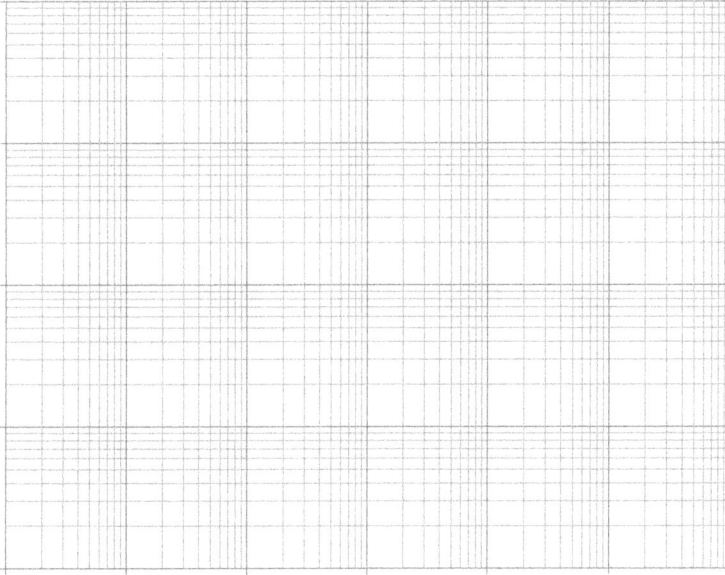

FIGURE B.2 Blank log–log paper.

This paper should be used in conjunction with the $W(u)$ plot provided in Appendix C. Both papers are drawn to the same scale, which is required for this technique. The horizontal and vertical axis increments should be labeled according to the range of analyzed drawdown versus time data. This paper can be also used for plotting x–y variables that change over several orders of magnitude.

3. **Semi-Log Graph Paper (Figure B.3)**: This paper is used for applying the semi-log curve matching method (see Chapter 6, Section 6.11.2). The axis increments should be labeled according to the range of analyzed drawdown versus time data. This paper is also suitable for other plots where a semi-log representation fits the data. For example, the logarithmic axis is appropriate for variables such as stream discharge, which can vary over several orders of magnitude.

4. The probability paper in Figure B.4 can be used in statistical analysis to plot cumulative distribution functions and perform probability plotting. The paper features a non-linear y-axis representing cumulative probability (ranging from near 0% to near 100%). The horizontal x-axis represents the values of the random variable being analyzed, such as river discharge. This axis is typically linear, as shown in Figure B.4, where a normally distributed variable will result in a straight line when plotted.

    Section 5.2.4 of Chapter 5 describes the use of such paper in creating river duration curves, which are a tool for estimating the probability that a specific river discharge will be equaled or exceeded. The plot can also be

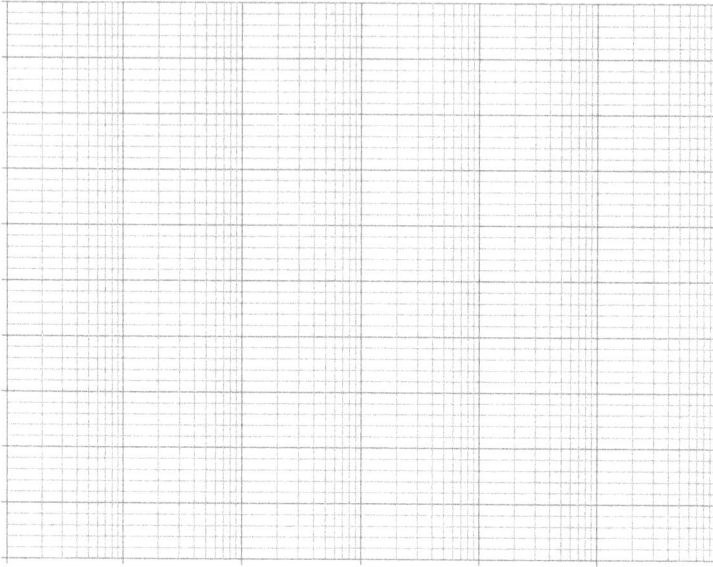

FIGURE B.3 Blank semi-log paper.

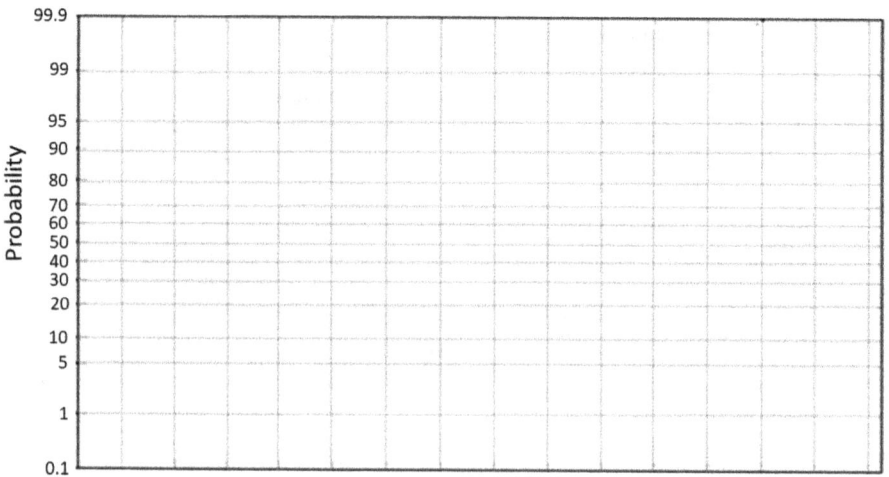

FIGURE B.4 Normal probability paper.

used to test whether the data follow a normal distribution. Another important application is to estimate percentiles, reliability, or other statistical characteristics. For example, such a curve can be used in dam design by estimating flood discharge at a certain probability. A typical design would use a hundred-year flood, where the probability of occurrence is 1%.

Appendix C
The Well Function

Appendix C presents the well function, $W(u)$, in both tabular (Table C.1) and graphical (Figure C.1) formats. This function is used to calculate aquifer response to pumping as well as in analyzing well-pumping data, as detailed in Chapter 6 (Section 6.9). To estimate the transient response of an aquifer to pumping, the table and figure are used to find the value of W for a given u. For well-pumping data analysis, the plot in Figure C.1 is used alongside the log–log plot in Appendix B (Figure B.2), both of which must be drawn to the same scale, a crucial element of the technique. While the axes in Figure C.1 are labeled, the axes in Figure B.2 should be labeled according to the range of analyzed drawdown versus time data.

TABLE C.1 The $W(u)$ (well) function

U	W	U	W	U	W	U	W	U	W
1.00E-14	31.66	1.00E-11	24.75	1.00E-08	17.84	1.00E-05	10.94	1.00E-02	4.04
2.00E-14	30.97	2.00E-11	24.06	2.00E-08	17.15	2.00E-05	10.24	2.00E-02	3.35
3.00E-14	30.56	3.00E-11	23.65	3.00E-08	16.74	3.00E-05	9.84	3.00E-02	2.96
4.00E-14	30.27	4.00E-11	23.36	4.00E-08	16.46	4.00E-05	9.55	4.00E-02	2.68
5.00E-14	30.05	5.00E-11	23.14	5.00E-08	16.23	5.00E-05	9.33	5.00E-02	2.47
6.00E-14	29.87	6.00E-11	22.96	6.00E-08	16.05	6.00E-05	9.14	6.00E-02	2.3
7.00E-14	29.71	7.00E-11	22.81	7.00E-08	15.9	7.00E-05	8.99	7.00E-02	2.15
8.00E-14	29.58	8.00E-11	22.67	8.00E-08	15.76	8.00E-05	8.86	8.00E-02	2.03
9.00E-14	29.46	9.00E-11	22.55	9.00E-08	15.65	9.00E-05	8.74	9.00E-02	1.92
1.00E-13	29.36	1.00E-10	22.45	1.00E-07	15.54	1.00E-04	8.63	1.00E-01	1.82
2.00E-13	28.66	2.00E-10	21.76	2.00E-07	14.85	2.00E-04	7.94	2.00E-01	1.22
3.00E-13	28.26	3.00E-10	21.35	3.00E-07	14.44	3.00E-04	7.53	3.00E-01	0.91
4.00E-13	27.97	4.00E-10	21.06	4.00E-07	14.15	4.00E-04	7.25	4.00E-01	0.7
5.00E-13	27.75	5.00E-10	20.84	5.00E-07	13.93	5.00E-04	7.02	5.00E-01	0.56
6.00E-13	27.56	6.00E-10	20.66	6.00E-07	13.75	6.00E-04	6.84	6.00E-01	0.45
7.00E-13	27.41	7.00E-10	20.5	7.00E-07	13.59	7.00E-04	6.69	7.00E-01	0.37
8.00E-13	27.28	8.00E-10	20.37	8.00E-07	13.46	8.00E-04	6.55	8.00E-01	0.31
9.00E-13	27.16	9.00E-10	20.25	9.00E-07	13.34	9.00E-04	6.44	9.00E-01	0.26

(Continued)

TABLE C.1 (*Continued*)   The *W*(*u*) (well) function

U	W	U	W	U	W	U	W	U	W
1.00E-12	27.05	1.00E-09	20.15	1.00E-06	13.24	1.00E-03	6.33	1.00E+00	2.20E-01
2.00E-12	26.36	2.00E-09	19.45	2.00E-06	12.55	2.00E-03	5.64	2.00E+00	4.90E-02
3.00E-12	25.96	3.00E-09	19.05	3.00E-06	12.14	3.00E-03	5.23	3.00E+00	1.30E-02
4.00E-12	25.67	4.00E-09	18.76	4.00E-06	11.85	4.00E-03	4.95	4.00E+00	3.80E-03
5.00E-12	25.44	5.00E-09	18.54	5.00E-06	11.63	5.00E-03	4.73	5.00E+00	1.10E-03
6.00E-12	25.26	6.00E-09	18.35	6.00E-06	11.45	6.00E-03	4.54	6.00E+00	3.60E-04
7.00E-12	25.11	7.00E-09	18.2	7.00E-06	11.29	7.00E-03	4.39	7.00E+00	1.20E-04
8.00E-12	24.97	8.00E-09	18.07	8.00E-06	11.16	8.00E-03	4.26	8.00E+00	3.80E-05
9.00E-12	24.86	9.00E-09	17.95	9.00E-06	11.04	9.00E-03	4.14	9.00E+00	1.20E-05

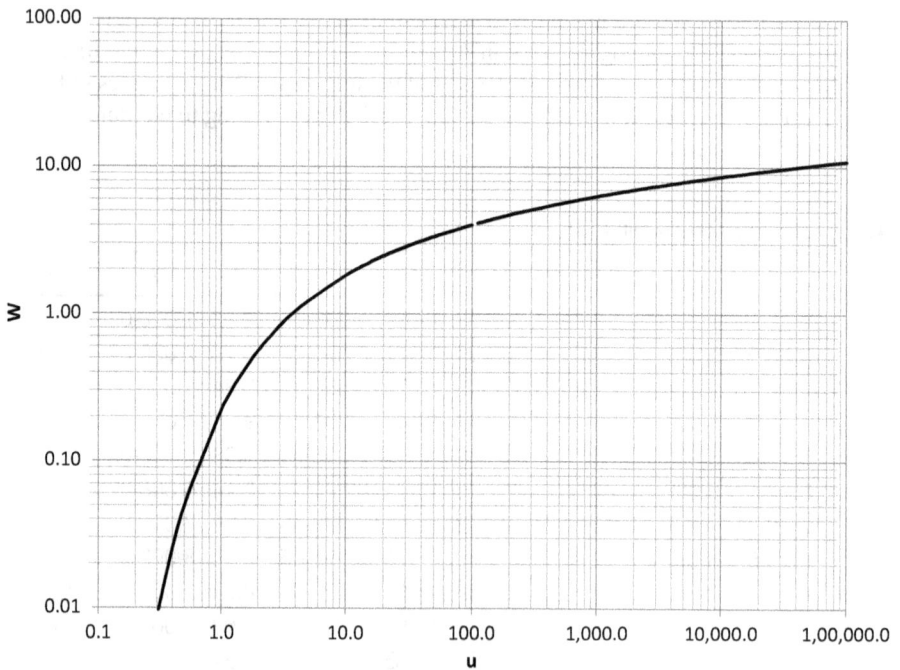

FIGURE C.1   The well function [*W*(*u*)] plot.

Index

Note: **Bold** page numbers refer to tables and *italic* page numbers refer to figures.

For Product Safety Concerns and Information please contact our EU
representative GPSR@taylorandfrancis.com
Taylor & Francis Verlag GmbH, Kaufingerstraße 24, 80331 München, Germany